Membranes for Energy Applications

Young Moo Lee

WILEY-VCH

Author

Prof. Young Moo Lee
Hanyang University
Department of Energy Engineering
Seoul 04763
South Korea

Library of Congress Card No.: applied for

British Library Cataloguing-in-Publication Data
A catalogue record for this book is available from the British Library.

Bibliographic information published by the Deutsche Nationalbibliothek
The Deutsche Nationalbibliothek lists this publication in the Deutsche Nationalbibliografie; detailed bibliographic data are available on the Internet at <http://dnb.d-nb.de>.

Print ISBN: 978-3-527-34764-3
ePDF ISBN: 978-3-527-34765-0
ePub ISBN: 978-3-527-34766-7
oBook ISBN: 978-3-527-82726-8

Typesetting Straive, Chennai, India

Contents

Preface

There are many books published in general membranes and processes, but there is no book published in membranes for energy applications. These books are fragmented; there are books and review papers on transport of gas, ions, and water separately, but there are no books touching these three very important small molecules for separation or their transportation. As there are shortages in energy sources from natural resources, many devices using alternative and renewable energy or energy storage are now a hot topic in recent years. Such fuel cell vehicles or smartphone/electrical vehicles should have fuel cell stacks or Li-ion batteries that equip with membranes to selectively separate ions and/or water. Here, membrane materials and membrane morphology or properties play an important role for electrochemical performance of the mentioned energy devices and provide a pivotal role in safety issues of these devices. Failure to block the transport of ions will cause explosion of devices such as smart phones. Fuels are also important for these energy devices. Such gases as hydrogen or oxygen can be used for these energy devices where membranes are also very important component in producing these fuels that need to be studied. Many processes involve membranes for fuel production. In addition to hydrogen gases as fuel for these devices, salinity in water can be used for energy generation as well if the devices are properly designed. Here, membranes should be mechanically strong enough to withstand very high applied pressure to overcome the osmotic pressure of saline water as a feed for membranes.

This book provides important information and recent advances on membranes for energy production. It starts from the history of membrane science and technology, fundamentals of membrane technology including principles of membrane formation and principle behind, will introduce on the gas separation using membrane technology, introduce membranes for ion transport or separation realized in energy generation and storage, and power generation, and conclude the future direction and outlook of the membrane technology in energy application and industry. Readers of this book will benefit not only on the application of membranes in energy industry, but also on the principles or theories behind this important application including the transport of small molecules such as gas, ions, and water. Also, the future direction of this topic will be covered at the end so that readers can advance the standard life of human beings in the future.

First chapter introduces an energy issue and a brief history of synthetic membranes and their technology and the existing concepts and phenomenon to commercial use. It will then describe how membrane separations compare with the other separation methods. Finally, it will bring the reader to the current state-of-the art membrane technology.

Chapter 2 is intended to bring-up the reader to what membranes are, how transport occurs through the membranes, what are synthetic polymer and inorganic membrane materials, how membrane and membrane modules are fabricated, how one selects which is the best membrane/membrane module for a given separation application, and finally how simple and complex membrane processes are developed.

In Chapter 3, the issue is how membrane processes can aid in energy and environmental issues. Membrane systems for air separation will be described, and enriched oxygen air combustion will be described and their advantages over the current systems in curbing greenhouse effects will be stated. Membrane processes for natural gas upgrading and carbon-dioxide capture and pervaporation for biofuel production will be described.

The problem included in Chapter 4 is to find membrane applications in the energy conversion using fuel cell devices. Electrochemical processes for the conversion of various fuels (hydrogen, methane, and methanol) will be described and concepts and equipment for the various fuel cells will be presented. Membranes as well as the importance of polymeric binders for catalysts will be discussed. Advantages and disadvantages of the various fuel cells and their applicability will be presented including the present status on fuel cell vehicles.

In Chapter 5, the problem addressed is the use of membranes in energy storage devices such as lithium-ion batteries. The basic principle of these devices will be described, and the needs associated with the membranes in these applications will be pointed out. Then, the various concepts and membranes and their use will be described. Membrane separators in new and emerging areas will also be covered.

The subject discussed in Chapter 6 is how to provide clean fuel for transportation and other applications for our future society. Hydrogen is regarded as a new energy source in future society. The key challenge is how membranes can play a role in production, purification, and storage and safe operation of hydrogen. Hydrogen is the main target fuels form the reaction between water and catalyst and this chapter provides insights how the membranes are playing a role in water electrolysis.

In Chapter 7, clean water supply is going to be a challenge for the coming generations. Here, the question posed is that how we can generate/recover energy while producing clean water from the saline and brackish waters. Followed by the importance of water-energy nexus in produced water, the concept of osmotic pressure and its exploitation for energy generation will be discussed.

April 2024

Young Moo Lee
Seoul, South Korea

Acknowledgments

It is difficult to credit all individuals who have contributed to the preparation of this book. Instead of including a long list of names, I would like to express our sincere thanks to colleagues and former PhD students from the Membrane Laboratory of Hanyang University, particularly, Drs. Jun Tae Jung, Chuan Hu, and Sun Ju Moon who helped coordinate our activities during editing, and friends in other universities and companies who provided valuable suggestions and discussions. I would like to thank the publisher for their dedication and hard work for their supporting role in realizing this ambitious project.

April 2024

Young Moo Lee
Seoul, South Korea

1

Introduction

1.1 Energy and Membranes

Energy crisis, water and food shortage, environmental pollution, and so on are important problems of human beings for the next 50 years. Energy crisis is the single most important problem because of an increase in energy usage from the technological development and its influence on energy toward water, food shortage, and environmental pollution. According to International Energy Outlook 2021 (International Energy Association 2021) from the United States Energy Information Administration (EIA), the world energy consumption will grow by nearly 50% between 2020 and 2050. Most of this growth comes from regions where strong economic growth is driving demand, particularly in Asia. Demand for all fuels increased, with fossil fuels meeting nearly 70% of the growth for the second year running. As a result, global energy-related CO_2 emissions rose by 1.7% to 33 gigatons (Gt) in 2018. Coal use in power generation alone surpassed 10 Gt, accounting for a third of total global CO_2 emission. Most of that came from newly built coal-fired power plants in developing countries [1]. As the amount of energy consumption increased enormously, in particular, after the Second World War, there was a worldwide issue on energy consumption and production that raised intensive environmental issues (see Figure 1.1). The largest portion of the energy sources are fossil fuels such as coal, oil, and natural gas that cover more than 80% of all the energy sources.

To solve the energy crisis, various efforts have been made to economically produce energy and use it efficiently. From fossil fuels to renewable energy, various kinds of resources have been utilized to produce energy. Fossil fuels are hydrocarbons, primarily coal, petroleum, or natural gas, generated from buried combustible geologic deposits of organic materials that have been converted to petroleum, coal, and natural gas by heat and pressure in the earth's crust over hundreds of millions of years. Because of their origin, fossil fuels have a high carbon content, resulting in a high heat content. According to the Statistical Review of World Energy from British Petroleum in 2020, the main primary energy sources worldwide consisted of coal (24%), petroleum (33%), and natural gas (27%), amounting to an 84% share for fossil fuels in primary energy consumption in the world [2]. Fossil fuel is the most economic resource to produce energy; however, it is required to reduce the share

Membranes for Energy Applications, First Edition. Young Moo Lee.

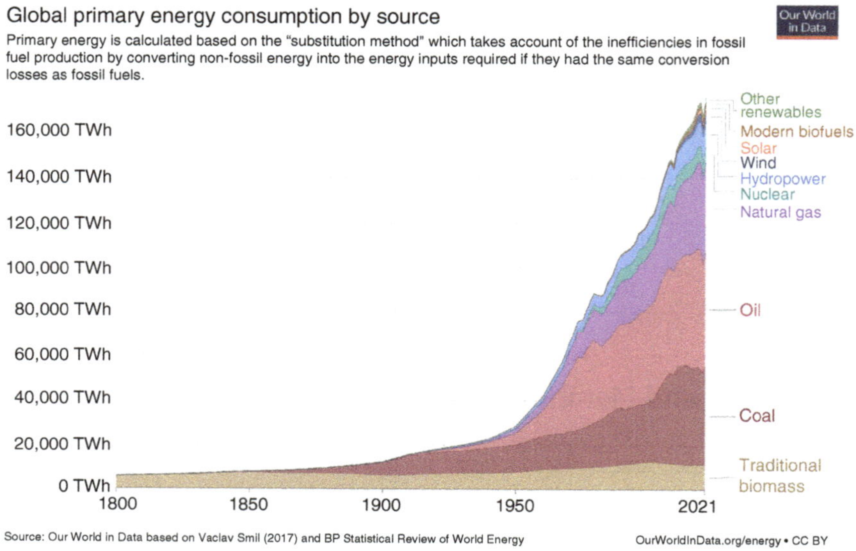

Figure 1.1 Global primary energy consumption by source.

of fossil fuel due to emission of CO_2 inducing global climate change. Currently, technology innovation is applied as a means to decrease the environmental impact of coal combustion by increasing the boiler efficiency, co-combustion with biomass or carbon capture, utilization, and storage [3]. Nuclear power is one of the candidates to replace fossil fuels owing to its low price for energy production. However, it is hard to completely replace fossil fuels with nuclear power due to its drawbacks such as uncertain accidents and the production of radioactive nuclear wastes. Especially, the three nuclear disasters (that is, Three Mile Island in 1979, Chernobyl in 1986, and Fukushima in 2011) caused by human mistakes and/or natural disasters such as flood and earthquake make it difficult to expand the number of nuclear power plants. Renewable energy including solar, wind, and geothermal energy has been developed during the last decades with an infinite potential; however, they require further research to overcome their limitations such as low energy density, intermittent energy production, and so on. As such, each energy resource has strengths and possibilities; however, it also has weaknesses to be solved.

Let us briefly look at the conversion of fossil fuels to electrical energy. Thermal power plants use coal, petroleum, and natural gas as fuel. According to the species and phases of the fuel, transportation of the fuel to the plant and post- and pre-treatment of the fuel are different resulting in the application usage. However, the principle of generating electricity from fossil fuels is the same. The steam is produced at a high pressure in the steam boiler after burning fuel in boiler furnaces. Superheated steam then enters into the turbine and rotates the turbine blades. The turbine is mechanically coupled with alternator that its rotor will rotate with the rotation of turbine blades. After entering into turbine, the steam pressure suddenly falls and the corresponding volume of the steam increases. After imparting energy to the turbine rotor, the steam passes out of the turbine blades into the condenser. In the condenser, cold water is circulated with the help of a pump which condenses the low-pressure wet steam. This condensed water is further supplied to a low-pressure water heater where the low-pressure steam increases the temperature of this feed water and it is again heated at high pressure. The chemical energy in fossil fuels is converted to thermal energy through carbon and hydroxide combining with oxygen during the combustion. Heat energy boils pressurized water to steam, and the steam travels through a mechanical turbine, causing it to rotate converting thermal energy to electric energy, i.e. electricity.

Pulverized coal (less than 5 cm) is burned in a boiler where water boils to steam. Converted steam is used to operate turbines for electronic generators. Compared to thermal power plants using other fuel types, coal requires a specific fuel processing and ash disposal. Flue gas from the combustion of fossil fuels contains carbon dioxide (CO_2) and water vapor, as well as pollutants such as nitrogen oxides (NOx), sulfur oxides (SOx), and, for coal-fired power plants, mercury, traces of other metals, and fly ash. Solid waste ash from coal-fired boilers should be removed. Gas-fired power plants, which burn natural gas (mainly methane) to generate electricity, produce a quarter of the world's electricity; however, they are also a significant part of global greenhouse gas emission resulting in global warming.

As the majority of current global primary energy relies on fossil fuels, the energy system is the source of approximately two-thirds of global CO_2 emissions. As methane and other short-lived climate pollutant emissions are believed to be severely underestimated, it is likely that energy production and use are the source of an even greater share of emissions.

Meanwhile, membrane technology, which has been able to be utilized with various kinds of energy application, has been developed to solve the water issues and the problems of energy production. For example, in energy production using fossil fuels, membrane technology can be utilized to reduce the environmental impact from the emission of carbon dioxide. Membranes can separate the carbon dioxide from the ash inducing environmental issues and prevent them from being discharged to the outside. Moreover, membrane technology is also used as an integral part of the energy production in various ways. For example, in a fuel cell application, membrane performance determines the overall system performance including power density and long-term stability of the fuel cells. In energy production using salinity gradients and osmotic pressure, membranes enable the harvesting of blue energy from the ocean, induced by the chemical potential difference using the diffusion difference of water and salt.

1.2 Brief History of Membrane Technology

Membrane technology has been known for a long time. The first report on gas separation was made by Mitchell [4], who observed the separation of gases through membranes. He noticed that gases had different permeabilities in experiments using rubber balloons exposed to various gas atmospheres. Graham [5, 6] reported in 1829 about different rates of permeation in experiments where a wet pig bladder inflated when stored in CO_2 environment. Fick established the theoretical equations for diffusion through membranes. Graham proposed the concept of solution–diffusion as the permeability process in polymeric membranes in 1866 [7]. Daynes [8] established a time lag method to determine diffusion coefficients employing the unsteady-state solution of Fick's second law to mathematically determine the diffusion coefficient from the extrapolation of the steady-state flux to the time axis. Barrer and Rideal [9] noted the temperature-dependent Arrhenius equation of gas permeability and diffusion of gases. Although most of these scientific advances have enlightened the membrane community, historic advances in gas and liquid separation membranes occurred in 1960 with the discovery of asymmetric cellulose acetate membranes with a selective layer thickness <1 μm for reverse osmosis by Loeb and Sourirajan [10–12]. Before this discovery, dense cellulose acetate membranes separated water and salts but were too thick to produce sufficient water flux to meet the commercial needs [13]. The production of asymmetric cellulose acetate membranes spurred the commercialization of reverse osmosis membrane and later led to gas separation membrane [14, 15]. A thin polyamide layer of 0.2 μm thickness can be coated on top of a microporous polysulfone support layer with a thickness of 200 μm to produce a thin-film composite membrane for reverse osmosis and nanofiltration [16]. The same concept of thin-film composite membrane can be applied to produce integrally skinned asymmetric membrane by depositing or coating selective dense layer on top of a porous sublayer [17].

The first commercial gas separation membrane system was introduced by Hennis and Tripodi [18, 19] at Monsanto to produce hydrogen in an ammonia synthesis plant. Membrane-based gas separation is being considered for carbon capture and storage of flue gas evolved from coal-fired power plants to meet the requirements of global climate change [20, 21].

The market for polymer membranes for gas and liquid separation has been expanding annually. The sales of membrane gas separation systems were approximately US$ 846 in 2019 and projected to be US$ 1132 million in 2024 at a CAGR of 6%. The worldwide sales of all synthetic membranes are estimated to be over US$ 5.4 billion as of 2019 and projected to grow to US$ 8.5 billion by 2024, at a CAGR of 9.0%. These numbers exclude the membranes for energy generation applications such as lithium-ion batteries and fuel cells. The main reasons for the fast growth of membrane markets and sales include consumer demand for higher quality products, increased regulatory pressures, deteriorating natural resources, and the need for environmental and economic sustainability. The major drivers for the membrane market include increasing population, raising awareness about wastewater reuse, and rapid industrialization. Also, the shift from chemical treatment of water to physical treatment of water, strict regulations regarding water treatment and discharge, and changing climate dynamics in terms of the amount of precipitation are also driving the membrane market. The future market is expected to expand, and further growth of this technology is anticipated for the next 10 years or so.

It is interesting to note that energy can be produced using membranes, for example, by using hydrogen as a fuel. Fuel cell concept was first demonstrated by

William Grove in 1839 using an experiment applying hydrogen and oxygen [22]. Water is electrolyzed into hydrogen and oxygen by passing an electric current through it. A fuel cell acts as an engine as it is a device that converts chemical energy into useful work, whereas a combustion engine converts the chemical energy of the fuel and oxidizer into mechanical work. The first fuel cell was developed by General Electric in the United States for the US Gemini space program in the early 1960s [23]. The fuel cell was expected to provide electricity and water for an animal orbiting in space in a satellite. The development of the space program continued with the incorporation of a new polymer, Nafion. DuPont and General Electric looked into the application of Nafion that received much attention due to its use as an electrolyte in the proton exchange membrane fuel cell (PEMFC) [24].

The same Nafion membrane was used in proton exchange membrane water electrolysis (PEMWE). The first PEMWE using this polymer began to be commercialized in 1978 [25]. Because of the use of expensive catalyst and membrane, an alternative to PEMWE was proposed, i.e. anion exchange membrane water electrolysis (AEMWE). Although many membrane materials have been tested, large-scale commercial AEMWE needs more time to appear.

We are living in an age of electrical vehicles. About 10% of the new cars are electrical vehicles equipped with lithium-ion batteries. European countries will ban the cars equipped with internal combustion engines by 2030, and this trend is moving forward to other countries. The proportion of electrical vehicles in total automobile market will exceed 30% by 2030, where LIB demand accounts to be 1293 GWh [26]. Batteries are made of many components such as cathode, anode, electrolyte, and separator. Separator is a porous membrane made mainly of polyolefins. Separator production in 2025 is estimated to be 2.7 billion m^2 with a CAGR of 12% [27]. Separators are made by dry process and wet process.

1.2.1 Current State-of-the-Art Membrane Technology

Membrane-based gas separation is currently being used to separate hydrogen from ammonia plants, oxygen from air, nitrogen from air, CO_2 from nitrogen stream or methane mixture, volatile organic compounds from petrochemical plants, water from organic vapors, and isopropanol from semi-conductor plants. For these applications, membranes are in hollow fiber or flat sheet forms that are constructed to make modules. Polysulfone, polyimide, and cellulose acetate are the most widely used membrane materials in industry while many microporous materials are under investigation mainly in academia.

PEMFC is currently utilized in fuel cell automobiles or vehicles while direct methanol fuel cell is used in forklifts. These membranes use perfluorosulfonic acid (PFSA) membranes. Solid oxide fuel cell or metal ceramic fuel cell is used in electricity generation for households or small residential areas.

PEMWE is currently being used in generating hydrogen. This technology also uses PFSA as a membrane. Proton exchange membranes are incorporated to fabricate membrane electrode assembly where catalysts are coated on each side of the membrane. To reduce the cost of membrane and catalyst in PEMWE, AEMWE is evolved these days. A lot of progress has been made in AEMWE which will soon be commercially available.

Separator is an exploring market and has been widely used in LIB manufacturing. Polyethylene or polypropylene has been used to manufacture LIB separators. Ceramic coating technology is widely used to increase the thermal stability of the separator. Other high-performance polymers are considered for separator materials, but the cost of the separator hinders the widespread of these materials for separator. Dry extension and wet manufacturing technology are the two major technologies widely used in the separator manufacturing industry.

Fuel cells and battery separators use membranes in flat sheet type which can be fabricated in stack or module form.

Throughout this book, we will introduce various kinds of membrane technologies for energy application, from their history to the recent development. We will also describe how membrane separations compare with other separation methods, and bring the readers to the current state of the art.

References

1 IEA (2018). Renewables 2018. International Energy Agency (IEA), Paris, France, pp. 1–10.

2 Looney, B. (2020). bp Statistical Review of World Energy June 2020.

3 Barnes, D.I. (2015). Understanding pulverised coal, biomass and waste combustion–a brief overview. *Applied Thermal Engineering* 74: 89–95.

4 Mitchell, J.K. (1995). On the penetrativeness of fluids. *Journal of Membrane Science* 100: 11–16.

5 Graham, T. (1995). On the law of the diffusion of gases. *Journal of Membrane Science* 100: 17–21.

6 Graham, T. Notice of the Singular Inflation of a Bladder (Reprinted from Quarterly Journal of Science, No II, Pg 88, 1829). *Journal of Membrane Science* 100 (1995): 9–9.

7 Graham, T. (1867). On the absorption and dialytic separation of gases by colloid septa. *Journal of the Franklin Institute* 83: 39–41.

8 Daynes, H.A. (1920). The process of diffusion through a rubber membrane. *Proceedings of the Royal Society of London Series A, Containing Papers of a Mathematical and Physical Character* 97: 286–307.

9 Barrer, R. and Rideal, E.K. (1939). Permeation, diffusion and solution of gases in organic polymers. *Transactions of the Faraday Society* 35: 628–643.

10 Loeb, S. and Sourirajan, S. (1963). Sea Water Demineralization by Means of an Osmotic Membrane. In: *Saline Water Conversion—II*, 117–132. ACS Publications.

11 Riley, R., Gardner, J., and Merten, U. (1964). Cellulose acetate membranes: electron microscopy of structure. *Science* 143: 801–803.

12 Loeb, S. (1965). Desalination research in California. *Science* 147: 1241–1242.

13 Reid, C. and Breton, E. (1959). Water and ion flow across cellulosic membranes. *Journal of Applied Polymer Science* 1: 133–143.

14 Glater, J. (1998). The early history of reverse osmosis membrane development. *Desalination* 117: 297–309.

15 Strathmann, H., Kock, K., Amar, P., and Baker, R.W. (1975). The formation mechanism of asymmetric membranes. *Desalination* 16: 179–203.

16 Cadotte, J.E., Petersen, R.J., Larson, R.E., and Erickson, E.E. (1980). A new thin-film composite seawater reverse osmosis membrane. *Desalination* 32: 25–31.

17 Shenvi, S.S., Isloor, A.M., and Ismail, A.F. (2015). A review on RO membrane technology: developments and challenges. *Desalination* 368: 10–26.

18 Henis, J.M. and Tripodi, M.K. (1980). Multicomponent membranes for gas separations, Google Patents.

19 Henis, J.M. and Tripodi, M.K. (1983). The developing technology of gas separating membranes. *Science* 220: 11–17.

20 Wilberforce, T., Baroutaji, A., Soudan, B. et al. (2019). Outlook of carbon capture technology and challenges. *Science of The Total Environment* 657: 56–72.

21 Favre, E. (2011). Membrane processes and postcombustion carbon dioxide capture: challenges and prospects. *Chemical Engineering Journal* 171: 782–793.

22 Grove, W.R. (1839). XXIV. On voltaic series and the combination of gases by platinum. *The London, Edinburgh, and Dublin Philosophical Magazine and Journal of Science* 14: 127–130.

23 Barbir, F. (2012). *PEM Fuel Cells: Theory and Practice*. Academic press.

24 Banerjee, S. and Curtin, D.E. (2004). Nafion® perfluorinated membranes in fuel cells. *Journal of Fluorine Chemistry* 125: 1211–1216.

25 Millet, P., Dragoe, D., Grigoriev, S. et al. (2009). GenHyPEM: a research program on PEM water electrolysis supported by the European Commission. *International Journal of Hydrogen Energy* 34: 4974–4982.

26 Berckmans, G., Messagie, M., Smekens, J. et al. (2017). Cost projection of state of the art lithium-ion batteries for electric vehicles up to 2030. *Energies* 10: 1314.

27 Lebedeva, N., Di Persio, F., and Boon-Brett, L. (2016). Lithium ion battery value chain and related opportunities for Europe, European Commission, Luxembourg.

2

Fundamentals of Membrane Technology

2.1 Introduction

Membranes applied in energy application is associated with synthetic membrane processes such as gas separation, ion transport processes, fuel cell, and energy storage differ widely as far as the applied driving forces for mass transport are concerned. In membrane science, fundamental aspects of mass transport can be described with thermodynamic and kinetic relations at the membrane/bulk solution interface. Therefore, in this chapter, before discussing further on the membranes in energy applications, the definition of terms used in membrane science are provided, such as permeability, perm-selectivity, rejection, and so on. Then, membrane structures such as porous and homogeneous or symmetric and asymmetric membrane and their function in membrane processes are described. Finally, the basic thermodynamic, kinetic, and electrochemical relationships relevant to the description of mass transport phenomena in membranes and membrane processes are treated.

2.2 Definition of Terms

Membrane performance in each application is determined by a number of components and processes such as membranes, modules, and operation parameters or engineering aspects of a process. These components and parameters and their functions are described and defined.

2.2.1 The Membrane and Its Function

It is difficult to define a precise and complete definition of a membrane that covers all aspects of membranes. In a general term, a membrane separates two phases and controls the transport of various components from one phase into the other in a specific manner. A membrane may have a dense and homogeneous structure or can have pores. Membrane may be formed from organic, inorganic, carbon, or metallic materials and their hybrid materials. It may be neutral or it may give positive or negative

Membranes for Energy Applications, First Edition. Young Moo Lee.

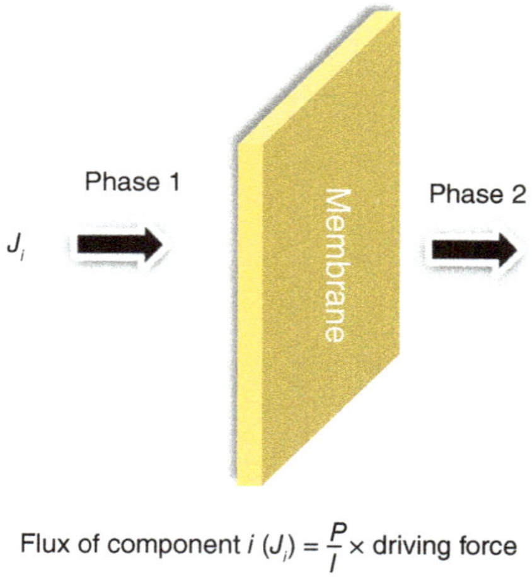

Figure 2.1 The mass transport across a membrane as a function of the membrane permeability (P), the driving force, and the membrane thickness (l).

charges, or functional groups with specific binding capabilities. Its thickness can be less than 100 nm or more than several millimeters. The electrical area membrane resistance of a membrane may vary from more than $10^6\ \Omega\,cm^2$ to less than $\Omega\,cm^2$.

The separation of a mixture in a membrane process is resulting from different transport rates of different components through the membrane. The transport rate of a component through a membrane is determined by the driving force or forces acting on the component in question and the permeability of the membrane matrix. The transport rate is inversely proportional to the thickness of the membrane as described in Figure 2.1 which shows the flux J of a component i from phase 1 into phase 2 through a membrane as a function of a driving force, permeability, and the membrane thickness.

The flux of a component through a membrane can be resulted from a convective bulk flow of matter through defined pores in the membrane or diffusion of individual components through a dense membrane. The permeability of a membrane for a given component is determined by the mobility and concentration of the component in the membrane matrix. Driving forces in membrane processes are gradients in the electrochemical potential of a component. Thus, activity, hydrostatic pressure, electrical potential, and temperature gradients can lead to a diffusive transport of individual molecules through the membrane. In porous membranes, a hydrostatic pressure difference between membranes can also result in convective mass transport.

2.2.2 Membrane Materials and Structure

Synthetic membranes used in energy applications show various physical structure and can be classified into the following three categories.

- Porous membrane
- Homogeneous and dense membrane
- Dense membranes carrying electrical charges or selective functional groups

Structure of membranes may be

- Isotropic or symmetric
- Anisotropic or asymmetric

The materials used for the preparation of membranes can be

- Polymers
- Ceramics
- Carbons
- Metals
- Hybrids

The membranes can be manufactured as

- Flat sheets
- Hollow fibers

2.2.2.1 Symmetric and Asymmetric Membranes

Note that symmetric membranes are used mainly in energy applications such as in battery separator, fuel cell, water electrolysis, and so on. In symmetric and isotropic structure, the transport properties and the structure are identical over the entire cross-section of the membrane and its flux is directly proportional to the membrane thickness (see Figure 2.2). The thickness of symmetric membranes is typically between 20 and 200 μm. An asymmetric membrane consists of a 0.1–5 μm thick skin layer on a highly porous 100–300 μm thick substructure. The skin layer is the actual selective barrier of the asymmetric membrane. The nature of the material and/or the size of pores in the skin-layer determines the separation characteristics. The porous sublayer is only a support for the thin and fragile skin and has little effect on the separation characteristics or the mass transfer rate of the membrane. Asymmetric membranes are used primarily in pressure-driven membrane processes such as gas separation and reverse osmosis or nanofiltration. High flux and good mechanical properties are the advantages of asymmetric membranes. To prepare

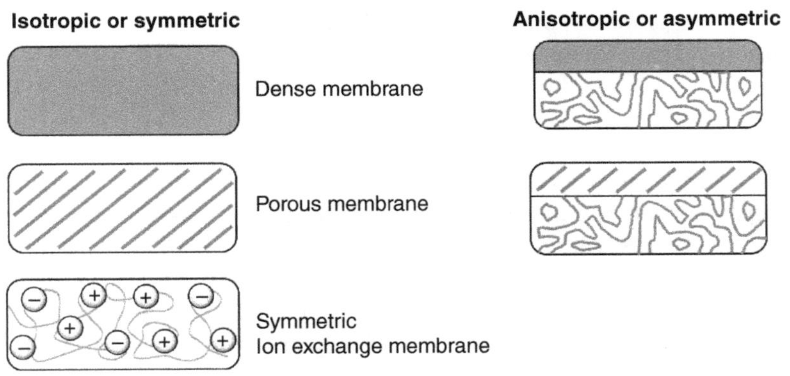

Figure 2.2 The structure of the various membranes.

Table 2.1 Categories of membranes.

Type	Pore size (nm)
Macroporous	>50
Mesoporous	2–50
Microporous	<2
Dense	—

Source: Adapted from Koros et al. [1].

the asymmetric membranes, two techniques are typically used. One is based on a phase inversion process that leads to the integral skin with asymmetric structure and porous substructure. The other is a composite membrane structure where a thin layer is cast on a porous sublayer in a two-step process. In this case, skin layer and support structures are typically made from different materials.

2.2.2.2 Porous Membranes

A porous structure is a very simple form of a membrane, which resembles the conventional filter. These membranes consist of a solid matrix with defined pores with diameters ranging from less than 1 nm to more than 10 μm. The size of the species to be separated is the key to determine the pore size of the membrane to be used and the related membrane process. As listed in Table 2.1, porous membranes with average pore diameters larger than 50 nm are classified as macroporous, and those with average pore diameters in the intermediate range between 2 and 50 nm are classified as mesoporous. Membranes with average pore diameter between 0.2 and 2 nm are classified as microporous. Some of the membranes used in energy application such as in battery separator are microporous.

Porous membranes can be made from various materials such as polymers, ceramics, graphite, and metal or metal oxides. Their structure may be symmetric or asymmetric depending on the application. For example, the structure of a typical battery separator is porous, whereas that of a gas separation membrane is asymmetric with a skin layer with a thickness of 100 nm. The techniques for the preparation of porous membranes can vary and include phase inversion method and sol–gel conversion techniques.

2.2.2.3 Homogeneous Dense Membranes

A homogeneous membrane is a dense membrane through which a mixture of molecules is transported by a pressure, a concentration, or an electrical potential gradient. Separation of various components of a mixture is directly dependent on their transport rates within the membrane, which is determined by their diffusivities and concentrations in the membrane matrix and the driving forces acting on the individual components. Therefore, dense membranes are often called as solution–diffusion type membranes if they are used in gas separation. Dense membranes are typically made from various materials such as polymers, metals, ceramics, and their hybrid. Dense membranes may carry positive or negative electrical charges or functional groups. Since the mass transport in dense membranes is based on diffusion, their permeability is rather low. They are usually used to separate components with similar size. Gas separation is a typical membrane process using dense asymmetric membranes with small thickness. Dense asymmetric structure on the porous substructure provides high fluxes of small gas component.

2.2.2.4 Ion Exchange Membranes

Ion exchange membranes carries positive or negative electrical charges or functional groups in dense structures. Two different types of ion exchange membranes are present: (i) anion-exchange membranes which carry positively charged groups fixed to the polymer matrix and (ii) cation exchange membranes that contain negatively charged groups fixed to the polymer matrix. In a cation exchange membrane, the fixed anions are in electrical equilibrium with mobile cations in the interstices of polymer. In contrast, the mobile anions are more or less completely excluded from the cation exchange membrane due to the electrical charge, which is the same as that of the fixed ions. This exclusion of the anions means that a cation exchange membrane only permits transfer of cations. Anion exchange membranes carry positively charges fixed on the polymer matrix. Therefore, they exclude all cations and are permeable only to anions. Polymeric ion exchange membranes are mainly used in fuel cell and water electrolysis.

2.2.2.5 Membrane Shapes

Industrial membrane process requires membranes to be in a certain geometrical shape. The most suitable membrane geometry in each separation process depends on the application and is determined by the performance as well as by manufacturing and operating costs. The most used membrane geometries in energy application are illustrated in Figure 2.3.

Flat sheet membranes which are 0.5–2 m wide with any length can be in modules which are used in membrane separation processes. Their structure can be symmetric or asymmetric with the actual separation barrier on the surface of the membrane. Flat sheet membranes are laminated to a porous polyester nonwoven fiber for additional mechanical strength. In laboratory and medical applications, small-scale membrane modules are often used without any support structure.

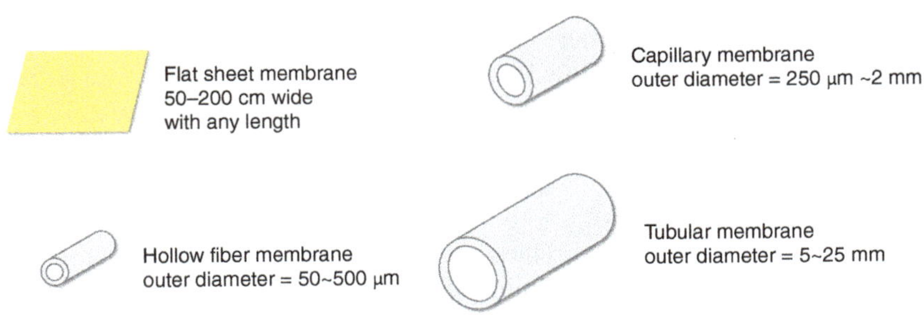

Figure 2.3 The shapes of membranes.

Hollow fiber membranes have an outer diameter of 0.05–0.5 mm and any length and a wall thickness of 0.01–0.1 mm. Hollow fiber membranes have an asymmetric structure with the selective barrier layer. Skin layer may be on the outside or inside of hollow fibers depending on the applications. Hollow fiber membranes have the advantage of providing an extremely large membrane area per unit module volume and their production costs are relatively low.

2.2.3 Mass Transport in Membranes

Membrane separation processes are resulted from the differences in the fluxes of individual chemical species through the membrane. The transport rate is determined by the driving forces or forces applied on the various components and their mobility and concentration in the membrane. In symmetrical ion exchange membranes, the driving forces are differences in electrochemical potentials of the different components in the two phases separated by the membrane. The synthetic membrane itself does not provide any energy for the transport of mass or energy. This is the difference between biological membrane and synthetic membrane. Biological membrane participates actively in the transport of various components providing required energy by a chemical reaction within the membrane structure. Therefore, biological membranes are capable of active transport whereas synthetic membranes are only capable of passive transport.

The size and the physical structure of the membrane material determine the mobility of a component in the membrane, while the chemical compatibility between the permeating component and the membrane material determine the concentration of the solute in the membrane. In the membrane separation processes, different forms of mass transport occurs. The most general form of mass transport through a membrane is illustrated in Figure 2.4. The membrane separates two phases and acts as barrier that reduces the flux of various components according to their chemical nature, size, and their electrical charges. The driving force for the transport of a component A across the membrane is a difference in the Gibb's free energy of component A (ΔG_A) in the two phases separated by the membrane, or the difference in the electrochemical potential of the component in the two phases.

The mass transport in a dense and a porous membrane is shown in Figure 2.5a,b. Figure 2.5a illustrates the transport of two components of the same or similar size, such as two small gas molecules, by diffusion through a dense membrane. One component permeates through the membrane under a driving force of a chemical potential while the other is retained. The transport of a component in a dense membrane is determined by two terms, that is, the mobility and the concentration of the component in the membrane matrix. The mobility of a component is a kinetic factor which is inversely proportional to its size whereas the concentration is a thermodynamic factor which is directly proportional to its solubility in the membrane material. When two components have the same size, their mobility is the same and a difference in the membrane permeability of the two components is the result of a difference in their solubility in the membrane material. Thus, two

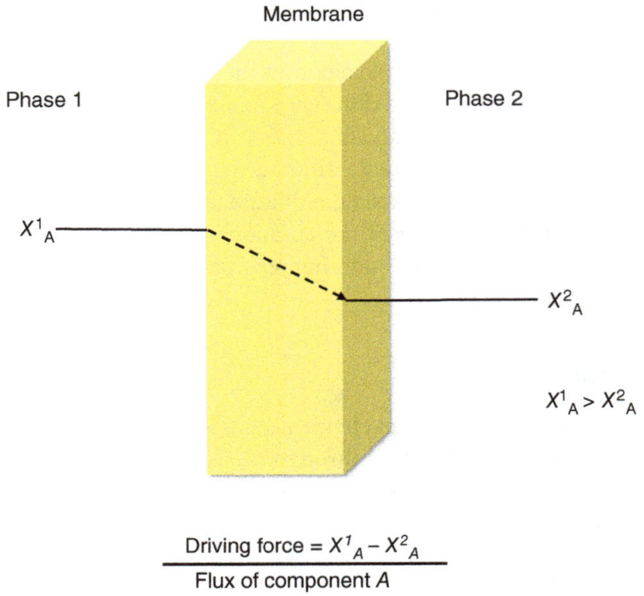

Figure 2.4 The mass transport through a synthetic membrane. X^1_A and X^2_A represent the Gibbs free energy levels of a component A in phase 1 and phase 2.

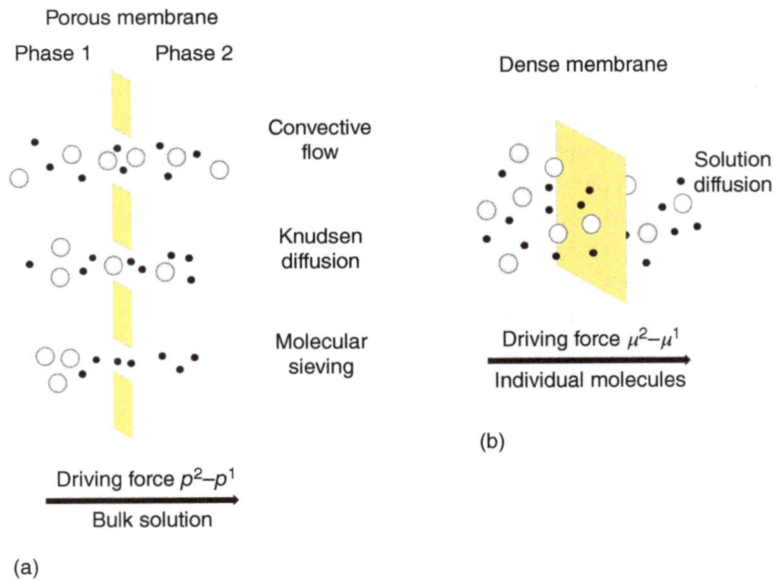

Figure 2.5 The mass flux through porous membranes and a dense membrane. (a) The viscous flow of two components at different size through porous membrane with a different pore size. (b) The diffusive flux of two components of identical size but different solubility a dense membrane.

components with the same size can be separated by a dense membrane if they have different solubility in the membrane matrix.

Figure 2.5b shows the flux of a solution through a porous membrane. In this case, the flux is based on the viscous flow due to the difference in a hydrostatic pressure across the membrane, and only particles that are smaller in diameter than the pores can permeate through the membrane while all other particles are retained. Therefore, the separation of a mixture in a porous membrane is determined by the pore size of the membrane and the diameter of the components. The flux of the porous membranes is several orders of magnitude higher than that of the dense membrane of identical membrane thickness.

2.2.4 Separation Properties

The perm-selectivity of a membrane for different chemical components in a mixture is determined by the fluxes of these components through the membrane. The flux of a component through the membrane is proportional to a permeability coefficient and a driving force and is inversely proportional to the membrane thickness. The flux through a membrane is given by:

$$J_i = \frac{P_i}{\Delta l} X_i \tag{2.1}$$

where J is the transmembrane flux, P is the permeability coefficient of the membrane, X is the driving force, and Δl is the membrane thickness and the subscript i refers to a component in the mixture.

Membrane separates two phases which may consist of two liquids or two gases or a liquid and a gas, as illustrated in Figure 2.6. Assuming that one phase, which is referred to as feed, contains a mixture of two components A and B and assuming also that the flux of the component A across the membrane is much higher than that of component B, a separation of the two components is achieved through the membrane, and a permeate enriched in the component A and a retentate enriched in the component B are obtained.

The separation capability of a membrane can be expressed by the perm-selectivity, the rejection, or the separation factor.

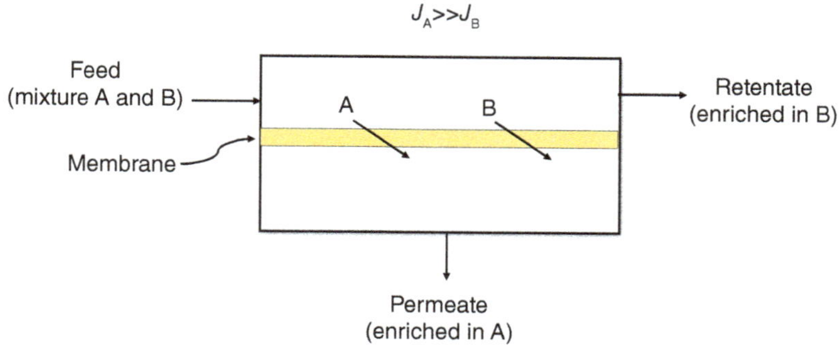

Figure 2.6 The flux and the separation of the components A and B in a membrane process.

Considering a two-component mixture A and B, then membrane perm-selectivity is given:

$$S_{A/B} = \frac{P_A}{P_B} \tag{2.2}$$

where $S_{A/B}$ is the perm-selectivity of a membrane toward components A and B, and P_A and P_B are the permeability of the membrane for the components A and B.

The separation capability of a membrane process is usually expressed by a separation factor α. For a mixture composed of components A and B, the separation factor is given by:

$$\alpha_{A/B} = \frac{X_A^p X_B^f}{X_A^f X_B^p} \tag{2.3}$$

where α is the separation factor, X is the mole fraction, the superscript p and f refer to permeate and feed, respectively, and the subscripts A and B refer to components in a mixture. The separation factor α can be expressed by mole fraction (mol mol^{-1}), volume fraction (m^3 m^{-3}), or mass fraction (kg kg^{-1}).

The perm-selectivity and the separation factor are dimensionless parameters which vary between 1 and ∞. Thus, if the separation factor $\alpha_{A/B}$ is >1, the permeation rate of component A is higher than the one of component B. If $\alpha_{A/B} = 1$, the membrane does not have any selectivity and no separation occurs. If $\alpha_{A/B}$ approaches ∞, only the component A will permeate the membrane while component B is completely retained. The same is true for the perm-selectivity of a membrane.

For dilute solutions containing a solvent and a solute, where the solute is partly or completely retained and the solvent passes through the membrane, it is convenient to express the selectivity in terms of the rejection of the membrane for a component. The rejection of a membrane is given by:

$$R_A = \left(1 - \frac{C_A^p}{C_A^f}\right) \tag{2.4}$$

where R_A is the rejection of the component A by the membrane, C_A^p and C_A^f are the concentrations of the component A in the permeate and the feed solution, respectively. The rejection is also a dimensionless parameter. It varies between zero for a membrane with no perm-selectivity, and one for an ideal semipermeable membrane.

2.3 Membrane Materials

As can be seen in Figure 2.7, membrane technology consists of (i) development of membrane materials, (ii) fabrication of membranes and modules, and (iii) membrane process design and plant engineering. In this section, we will deal with the membrane materials followed by the module fabrication.

The choice of a suitable material and preparation method depends on the application of the membrane. For example, membrane material for gas separation act as a barrier layer and is of significant importance for the performance of the membrane.

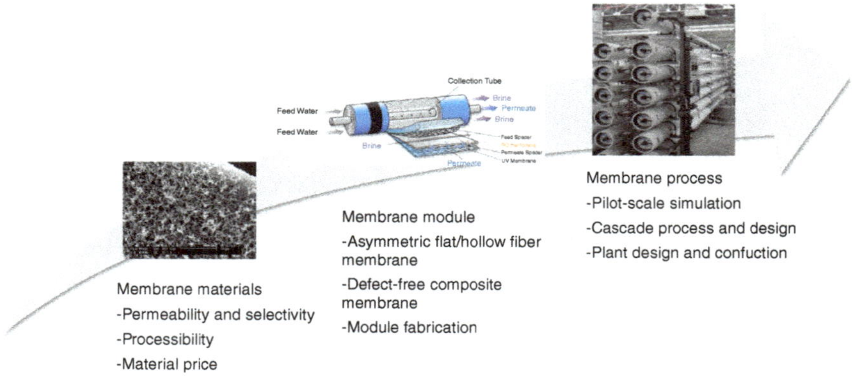

Figure 2.7 Membrane development consist of (i) development of membrane materials, (ii) fabrication of membranes and modules, and (iii) membrane process design and plant engineering. Source: Warsinger et al. [2]/with permission from Elsevier.

In other application such as fuel cell and water electrolysis, ion conductivity, and mechanical properties are importance for the performance of the membrane. However, the most important property of a membrane is its permeation selectivity and its transport rate. The permeability and the selectivity are determined by the intrinsic chemical properties of the membrane material and its physical structure such as the glass transition temperature, hydrophobic and hydrophilic nature of the basic material, or pore size and pore size distribution of the membrane.

Most of the commercially available membranes are made of polymers. Metals and ceramics or glasses are used for special applications such as the production of very pure hydrogen or the separation of isotopes at elevated temperature.

2.3.1 Polymer Materials

Polymers are defined as high-molecular-weight materials built from many low-molecular-weight compounds (or monomers) by addition or condensation reaction. The molecular weight and the chemical structure of the building blocks that are linked by a covalent bond determine the chemical and mechanical properties of the polymer. The interaction of the polymer chain also determines the sorption of low-molecular-weight organic or inorganic components such as water, salts, gases, or organic solvents or vapors. Crosslinked polymers are linked by the branches between building blocks and are liked by a covalent bond. Crosslinked polymers are generally not soluble in organic solvents. Some polymers are linked by a secondary bonding force such as hydrogen bonding forces and dipole–dipole interaction or dispersion forces. Covalent bonding forces range from 300 to 600 kJ mol^{-1} while the secondary bonding forces are much lower than covalent bonding forces, ranging from 10 to 40 kJ mol^{-1}. However, the presence of the secondary bonding forces influences the molecular interactions between or within the polymer chains and usually make the polymer much stronger. For example, polyamide has hydrogen bonds between amide links and make the membrane mechanically strong. It is soluble in organic solvents.

2.3.1.1 Physical State and Properties of Polymer

Polymers exist in a solid and a liquid phase. Solid polymer exists in crystalline, semicrystalline, amorphous, or rubbery state. The temperature at which glassy state of polymer becomes rubbery is called the glass transition temperature (T_g). Further increase of the temperature leads to a point where the polymer becomes liquid. The temperature at this point is referred to as the melting temperature (T_m). At both T_g and T_m, the physical properties of a polymer change drastically. As most of the membrane materials are useful in the solid state, the T_m of a polymer is regarded as the maximum temperature of operation in a practical application. The T_g is the temperature where the mechanical properties of a polymer change drastically. At T_g, free volume of a polymer also changes. The free volume exists in any polymer materials and influences the permeability of small molecules such as gases. The T_g is regarded as an important factor for the selection of a polymer for the preparation of membranes because at T_g, the mechanical properties of a polymer change drastically.

Free volume (V_f) is defined as the unoccupied volume in the form of holes in a polymer in contrast to the term, occupied volume (V_o), which is the volume occupied by the polymer molecules. As temperature increases, the polymer expands and both the occupied volume and the free volume increase. In the glassy state, both the free volume and the occupied volume increase linearly with the temperature according to the expansion coefficient.

$$V(T) = V_o(T) + V_f \tag{2.5}$$

Below T_g,

$$V(T < T_g) = V_o(T = 0) + V_f(T = 0) + \left(\frac{dV_o}{dT}\right)_g T \tag{2.6}$$

Above the T_g,

$$V(T > T_g) = V(T_g) + \left(\frac{d(V_o + V_f)}{dT}\right)_g (T - T_g) \tag{2.7}$$

where V, V_o, and V_f are the volume, the volume occupied by the polymer, and the free volume, respectively, T is the temperature, and the subscript g refers to the glass state.

Figure 2.8 illustrates the volume change of a polymer as a function of temperature at temperatures below and above the T_g. Table 2.2 lists the type of polymers – rubbery or glassy.

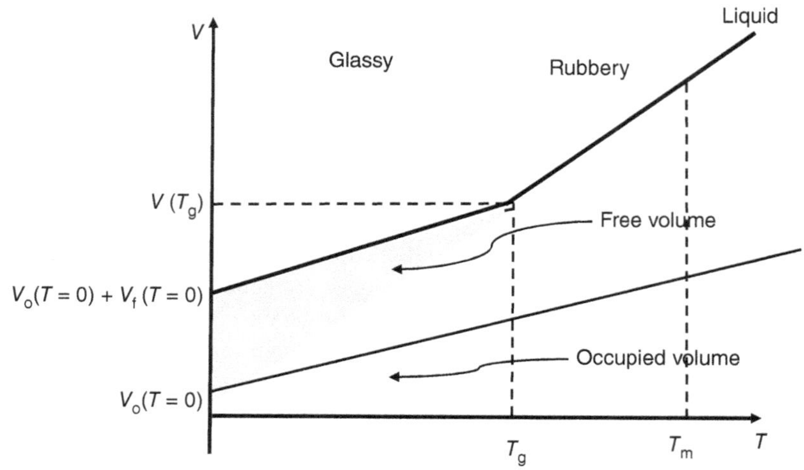

Figure 2.8 The change of the free volume in a polymer.

Table 2.2 List of rubbery polymers and glassy polymers.

	Rubbery polymers	**Glassy polymers**
Polymers	Fluoro elastomers Neoprene Nitrile rubber Poly(dimethylsiloxane) Polyethylene chlorine sulfonated Polyvinilchloride plasticized Polyuretane	Aromatic polyamides Cellulose acetate Perfluorodioxoles Polyacetilenes Polycarbonates Polyimides Poly(phenileneoxide) Polysulfone Poly(1-trimethysilyl-1-propyne)
Properties	High permeability Low selectivity	Moderate permeability High selectivity

2.3.2 Inorganic Materials

Inorganic membranes are based on metal or metal-oxide (or ceramic). Ceramic-based membranes are made by inorganic materials, such as Al_2O_3, TiO_2, and ZrO_2. Compared with the organic membranes, ceramic membranes perform better in harsh environments, e.g. at temperatures above 200 °C or during solvent separation, due to the superior thermal, chemical, and mechanical stabilities of ceramics.

Preparation of ceramic membranes commonly comprises of three essential steps: (i) formation of suspension/slurry/sol containing the target ceramics, (ii) formation of the membrane precursor from the suspension/slurry/sol; and (iii) consolidation of the particles in the precursor at elevated temperature. The procedure for synthesizing composite membranes with a pore-size gradient over multiple layers can be

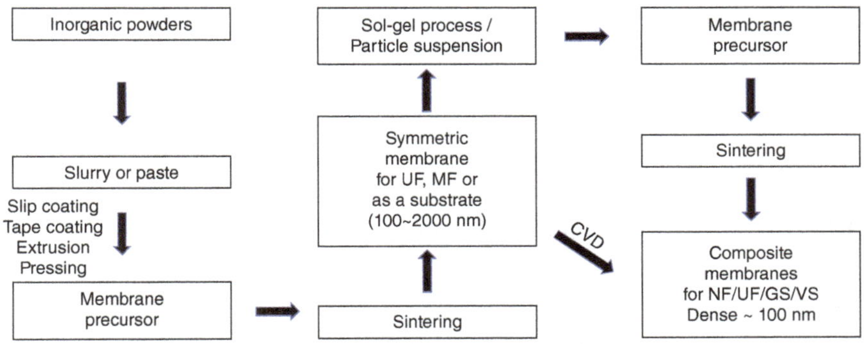

Figure 2.9 A flow chart illustrating preparation processes of inorganic membranes with different pores and their application.

more complicated. Figure 2.9 is a flow chart describing the sophisticated processes for preparing ceramic membranes with different selective pore sizes [3]. The process described in left is generally used for macro/mesoporous ceramic membranes for microfiltration (MF) or supports. The process described in the middle is usually used for mesoporous ceramic membranes for ultrafiltration (UF), nanofiltration (NF), or supports. Composite membranes with ultra-thin dense/microporous/mesoporous layers on the top of supports prepared using the aforementioned processes.

2.3.2.1 Preparation of Ceramic Membranes

Generally, ceramic membranes for MD processes are mainly multilayer structures (composite) consisting of a supporting layer (porous substrate), intermediate layer, and a functional top layer (selective layer). Figure 2.10 illustrates the cross-sectional structure of a composite ceramic membrane made from different materials [5]. The major raw materials for composite ceramic membranes include alumina (Al_2O_3),

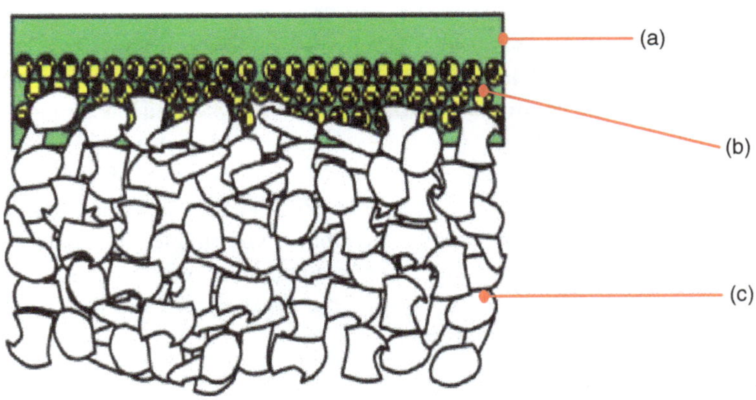

Figure 2.10 Schematic diagram of a composite ceramic membrane. (a) Top layer (selective layer). (b) Intermediate layer. (c) Supporting layer (porous substrate). Source: Reproduced from Tseng et al. [4]/with permission from Elsevier.

silica (SiO_2), zirconia (ZrO), titania (TiO_2), or a combination of these oxides [6]. Recently, many studies have focused on reducing the preparation cost by using low-cost materials, such as clay, fly ash, chamotte, kaolin, and quartz sand.

2.4 Basic Principles of Membrane Preparation

2.4.1 Thermodynamics of Phase Separation

Phase separation (or inversion) is a demixing process where a polymer dope solution is first phase separated into polymer-rich and polymer-lean phases. Starting from a homogeneous mixture, a change in the composition or conditions induces the phase separation into the polymer-rich and polymer-lean phases. All phase-separation processes can be described using the same thermodynamic principles, where a thermodynamically stable polymer solution is exposed to an unstable environment and ultimately precipitates to form a membrane. Hence, the compatibility between the polymer, solvent, and nonsolvent, or the thermodynamic stability, is a critical factor for the phase-separation processes. The thermodynamic state of a system is described by the Gibbs free energy, as follows:

$$G = H - TS \tag{2.8}$$

where G is the Gibbs free energy, H is the enthalpy, S is the entropy, and T is the temperature. The partial Gibbs free energy at constant temperature and pressure is referred to as the chemical potential of a component and is given by:

$$\left(\frac{\partial G}{\partial n_i}\right)_{p,T,nj} = \mu_i \tag{2.9}$$

where μ_i is the chemical potential of the component i, n is the number of moles in the mixture, p is pressure, and the subscripts i and j refer to the components of the mixture. The Gibbs free energy of a two-component mixture that is not completely miscible over the entire compositional range does not decrease over this range. There are three such compositional ranges for a two-component mixture. (i) A range where the two components are completely miscible and the Gibbs free energy is negative; (ii) A range where the Gibbs free energy of mixing is positive and the composition of the mixture. The mixture is unstable and decomposes spontaneously into two stable phases which are characterized by two minimized Gibbs free energy values; (iii) A region where the Gibbs free energy of mixing is negative. In this region, the composition is stable. However, on the microscopic scale, the composition of the mixture starts to precipitate with nucleation. Further information regarding the phase separation of a two-component mixture within the miscibility gap can be obtained from the first and second derivatives of the Gibbs free energy as a function of the composition:

$$\left(\frac{\mathrm{d}\Delta G_m}{\mathrm{d}X_i}\right)_{T,p,j} = \Delta\mu_i \tag{2.10}$$

$$\left(\frac{\mathrm{d}^2\Delta G_m}{\mathrm{d}X_i^2}\right)_{T,p,j} = \frac{\mathrm{d}\Delta\mu_i}{\mathrm{d}X_i} \tag{2.11}$$

For a stable mixture, the Gibbs free energy of mixing is negative and the second derivative is positive:

$$\Delta G_m < 0 \quad \text{and} \quad \frac{d^2\Delta G_m}{dX^2} > 0 \tag{2.12}$$

In the unstable region, the Gibbs free energy of mixing is positive and the second derivative is negative:

$$\Delta G_m > 0 \quad \text{and} \quad \frac{d^2\Delta G_m}{dX^2} < 0 \tag{2.13}$$

For a composition on the spinodal, the first derivative of Gibbs free energy of mixing is a maximum and the second derivative is zero:

$$\frac{d^2\Delta G_\text{m}}{dX^2} = 0 \tag{2.14}$$

For a composition on the binodal, the Gibbs free energy of mixing is a minimum and the first derivative is zero:

$$\frac{d\Delta G_\text{m}}{dX} = 0 \tag{2.15}$$

Any composition in the unstable region, which is enveloped by the spinodal decomposition, will spontaneously demix. Any composition in the metastable region, which is located between the binodal decomposition and the spinodal decomposition, will be stable and demixing can only be achieved by nucleation and growth.

Figure 2.11 illustrates the phase separation and precipitation in three component system – polymer, solvent, and nonsolvent. To obtain a porous structure,

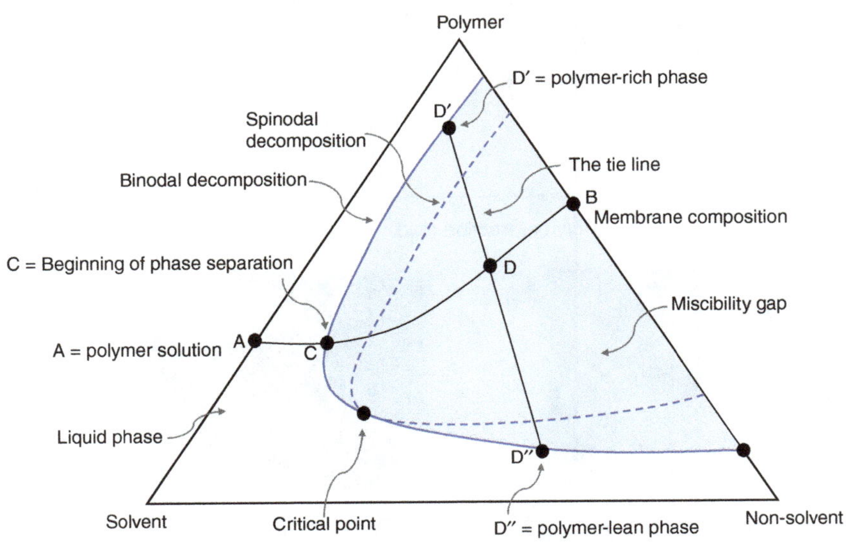

Figure 2.11 Three-component phase diagram showing the formation of a porous membrane by addition of a nonsolvent to a polymer solution.

nonsolvent-induced phase separation (NIPS) and thermally induced phase separation (TIPS) are widely used membrane preparation methods. During NIPS process, the casting solution (or dope solution) is immersed in a coagulation bath containing nonsolvent, resulting in solvent–nonsolvent exchange.

Immersing the polymer solution in a nonsolvent is one of the major strategies used to induce phase separation. Phase separation can be achieved using a polymer solution containing three components (i.e. polymer, solvent, and nonsolvent, which are completely miscible over a certain compositional range) and has a miscibility gap over another compositional range. The polymer solution is composed of a polymer dissolved in a good solvent, generally an organic solvent, and a poor solvent, generally water. As illustrated in Figure 2.12, the three-component mixture

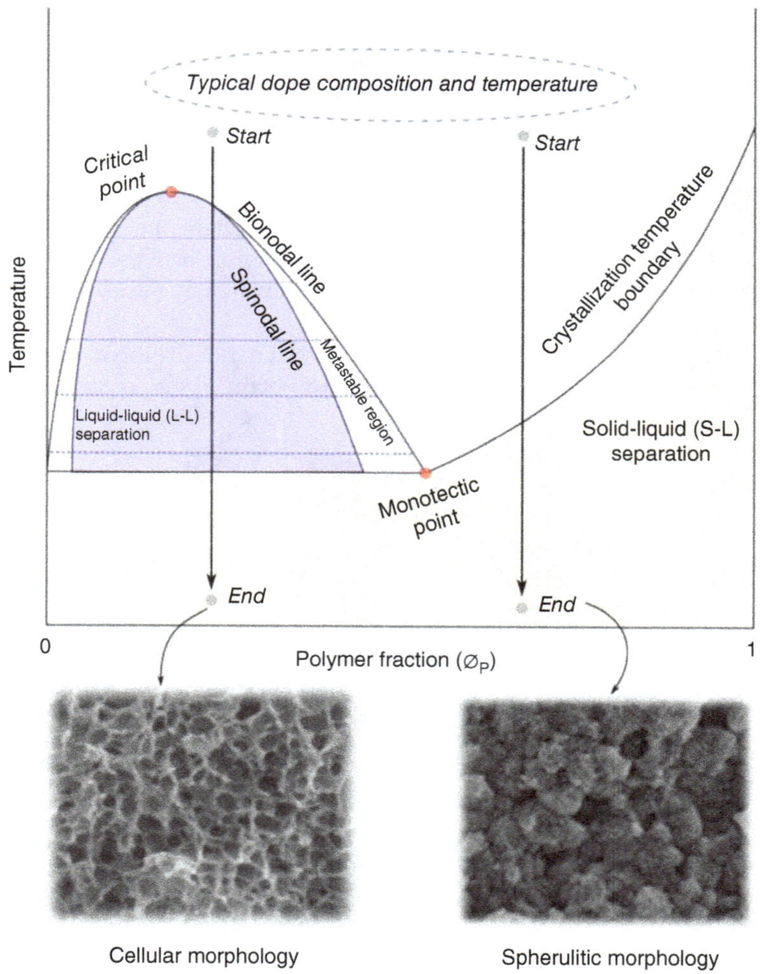

Figure 2.12 Characteristic phase diagram of thermally induced phase separation. Source: Kim et al. [7]/John Wiley & Sons, Inc.

is completely miscible over a certain compositional range, but exhibits a miscibility gap over another range. The corners of the triangle represent the pure components. Boundary lines between any two corners represent mixtures of two components, and any point within the triangle represents a mixture of all three components. If a nonsolvent is added to a polymer solution, the composition is indicated by point A on the boundary lines, i.e. solvent–polymer lines. If the solvent is exchanged to the nonsolvent from the polymer solution, the composition of the mixture will change following the line A–B. At point C, the composition of the system will reach the miscibility gap and the polymer solution will begin to precipitate. At a certain composition of the three-component mixture, the polymer-rich phase is considered a solid, represented by point D in Figure 2.11. The polymer-rich phase indicated by point D′ and the polymer-lean phase indicated by point D″ are connected to point D by the tie line.

2.4.2 Nonsolvent-induced Phase Separation

The basic procedure of the NIPS method is as follows:

i. Dissolve the polymer solution, which is composed of at least one polymer and at least one good solvent, and may contain additives. The polymer solution is mixed to obtain a homogeneous mixture.
ii. Cast the dope solution into the desired shape, e.g. flat sheet or hollow fiber.
iii. Immerse the cast solution in a nonsolvent coagulation bath (mainly water) to induce phase separation and precipitation of the polymer.
iv. Extract the solvent and additives, often via solvent extraction to yield a membrane.

As described above, the membrane is prepared by exchange of solvent and nonsolvent; hence, the solvent–nonsolvent system must be miscible. The key parameters influencing the formation of phase-inversion membranes are the selection of solvent–nonsolvent system, composition of the polymer solution, composition of the coagulation bath, and film casting or hollow fiber spinning conditions. Particularly, Strathmann et al. reported two fundamentally different structures depending on the rate of polymer precipitation by NIPS [8]. The precipitation rate is measured as the time between immersing the casting solution into a coagulation bath and the solution turning opaque or the membrane separating from the glass plate. They explained that in general, low precipitation rates produced membranes with "sponge-like" morphologies. Such morphologies tended to form a dense top layer. High precipitation rates generally produced membranes with large "finger-like" macrovoids in the membrane structure. Such morphologies usually exhibited a highly porous structure and thin skin layers. Therefore, membranes with finger-like morphology usually have high water flux; however, membrane failure can occur due to their low mechanical strength [9, 10].

As mentioned above, several parameters influence the diffusion and demixing processes and hence, the final membrane morphology. Several representative parameters are introduced below. One of the main variables in phase-inversion membrane formation is the choice of the solvent–nonsolvent system. The polymer should be soluble or easily dispersed in the solvent, and the solvent and nonsolvent must be miscible. Water is frequently used as a nonsolvent, but other nonsolvents can also be used. Higher mutual affinity between solvent and nonsolvent usually results in instantaneous demixing and more porous membranes [11].

2.4.2.1 Type of Polymer

The type of polymer affects the sorption of solutes, membrane hydrophilicity, and thermal and chemical stability of the membrane. Many different polymers have been used for energy applications. Hydrocarbon polymers such as polysulfone (PSf), polyethersulfone (PES), polyacrylonitrile (PAN), and polyimide (PI) are widely fabricated due to their good solubility in many organic solvents. Fluorocarbon polymers such as PVDF and poly(vinylidene fluoride-hexafluoropropylene) (PVDF-HFP) are widely used for membrane distillation because of their intrinsic hydrophobicity. PTFE has the highest hydrophobicity among the fluorocarbon polymers; however, PTFE is not prepared by phase inversion because of its poor solubility in organic solvents. PTFE and PVDF can be used as binders for energy applications such as battery separators and fuel cells.

2.4.2.2 Polymer Concentration

The polymer concentration of the dope solution is another parameter determining the membrane morphology. Typical polymer concentrations range from 15 to 30 wt% [12]. Increasing the polymer concentration in the casting solution leads to a much higher polymer concentration at the nonsolvent interface, implying that the volume fraction of polymer matrix increases at the nonsolvent interface; hence, a lower porosity is obtained [13]. Increasing the polymer concentration can result in a reduction of the pure water flux to zero, although demixing occurs instantaneously [13]. Strathmann et al. reported the effect of polymer concentration on membrane morphology [8]. They demonstrated that the initial polymer concentration determined the path to precipitation in the ternary phase diagram.

2.4.2.3 Additives

Additives in the casting solution have a considerable effect on membrane morphology. Most casting solutions contain additives for improving the morphology and function of phase-inversion membranes, which makes analysis of solvent–nonsolvent systems more complicated. The addition of these additives can form "sponge-like" membrane structures by suppressing the "finger-like" macrovoid formation, which enhances pore formation and improves pore interconnectivity. However, additives in casting solutions generally increase hydrophilicity. Typical additives include polyvinylpyrrolidone (PVP), poly(ethylene glycol) (PEG), poly (methy methacrylate) (PMMA), propionic acid (PA), alcohols, water, polyethylene oxide (PEO), and LiCl.

2.4.2.4 Casting Temperature

The casting temperature is also an important parameter affecting the membrane morphology. As the casting temperature increases, the viscosity of the casting solution decreases. The casting solution viscosity affects the exchange rate of the solvent and nonsolvent during phase separation. Zheng et al. reported that increasing the temperature of the casting solution decreased its viscosity, inducing miscibility between the solvent and nonsolvent [14, 15]. They explained the details of the kinetic parameters of membrane formation, and described a relationship between the membrane thickness and coagulation time. As the casting temperature increased, the kinetic parameter of membrane formation increased. Tsai et al. reported that the coagulation bath temperature had tremendous influence on the membrane morphology [16]. They demonstrated that the temperature of the coagulation bath affected the viscosity of the fluid in the bath and hence, the size of the formed macrovoids.

As summarized above, there are many relevant parameters when fabricating membranes using the NIPS method. To obtain the optimized membrane morphology for energy applications, it is possible to predict the factors that favor the formation of membranes. After preparing the phase-inverted membrane, surface modification needs to be conducted to enhance the membrane hydrophilicity or hydrophobicity.

2.4.3 Thermally Induced Phase Separation

In addition, according to Eq. (2.8), increasing the temperature decreases the Gibbs free energy of mixing as the entropy of mixing is always positive. Once a polymer solution is dissolved at high temperature, a homogeneous and thermodynamically stable dope solution is formed. If a polymer solution is cooled at a controlled rate, it follows the path shown by the arrows in Figure 2.12 [7]. This phase diagram can be described by temperature and polymer concentration as shown in Figure 2.13. At high polymer concentrations, the membrane passes through the crystallization temperature boundary, and phase separation proceeds via solid–liquid (S–L) separation where the polymer crystallizes directly from the dope solution. Conversely, at lower polymer concentrations, the polymer solution enters the metastable region, where liquid–liquid (L–L) demixing occurs. In the L–L region, the polymer solution separates into a polymer-lean phase and polymer-rich phase along the tie line. The polymer-rich phase forms a continuous polymer phase, which forms the membrane, and the polymer-lean phase forms membrane pores.

During TIPS, phase separation of the polymer solution is controlled by the temperature. Owing to changes in the solubility of the polymer with temperature, the polymer solution can be precipitated at low temperature, while it is homogeneously mixed at high temperature. TIPS membranes usually have advantages for a certain application such as membrane distillation, such as a low tendency to form defects, high porosity, and the ability to form interesting microstructures with narrow pore size distributions [7, 11, 17].

The basic procedure for TIPS is as follows:

i. Dissolve a polymer in a high-boiling point, low-molecular weight solvent at an elevated temperature, typically near or above the melting point of the polymer to obtain a homogeneous mixture.

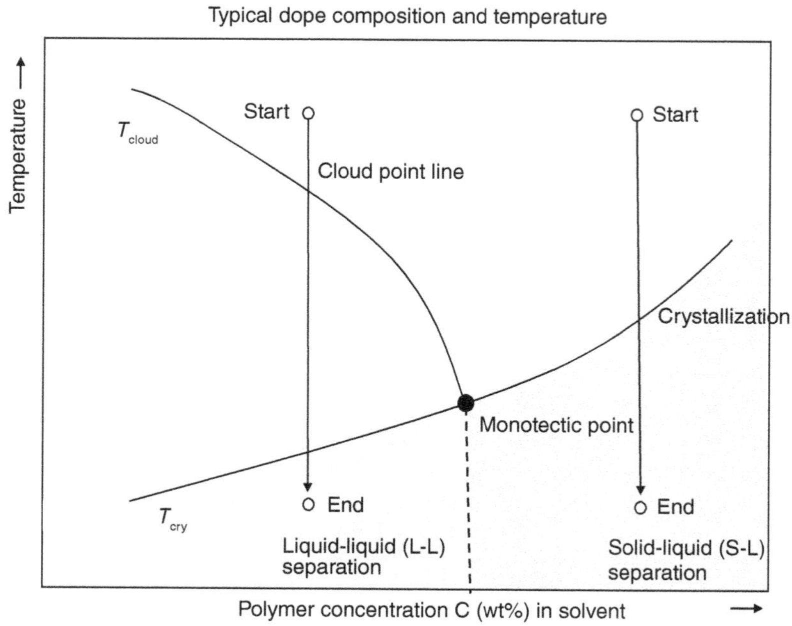

Figure 2.13 Phase diagram of thermally induced phase separation determined by cloud point (T_{cloud}) and crystallization temperature (T_{cry}).

ii. Cast the dope solution into the desired shape (flat sheet or hollow fiber).
iii. Cool the casting solution in a controlled procedure to induce phase separation and precipitation of the polymer.
iv. Extract the diluent, often *via* solvent extraction, to yield a membrane.

As described, the membrane is fabricated from a homogeneous dope solution by cooling to induce phase separation. The phase inversion process affects the polymer–solvent interaction, cooling rate, cooling media, and thermal gradient. As the TIPS process usually uses a binary system, it is inherently simpler (fewer variables to control) than the NIPS process, which is a ternary system. One of the distinct advantages of the TIPS process is the ability to fabricate semicrystalline polymers that are not usually soluble in organic solvents. Conversely, TIPS membranes typically have lower mechanical strength, and it is a more energy intensive process, than the NIPS process. For these reasons, the combination of NIPS and TIPS (N-TIPS) has been investigated to overcome the limitations of each process [18]. Similar to NIPS, representative variables for fabricating membranes by TIPS are summarized below.

2.4.3.1 Polymer–Solvent Interaction

The polymer–solvent interaction determines the overall macroscopic structure of the membrane (e.g. cellular, fibrillar, tubular, or spherulitic). A suitable solvent for the TIPS process should have a high boiling point, low molecular weight, low toxicity, low environmental impact, low cost, and be reasonably compatible with the polymer over wide temperature and concentration ranges [19]. Thermal conductivity and miscibility with the quenching media are also important to membrane morphology. Compatibility between the polymer and the solvent determines the binodal curve and polymer crystallization temperature of the mixture. Poor compatibility between the polymer and the solvent leads to the binodal curve being attained at higher temperatures. Membrane morphology is highly dependent on the choice of solvent. For example, Yang et al. prepared the PVDF membrane with diphenyl ketone (DPK) solvent using TIPS [20]; the monotectic point was approximately 30 wt%. The same group fabricated PVDF membranes using TIPS with a different solvent, diphenyl carbonate (DPC) [21].

2.4.3.2 Effect of Cooling Rate

The cooling rate usually determines the size of pores and spherulites, which affects the overall filtration performance and mechanical strength. The cooling rate is a kinetic factor in the overall membrane formation process. The cooling rate of the dope solution is mainly determined by the temperature difference between the dope and the cooling environment (or coagulation bath). According to the nucleation–growth (NG) model, faster cooling rates result in smaller spherulites with smaller pore size and higher mechanical strength. Conversely, slow cooling results in longer times for crystal growth and larger spherulites [22]. In addition, higher cooling rates result in membranes with slightly lower porosity than the slowly cooled membranes.

2.4.3.3 Effect of Additives

The presence of additives or nucleating agents in dope solutions can significantly affect the solution demixing phenomena, and hence, the morphology and performance of the membrane. Additives can act as a pore former that determines the porous structure and enhances the mechanical strength. Similar to NIPS, pore formers such as PVP, glycerol, PEG, PMMA, and LiCl can be used to produce porous membranes. Different spherulite sizes are formed depending on the additives [23]. Recently, inorganic additives such as zirconium dioxide (ZrO_2), aluminum oxide (Al_2O_3), and titanium dioxide (TiO_2) have been studied for TIPS to enhance the hydrophobicity and mechanical strength of the resulting membranes [24–26].

2.5 Membrane Fabrication

2.5.1 Asymmetric Membranes

Asymmetric membranes comprising of a thin dense layer (about 100 nm) and a porous support layer are commonly employed for gas separation and reverse osmosis. These membranes are produced by extrusion (in the form of hollow fiber or flat sheet) of a dope solution of polymer and solvent mixtures through an air gap (to allow for a thin dense layer) into a coagulation bath (usually water) via NIPS process. The polymer concentration is usually in the range of 30–35 wt% with solvent and other additives (such as pore formers) comprising the rest of the solution. To prevent the formation of void and obtain optimum membrane morphology, the polymer dope solution should be very close to the point of phase separation. This can be accomplished with addition of a non-solvent to approach the phase boundary. Polysulfone is one the major polymers employed for these membranes with *N*-methyl pyrrolidone (NMP) being a desired solvent. Other polymers typically used for gas separation asymmetric membranes include polyimides and cellulose acetate. Cellulose acetate was an early choice of asymmetric membranes for reverse osmosis applications and are recently cellulose acetate hollow fiber membranes are available for gas separation from Toyobo.

While the phase inversion process can produce very thin dense layers on a porous support membrane, the ability to produce defect free membranes is not an inherent feature of the process. Figure 2.14a depicts the inherent defects produced in selective layer. The solution of this problem involves surface defect repair methods. The most common method is the coating with a thin layer of a very permeable polymer coating. Silicone rubber coatings are a typical example (see Figure 2.14b). The defects (or pinholes) can drastically reduce the separation factor of the gas mixture due to hydrodynamic flow but coating with a highly permeable coating can eliminate the pin-holes without changing the overall permeability of the membrane

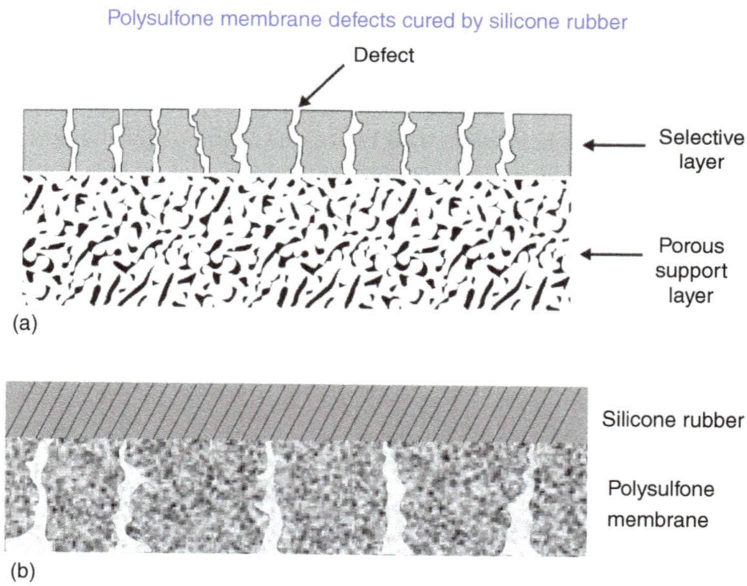

Figure 2.14 (a) Polymer membrane exhibits inherent defeats in selective layer when fabricated by non-solvent-induced phase separation. (b) These defects can be repaired by coating with a very thin layer of a very permeable coating such as silicone coating. Source: Defect repair methods were demonstrated by Henis and Tripodi [27]/American Association for the Advancement of Science – AAAS.

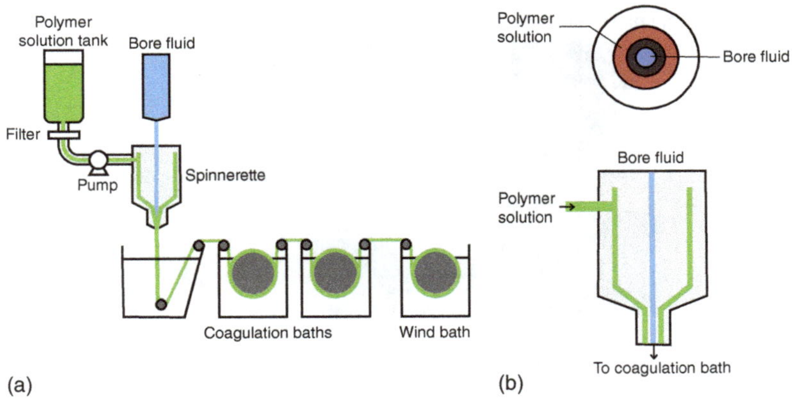

Figure 2.15 Schematic of (a) asymmetric hollow fiber spinning process using (b) dual spinneret via NIPS process. Temperature in coagulation baths may be set at different temperatures.

thus allowing polymer intrinsic separation and permeability values to be achieved. This concept was demonstrated by Henis and Tripodi and predicted to be the result from series-resistance model calculations [27–29].

The process for making an asymmetric hollow fiber using NIPS process is illustrated in Figure 2.15 and depends on the use of dual spinneret or triple spinneret (Figure 2.16).

The hollow fiber spinneret has two annular feed sections depending on the type of spinnerets. In dual spinnerets (Figures 2.15 and 2.17), the outer feed section delivers the polymer solution, and the inner feed section delivers a bore fluid; usually air or an inert gas. Water can also be employed to control the bore side skin layer. In triple spinneret, outer coating solution or second polymer coating solution is fed along with the polymer solution and bore fluid in the mid-section and inner section, respectively. In both cases, an air gap is provided between the exit of the spinneret and the coagulation bath to control the dense layer thickness on the hollow fiber surface. The coagulation bath is usually water but can be any appropriate coagulation liquid media containing solvent. Water is preferred as it is non-flammable, low cost, and environmentally benign. The fiber can be wound on spools and immersed in a water bath to remove residual solvent or can be allowed to wind back and forth in a water reservoir; a process referred to in the membrane field as "piddling."

Polymer surface (outer or inner) can be modified to meet the requirements of membrane application. As can be seen in Figures 2.16 and 2.17, triple spinneret can be utilized to produce porous outer surface when the appropriate solvents are used. Asymmetric hollow fiber can be spun using the triple spinneret to prepare porous surface or to coat the second polymer solution on top of the give polymer. Second polymer can be selective layer for gas separation.

Asymmetric membranes can be prepared via TIPS process using both dual spinneret (Figure 2.18) or triple spinneret (Figure 2.19). Asymmetric hollow fiber membranes prepared by TIPS process are usually stronger than those by NIPS process as thermal process induces crystal growth of polymers.

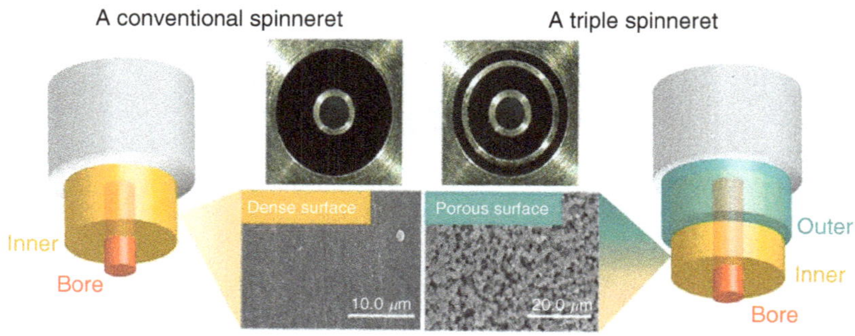

Figure 2.16 Dual spinneret and triple spinneret.

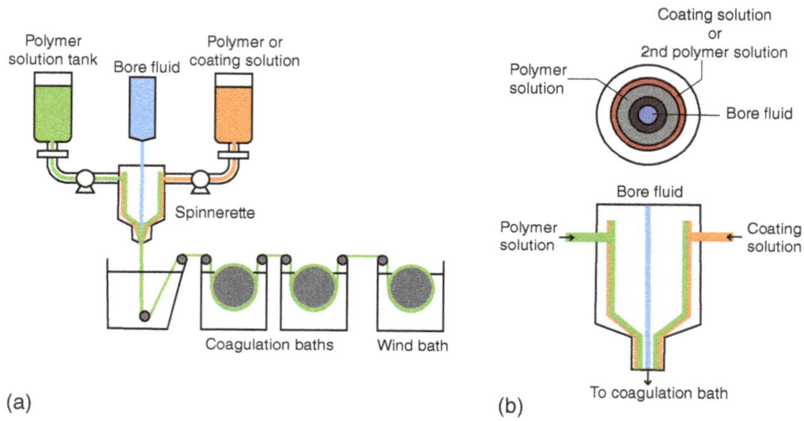

Figure 2.17 Schematic of (a) triple hollow fiber spinning process using (b) triple spinneret via NIPS process. Temperature in coagulation baths may be set at different temperatures.

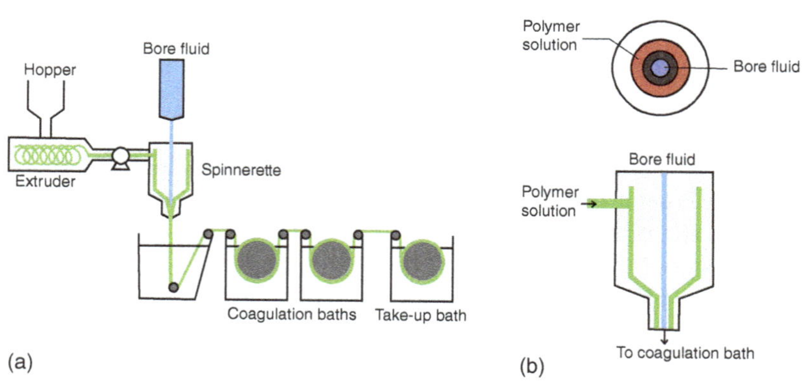

Figure 2.18 Schematic of (a) asymmetric hollow fiber spinning process using (b) dual spinneret via TIPS process. Temperature in coagulation baths may be set at different temperatures.

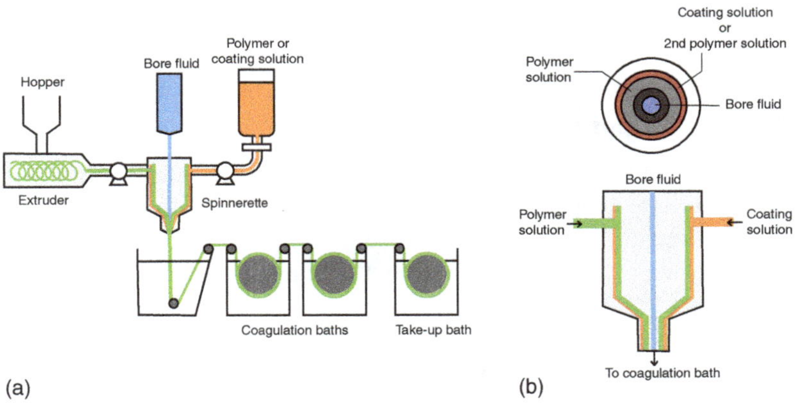

Figure 2.19 Schematic of (a) asymmetric hollow fiber spinning process using (b) triple spinneret via TIPS process. Temperature in coagulation baths may be set at different temperatures.

2.5.2 Flat Sheet Membranes

Typically flat sheet membranes are prepared from casting polymers solution on porous substrate (typically poly(ethylene terephthalate) (PET)) nonwoven fabrics. As seen in Figure 2.20a, continuously the polymer solution (i.e. polysulfone solution) is cast on PET surface and subsequently supplied to coagulation baths

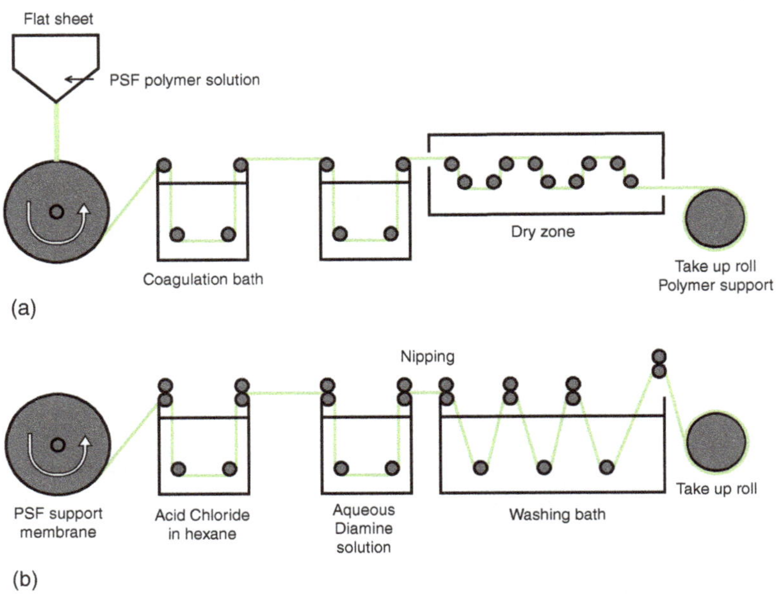

Figure 2.20 Schematic of (a) polymer supports membrane production and (b) thin-film composite membrane for reverse osmosis and nanofiltration.

with different temperatures to manufacture microporous membranes followed by drying and take-up.

2.5.3 Thin-Film Composite Membranes

One version of a thin film composite membrane involves the coating of a very thin layer of a polymer onto a porous support such as a porous polymer membrane or hollow fiber. This construction would be desired if the dense layer is a very expensive specialty polymer or a rubbery material not offering the capability to support pressure. A variation of this method is an asymmetric thin-film composite membrane produced by solution coating a porous membrane and subjecting the thin film coating to a non-solvent shortly after coating to yield an asymmetric layer.

Thin-film composite (TFC) membranes can be made by the interfacial polymerization of a polyamide at the surface of a microporous membrane (Figure 2.20a). This type of membrane is the well-known reverse osmosis membrane for water desalination. The interfacial polymerization of polyamides on a porous polysulfone membrane was initially reported by Cadotte and coworkers [30–32]. This technology was developed at North Star and FilmTec Corporation and later acquired by Dow Chemical Co. These membranes are commercially available as FilmTec™ membranes. To prepare TFC membranes for reverse osmosis and nanofiltration, PET nonwoven fabric is fed into the first bath containing acid solution in hexane followed by the second bath containing aqueous hexamethylene diamine solution as can be seen in Figure 2.20b and excessive washing monomers. Nipping rolls remove excessive monomers as well. For coating the second polymer solution on top of the microporous polymer support, Figure 2.20b can be modified. The initial membrane involved a microporous polysulfone membrane initially saturated in a water solution containing polyethyleneimine [32]. This was reacted with isophthaloyl chloride or toluene diisocyanate in hexane. The concentration of the reactants in their respective solutions was in the range of 0.5–1.0 wt%. The resultant membrane exhibited salt rejections of 99%. Many variations of this process have been reported in the literature with aromatic di- and triamines reacted with aromatic di- and triacid chlorides to produce a crosslinked aromatic polyamide membrane. This process produces dense interfacially polymerized membranes of the order of 100 nm thickness. Various modifications of this method are noted in additional references [33–35].

2.6 Membrane Module Fabrication

The two most common separation modules for gas and liquid separation processes are the hollow fiber and spiral wound constructions shown in Figures 2.21 and 2.22. The hollow fiber module consists of the hollow fibers placed in a cylindrical pressure chamber and encapsulated in a thermosetting polymer (e.g. epoxy) at both ends. The feed stream is usually on the shell side of the module and the permeate stream is collected from the open ends of the hollow fibers. Optionally, the feed stream can

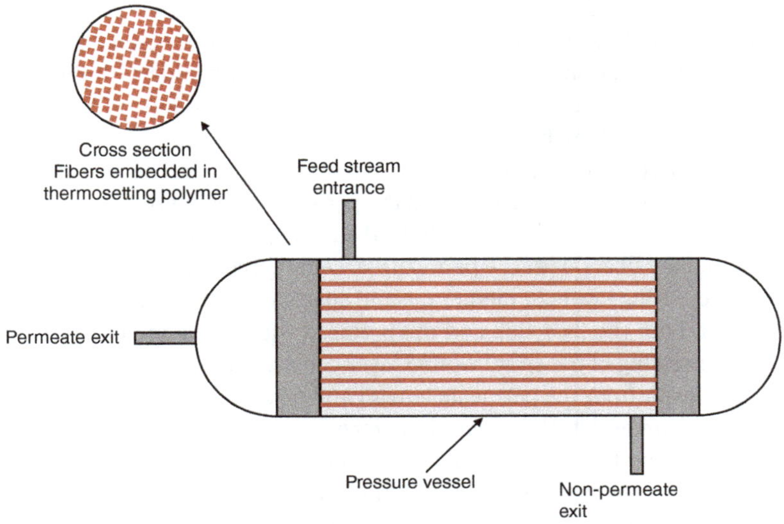

Figure 2.21 Schematic of hollow fiber membrane module.

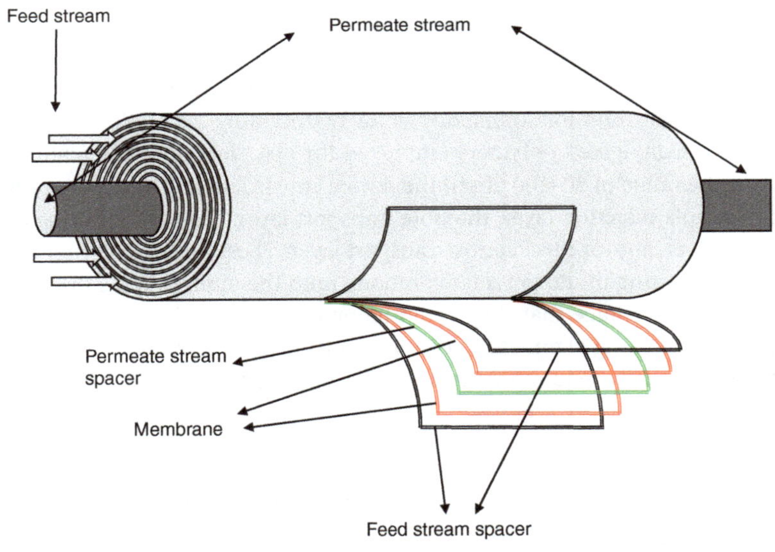

Figure 2.22 Schematic of spiral wound module.

be delivered to the bore of the hollow fibers with the permeate stream on the shell side. The spiral wound module is comprised of spiral wound sheets of alternating layers of the membrane, feed stream spacer, membrane, permeate stream spacer, and repeating layers of the noted construction. Proper sealing of the spiral wound layer constructions allows for separation of the feed stream from the permeate stream and flow of the permeate stream to the permeate stream outlet.

For membrane separation processes, two types of membrane modular units are generally employed: hollow fiber and spiral wound. The hollow fiber-based modules are comprised of bundles of hollow fibers encapsulated at both ends with the feed stream separated from the permeate stream by the hollow fibers. The hollow fiber module consists of the hollow fibers placed in a cylindrical pressure chamber and encapsulated in a thermosetting polymer (e.g. epoxy) at both ends. The feed stream is usually on the shell side of the module and the permeate stream is collected from the open ends of the hollow fibers. Optionally, the feed stream can be delivered to the bore of the hollow fibers with the permeate stream on the shell side. For spiral wound units, alternating layers of membrane and spacers (to allow gas flow across the membrane on both the feed side and the permeate side of the membrane) are employed. Proper sealing of the spiral wound layer constructions allows for separation of the feed stream from the permeate stream and flow of the permeate stream to the permeate stream outlet. Gas separation processes generally use hollow fibers whereas desalination and water filtration processes generally employ spiral wound constructions. For reverse osmosis and nanofiltration processes, spiral would modules are commonly used. For membrane bioreactors for sewage water treatment, hollow fiber constructions are common. In fuel cell applications, thin (~25 μm) dense films of a proton exchange membrane are positioned between the anode and the cathode in the membrane electrode assembly (MEA). These MEAs are stacked to prepare stacks for power generation. For lithium-ion batteries, microporous membranes (usually of polyethylene or polypropylene) are positioned between the electrodes allowing lithium-ion transport while maintaining protection against electrode contact. Conjugated polymers employed for optoelectronic applications are typically in the range of 50–100 nm in the device construction. These polymers are used for the hole injection layer, the hole transport layer, the light emitting or light harvesting layer, and/or the electron transport layer. These layers are generally produced by spin coating dilute polymer solutions unto the appropriate substrate. Polymers have the advantage that they could be processed by roll-to-roll printing or ink jet printing yielding significant economic advantages over their inorganic competitors which usually require vacuum deposition (Figures 2.23 and 2.24).

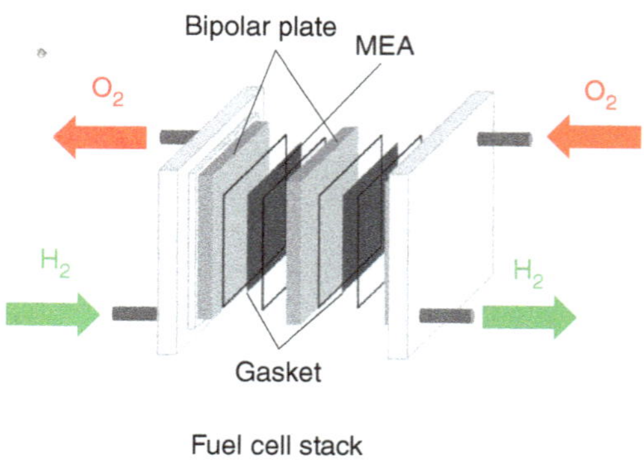

Figure 2.23 Schematic of fuel cell stack. Bipolar plate and gaskets are places in between MEAs. Air can be fed in place of O_2.

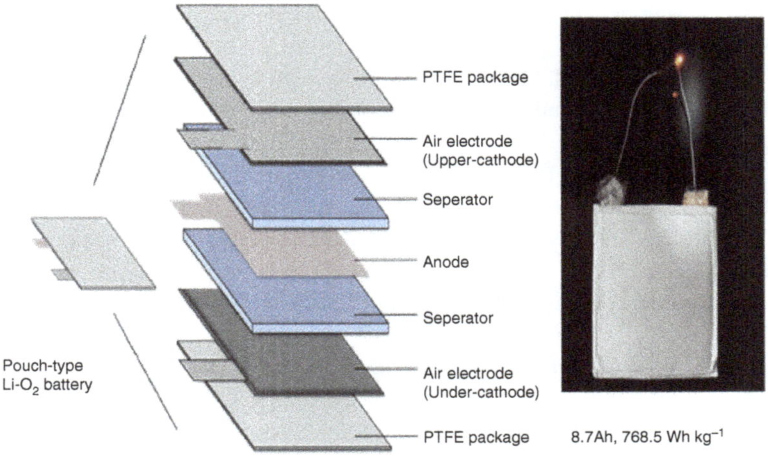

Figure 2.24 Schematic of lithium-air battery pouch for stacks. Source: Adapted from Shangqian Zhao et al. [36]/Elsevier.

References

1 Koros, W.J., Ma, Y.H., and Shimidzu, T. (2009). *Pure and Applied Chemistry* 68: 1479–1489.

2 Warsinger, D.M., Chakraborty, S., Tow, E.W. et al. (2016). *Progress in Polymer Science* 81: 209–237.

3 Li, K. (2007). *Ceramic Membranes for Separation and Reaction*. England: Wiley.

4 Tseng, H.-H., Wang, C.-T., Zhuang, G.-L. et al. (2016). *Journal of Membrane Science* 510: 391–404.

5 Wan, C.F., Yang, T., Lipscomb, G.G. et al. (2017). *Journal of Membrane Science* 538: 96–107.

6 Xing, W.H. (2017). *Membrane-Based Separations in Metallurgy* (ed. L.Y. Jiang and N. Li), 357–370. Amsterdam: Elsevier.

7 Kim, J.F., Kim, J.H., Lee, Y.M., and Drioli, E. (2016). *AIChE Journal* 62: 461–490.

8 Strathmann, H. and Kock, K. (1977). *Desalination* 21: 241–255.

9 García-Payo, M., Essalhi, M., and Khayet, M. (2009). *Desalination* 245: 469–473.

10 Silva, T.L., Morales-Torres, S., Figueiredo, J.L., and Silva, A.M. (2015). *Desalination* 357: 233–245.

11 Mulder, J. (2012). *Basic Principles of Membrane Technology*. Springer Science & Business Media.

12 Baker, R.W. (2012). *Membrane Technology and Applications*. Wiley.

13 Mosqueda-Jimenez, D., Narbaitz, R., Matsuura, T. et al. (2004). *Journal of Membrane Science* 231: 209–224.

14 Zheng, Q.-Z., Wang, P., and Yang, Y.-N. (2006). *Journal of Membrane Science* 279: 230–237.

15 Zheng, Q.-Z., Wang, P., Yang, Y.-N., and Cui, D.-J. (2006). *Journal of Membrane Science* 286: 7–11.

16 Tsai, H., Li, L., Lee, K. et al. (2000). *Journal of Membrane Science* 176: 97–103.

17 Lloyd, D.R., Kinzer, K.E., and Tseng, H. (1990). *Journal of Membrane Science* 52: 239–261.

18 Jung, J.T., Kim, J.F., Wang, H.H. et al. (2016). *Journal of Membrane Science* 514: 250–263.

19 Tang, Y.-H., He, Y.-D., and Wang, X.-L. (2015). *Journal of Membrane Science* 474: 196–206.

20 Yang, J., Li, D., Lin, Y. et al. (2008). *Journal of Applied Polymer Science* 110: 341–347.

21 Lin, Y., Tang, Y., Ma, H. et al. (2009). *Journal of Applied Polymer Science* 114: 1523–1528.

22 Hassankiadeh, N.T., Cui, Z., Kim, J.H. et al. (2014). *Journal of Membrane Science* 471: 237–246.

23 Lee, J., Park, B., Kim, J., and Park, S.B. (2015). *Macromolecular Research* 23: 291–299.

24 Yan, L., Li, Y.S., and Xiang, C.B. (2005). *Polymer* 46: 7701–7706.

25 Yan, L., Li, Y.S., Xiang, C.B., and Xianda, S. (2006). *Journal of Membrane Science* 276: 162–167.

26 Bottino, A., Capannelli, G., and Comite, A. (2002). *Desalination* 146: 35–40.

27 Henis, J.M.S. and Tripodi, M.K. (1983). *Science* 220: 11–17.

28 Henis, J.M.S. and Tripodi, M.K. (1981). *Journal of Membrane Science* 8: 233–246.

29 Henis, J.M.S. and Tripodi, M.K. (1980). US Patent 4,230,463.

30 Cadotte, J.E. (1981). US Patent 4,277,344.

31 Cadotte, J.E., Petersen, R.J., Larson, R.E., and Erickson, E.E. (1980). *Desalination* 32: 25–31.

32 Cadotte, J.E. (1985). *Materials Science of Synthetic Membranes*, ACS Symp Ser., vol. 269 (ed. D.R. Lloyd), 273–294.

33 Zhou, Y., Yu, S., Liu, M., and Gao, C. (2005). *Desalination* 180: 189–196.

34 Petersen, R.J. (1993). *Journal of Membrane Science* 83: 81–150.

35 Qiu, S., Wu, L., Zhang, L. et al. (2009). *Journal of Applied Polymer Science* 112: 2066–2072.

36 Zhao, S., Zhang, L., Zhang, G. et al. (2020). *Journal of Energy Chemistry* 45: 74–82.

3

Membranes in Gas Separation for Energy and Environment

3.1 Introduction

Membrane-based gas separations have been developed rapidly in the last two decades, competing with conventional separation technologies. Polymer membranes for gas separation have advanced from simple lab-scale experiments to large industrial applications such as hydrogen separation from streams in ammonia synthesis plants, carbon dioxide separation from natural gas processing, and nitrogen production or oxygen enrichment from air. The advantages of using polymer membranes for gas separation include lower energy requirements, easier installation or extension, and lower initial capital cost compared with other separation processes, such as adsorption, absorption, and cryogenic distillation. Superiority of polymer membranes over other separation methods is based on the premise that they have high separation capacity of membranes such as high selectivity and high permeability and, they must be competitive in cost and available for operating environments. However, among many available polymers, only a few polymer membrane materials have been used in practical applications of gas separation despite exhaustive efforts of the past decade of intensive activity to obtain superior gas separation performance of polymers and ceramics or carbons. Therefore, selection of suitable membranes for each separation process and specific gas mixtures is crucial. This chapter reviews the latest developments of polymer membrane materials for gas and vapor separation and their current and potential applications, particularly focusing on energy applications. Gas separation applications in energy application that use promising polymer membranes, such as hydrogen recovery or production and separation for energy devices are briefly reviewed.

To date, polymer membranes have served as key elements in many energy and environmental applications. The key property of membranes is their ability to control the permeation rate of a chemical component through the membrane with driving forces such as pressure, concentration, and electrical potential. Among these driving forces, pressure difference across the membrane is mostly used.

Membranes for Energy Applications, First Edition. Young Moo Lee.

Mitchell [1] first observed the separation of gases through membranes. He reported that gases had different permeabilities in experiments involving natural rubber balloons exposed to various gas atmospheres. Another early report by Thomas Graham in 1829 [2] reported different rates of permeation in his experiments that a wet pig bladder inflated when placed in CO_2 environment. In 1855, Fick established the basic equations for diffusion through membranes which are now termed as Fick's first and second law of diffusion [3]. Graham first proposed the concept of solution–diffusion as the permeability process in polymeric membranes in 1866 [4]. Gas permeability through a membrane was measured with vacuum on one side that became possible for natural rubber using current permeability methods. During the same period, these experiments exhibited the ability of membranes to separate oxygen from air. It was reported that increasing pressure and temperature should result in increased flux. Knudsen noted a specific diffusion in pores that dominated for pore diameters of 2–50 nm [5]. Knudsen diffusion happens when the gas mean free path is large compared to the pore diameter. Daynes established the time lag method to determine diffusion coefficients in 1920 employing the unsteady-state solution of Fick's second law to mathematically determine the diffusion coefficient from the extrapolation of the steady-state flux to the time axis [6]. Later, Barrer [7] noted that the Arrhenius equation was applicable to the permeability and diffusion of gases as a function of temperature. A major advance in commercial development of membrane for gas and liquid separation occurred with the discovery of the phase inversion method of Loeb and Sourirajan [8] producing asymmetric cellulose acetate membranes with very thin separating layers of $<0.5\,\mu m$. Before this historic discovery, typical cellulose acetate dense membranes had separating layer thicknesses several orders of magnitude too thick to produce the flux required for commercial applicability. The development of asymmetric cellulose acetate membranes advanced the commercialization of reverse osmosis membrane and later led to gas separation membrane.

In 1980, the first commercial gas membrane separations system (Prism) appeared on the gas separation market developed by Henis and Tripodi [9] at Monsanto (now Air Products Co.) to produce hydrogen from purge gas streams in ammonia synthesis plants. Membrane-based gas separation system was considered promising if feed was available at high pressure and permeate could be tolerated at lower pressures. Since then, gas separation processes using polymer membranes have continuously competed with other separation processes such as adsorption and cryogenic distillation in niche application markets. Membrane-based gas separations remain competitive because of their easy operation, small footprint, low energy cost, portability, reliability, and space efficiency [10].

Current membrane-based gas separation processes and potential applications include hydrogen separation and recovery from ammonia purge gas, refinery, and syngas stream in the petrochemical industry. Other applications include carbon dioxide enhanced oil recovery, natural gas processing, landfill gas upgrading, air separation (oxygen-enrichment or nitrogen production), air dehydration, helium recovery, and carbon dioxide capture from flue gas [11].

3.2 Basic Principles of Gas Separation in Polymer Membranes

3.2.1 Permeability and Selectivity

Gas permeation through a non-porous dense polymer membrane is described using a solution–diffusion model, in which the permeability coefficient (P_A) (cm^3 ((STP) cm/cm^2 s cmHg))$^{-1}$ of gas molecule A is the product of a diffusion coefficient (D_A) (cm^2 s^{-1}) and a solubility coefficient (S_A) (cm^3 (gas cm^3-*polymer* cmHg))$^{-1}$.

$$P_A = D_A \cdot S_A \tag{3.1}$$

Equation (3.1), first postulated by Graham in 1866, is a basic expression representing the solution–diffusion mechanism of gases through homogeneous dense polymer membranes. Diffusion is the process by which a small molecule (of gas, vapor, etc.) is moved in a system due to random molecular motions. Diffusion in a polymer matrix is classified into three categories depending on the relative mobility of the polymer and penetrant gas [12]. These categories are (i) Fickian diffusion (case I), (ii) non-Fickian diffusion (case II), and (iii) anomalous diffusion (see Table 3.1). Solubility is related to the local concentration (C) of the gas dissolved in the polymer at a given temperature and the solubility coefficient $S(C)$ is expressed as a function of pressure (p).

$$S(C) = C/p \tag{3.2}$$

$S(C)$ depends on gas condensability as well as polymer–penetrant interactions. Equation (3.2) can be expressed differently using various sorption modes in polymers. These modes, shown in Figure 3.1, are Henry's law sorption,

Table 3.1 Different diffusion mechanisms of gas through a polymer membrane.

Diffusion mode	Description
Case I: Fickian diffusion	The diffusion process has a rate much smaller than that of the relaxation modes of the polymeric matrix. The sorption equilibrium is rapidly reached. The boundary conditions are independent of time and swelling kinetics.
Case II: non-Fickian diffusion	This mode relates to a fast diffusion process compared to the simultaneous relaxation processes of the polymer. The sorption phenomena are complicated by a strong dependence with the swelling kinetics. These deviations from the Fickian behavior are generally found in the case of the sorption of organic vapors by solid polymers.
Anomalous diffusion	This mode relates to a process when the diffusion and the polymer relaxation rates are comparable. The sorption and the transport of molecules are affected by the presence of pre-existing microvoids in the matrix and the penetrant motion is affected by polymer microstructure.

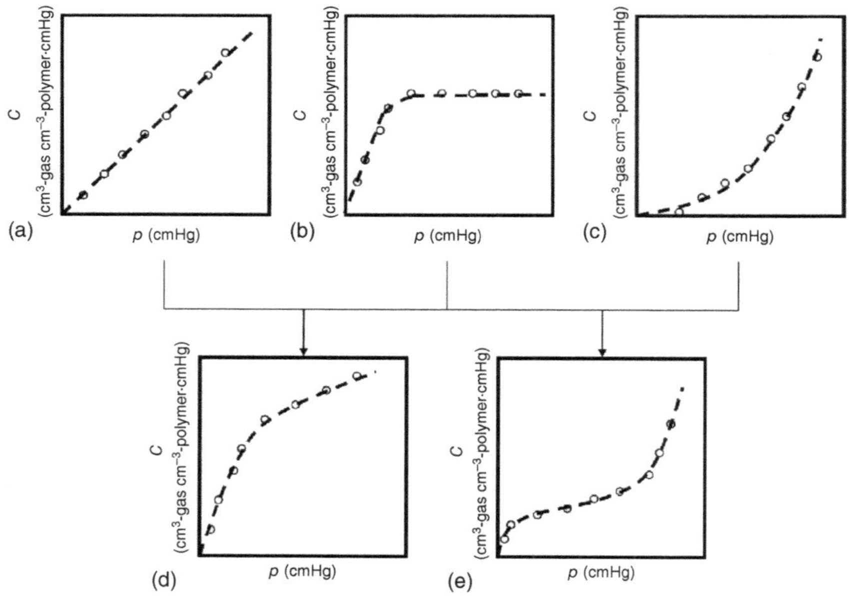

Figure 3.1 Representative sorption phenomena of gases in polymer membranes. (a) Henry's law sorption, (b) Langmuir-mode sorption, (c) Flory–Huggins mode sorption, (d) dual-mode sorption, and (e) BET mode sorption where C and p represent concentration (cm^3-(gas/cm^2-polymer.cmHg)$^{-1}$) and pressure (cmHg).

Langmuir-mode sorption, Flory–Huggins mode sorption, dual-mode sorption, and Brunauer–Emmett–Tellet (BET) mode sorption.

In general, gas transport phenomena in polymer membranes progresses through five successive steps as follows [13] as illustrated in Figure 3.2. (1) Diffusion through the boundary layer at the upstream side, (2) relative sorption of the gases by the polymer membrane, (3) diffusion of the gases inside the polymer membrane. The diffusion step is the slowest and becomes the rate-determining step in gas permeation, (4) desorption of gases at the permeate side (lower partial pressure), and (5) diffusion out of the boundary layer of the downstream side.

In membrane gas separation, the concentration polarization on the feed side is negligible and therefore the first and last steps are frequently excluded. However, membrane performance in the system generally decreases with time due to concentration polarization, resulting in the limited permeation of certain species. Therefore, in practice, membrane module and system designs must be thoroughly considered to prevent concentration polarization, fouling or both. Generally, the permeation rate is a dominating factor affecting concentration polarization, while the influences of separation factor is significant initially and levels off gradually. In addition, increasing feed gas velocity decreases the concentration polarization, while the effect of operation pressure is very limited, and the composition of feed gas is less significant.

Permeability and diffusivity of a dense membrane can be determined by a time-lag method. The development of mathematical relationships for the solution–diffusion

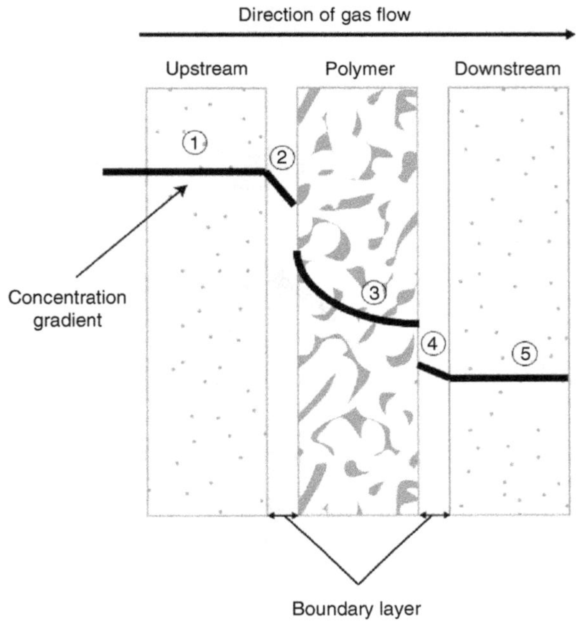

Figure 3.2 Schematic representative of gas permeation steps across polymer membranes according to solution–diffusion mechanism.

process start with Fick's first and second laws [13]. Fick's first law involves steady-state diffusion and is expressed in one dimension (through a dense polymer film) as:

$$J = -D\frac{\partial C}{\partial x} \tag{3.3}$$

where J is the flux, C is the concentration, and x is the dimension measured in the thickness of the film. The diffusion coefficient is negative as diffusion is in the opposite direction of increasing concentration. Fick's second law involves unsteady-state diffusion and is expressed in one dimension as:

$$\frac{\partial C}{\partial t} = D\frac{\partial^2 C}{\partial x^2} \tag{3.4}$$

Solution of Eq. (3.4) for the case involving unsteady diffusion for gas transport across the film yields the following expression:

$$\frac{Q_t}{lC_1} = \frac{Dt}{l^2} - \frac{1}{6} - \frac{2}{\pi}\sum_{n=1}^{\infty}\frac{(-1)^n}{n^2}\exp(-Dn^2\pi^2 t/l^2) \tag{3.5}$$

where Q_t is the quantity of gas permeating the film at time = t; C_1 is the gas concentration at the feed stream surface and the concentration at the permeate surface is 0.

At steady-state ($t \rightarrow \infty$), this equation reduces to:

$$Q_t = \frac{DC_1}{l}\left(t - \frac{l^2}{6D}\right) \tag{3.6}$$

Thus, for a plot of Q_t versus time, the intercept of the steady-state slope is:

$$\Theta = \frac{l^2}{6D} \tag{3.7}$$

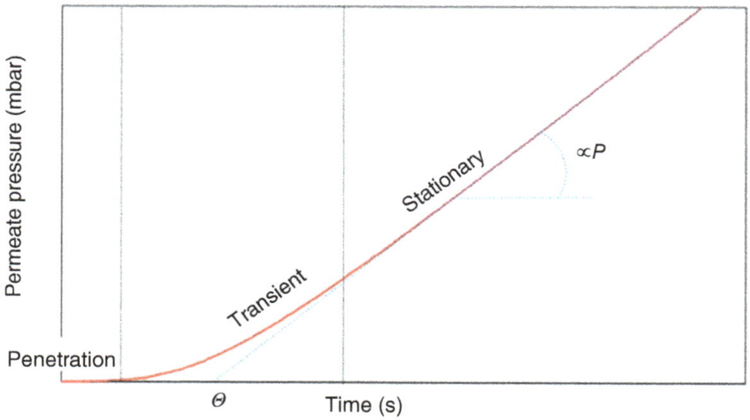

Figure 3.3 Graphical representation of the time lag determination from permeability data.

where Θ is typically referred to as the time lag, illustrated in Figure 3.3. Time lag measurement is a common method to determine the diffusion coefficient. Once the diffusion coefficient is determined the solubility constant, S, can then be calculated from the steady-state permeability coefficient ($P = DS$). Solutions to Eq. (3.4) for various geometries, initial and boundary conditions are tabulated in detail in a book by Crank [14]. These solutions are partly based on the similarity of Fick's Law for diffusion and Fourier's Law for heat conduction where a seminal text on solutions of Fourier's second law were presented in a book by Carslaw and Jaeger [12].

The diffusion coefficient is related to the physical state of the polymer (rubbery, crystalline, or glassy state). The free volume existing in the polymer is very important and correlations with fractional free volume have been shown in the literature [15–17]. The crystalline regions are generally considered impermeable with negligible diffusion through these regions. The specific gas diffusion rate in each polymer is a direct function of the gas diameter (specifically the kinetic diameter).

Selectivity ($\alpha_{A/B}$) of gas A to gas B is either due to differences in the solubility coefficient (solubility selectivity, S_A/S_B) or to differences in the diffusion coefficient (diffusivity selectivity, D_A/D_B).

$$\alpha_{A/B} = \frac{P_A}{P_B} = \left[\frac{D_A}{D_B}\right] \cdot \left[\frac{S_A}{S_B}\right] \tag{3.8}$$

Here, separation occurs because of the difference in solubilities and mobilities of different penetrants in the membrane.

3.2.2 Temperature Dependence of Gas Transport

Generally, at a given pressure, diffusion of gas through a polymer membrane increases with temperature. The temperature dependence of diffusion can be expressed by an activated process following a relationship like the Arrhenius equation:

$$D_A = D_{A0} \exp[-E_D/RT] \tag{3.9}$$

where D_A is the diffusion coefficient ($cm^2\ s^{-1}$), D_{A0} is the pre-exponential factor, E_D is the activation energy of diffusion ($kJ\,mol^{-1}$), R is a universal gas constant (8.31 J $(mol\,K)^{-1}$), and T is the Kelvin temperature (K). The activation energy is the energy used against cohesive forces between polymer segments in making the gaps through which diffusion occurs. Typically, gas diffusion coefficients increase over the temperature range of interest if the polymer does not undergo thermally induced structural rearrangements such as crystallization.

The temperature dependence of S_A of a given polymer–penetrant system can be described in the form of a van't Hoff relationship:

$$S_A = S_{A0} \exp[-\Delta H_s/RT] \tag{3.10}$$

where ΔH_s is the enthalpy of solution of the penetrant and S_{A0} is the pre-exponential factor. From Eqs. (3.1), (3.4), (3.5), it follows that over a narrow range of temperatures,

$$\begin{aligned} P_A &= D_{A0} \exp[-E_D/RT] \cdot S_{A0} \exp[-\Delta H_s/RT] \\ &= D_{A0} S_{A0} \exp[-(E_D + \Delta H_s)/RT] \\ &= P_{A0} \exp[-E_p/RT] \end{aligned} \tag{3.11}$$

where E_p, the activation energy of permeation, is:

$$E_p = E_D + \Delta H_s \tag{3.12}$$

Values of E_p, E_D, and ΔH_s depend upon the polymer morphology and thermal properties such as the glass-transition-temperature (T_g). For many polymer and gas pairs, these parameters can be found in the literature [18]. The heat of the solution, ΔH_s may be expressed as:

$$\Delta H_S = \Delta H_{Co} + \Delta H_{Pm} \tag{3.13}$$

where ΔH_{Co} is the molar heat of condensation (always < 0 and small for gases) [13] and ΔH_{Pm} is the partial molar heat of mixing (always > 0). These values can be estimated from the cohesive energy densities of the penetrant and the polymer using Hildebrand's theory [19].

3.2.3 Pressure Dependence of Gas Transport

Gas permeability of a polymer is affected by feed pressure variations. Koros and Chern [20] classified patterns observed in permeability versus pressure relationships into four characteristic patterns, illustrated in Figure 3.4. They are (a) linear, with a slope close to zero, representing the ideal case that meets the assumption of diffusion and solution being independent of gas pressure (e.g. low sorbing penetrants such as He or N_2 in rubbery or glassy polymers), (b) a nearly linear increase of permeability with increasing pressure, which often describes the permeability of an organic vapor into a rubbery polymer, (c) a decreasing trend of permeability with increasing pressure, which is typically observed for highly soluble gases such as CO_2 in glassy polymers, and (d) concave upwards, which can be perceived as a combination of (b) and (c) and is typical of a plasticizing penetrant such as an organic vapor in a glassy polymer.

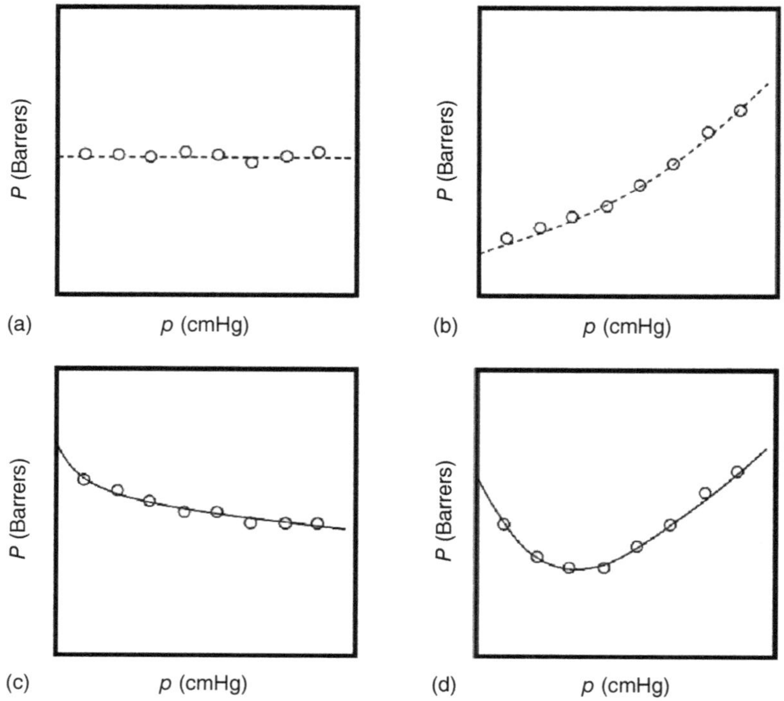

Figure 3.4 Changes in gas permeability (P, Barrer, 10^{-10} cm^3((STP) cm^2 s cmHg)$^{-1}$) of a polymer membrane as a function of pressure (p, cmHg). Adapted from [20] .

3.2.4 Unusual Sorption Behavior of Glassy Polymers

Polymers are classified into rubbery and glassy polymers where two polymers behave differently (see Figure 3.5). Rubbery polymers have more flexible chains in the backbone and side chain than glassy polymers. Rubbery polymers are at equilibrium at room temperature because their glass transition temperatures (T_g) commonly exist below room temperature. Therefore, gas transport behavior in rubbery polymers follows a simple solution–diffusion mechanism if the rubbery polymers do not contain any crystalline phase that acts as a complete barrier against the penetrant.

On the other hand, gas transport through glassy polymers is more complicated because the sorption sites of gases in glassy polymers are unique. In glassy polymers, changes of gas permeability as a function of pressure are often elucidated by the dual-mode and partial immobilization model [22–25] (see Figure 3.6). The basic assumptions applied to these models are: (i) Henry's and Langmuir's sorption modes co-exist in the glassy polymers, (ii) local equilibrium between the two modes is maintained throughout the glassy membranes, (iii) the gas sorbed on the Langmuir's sorption sites is completely immobilized, and (iv) diffusion occurs only in the Henry's sorption sites. The pressure dependence of the solubility coefficient for a single pure component is represented by:

$$S = \frac{C}{p} = k_D + \frac{C'_H b}{1 + bp} \tag{3.14}$$

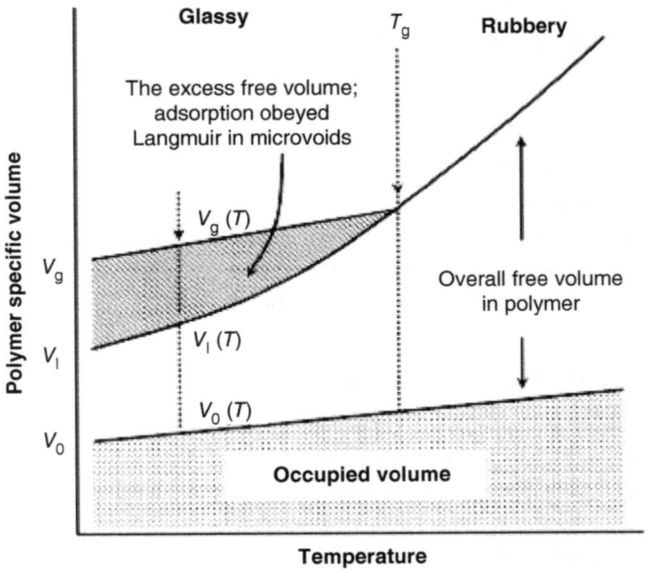

Figure 3.5 Schematic representation of the relationship between the polymer specific volume and temperature in amorphous polymers [21].

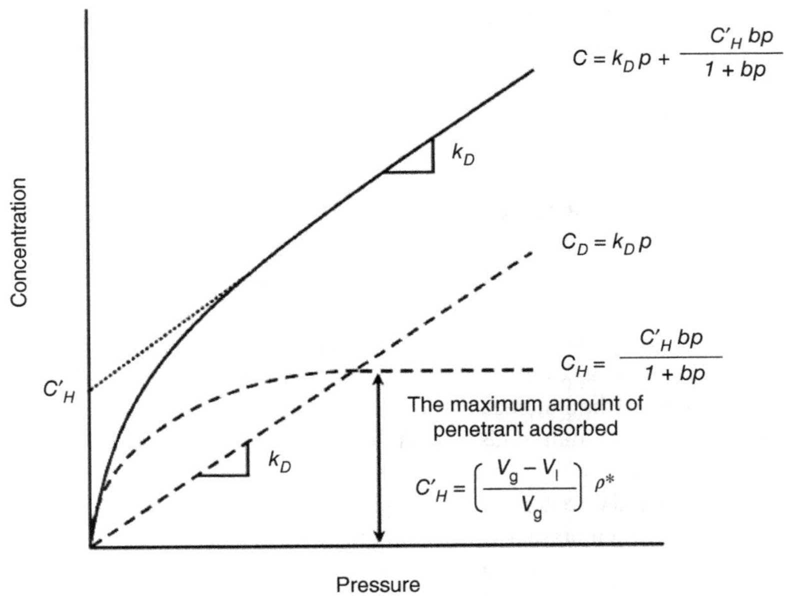

Figure 3.6 Schematic representation of a dual-mode sorption analysis using Eq. (3.14) [21].

where k_D is the Henry's law constant, C'_H is the Langmuir capacity constant, and b is the Langmuir affinity constant. The C'_H term indicates the amount of unrelaxed free volume in the glassy polymer. This term is used to determine the non-equilibrium nature of glassy polymers. The affinity constant, b, describes the specific nature of a given penetrant to sorb into the excessive unrelaxed volume in the non-equilibrium glassy polymer matrix. In these models, the true diffusion coefficient of a given penetrant is assumed to be constant regardless of concentration or position in the glassy membrane. The non-equilibrium state of glassy polymers can be affected considerably by process and operation parameters such as preparation conditions, pressure, temperature, and interactions between the polymer and penetrant [21].

3.2.5 Criteria for Membrane Material Selection

To select a suitable polymer membrane for each gas separation application, the nature of the polymer membrane (glassy or rubbery) must be considered as well as the target separating gas mixtures (permanent gas/permanent gas or condensable vapor/permanent gas) [26]. The main factors contributing to selectivity (solubility-selective or diffusivity-selective materials) can then be determined and used to improve separation efficiency with minimal loss of productivity. The most ideal membrane material design is to improve both diffusivity–selectivity and solubility–selectivity, but it is difficult to achieve both simultaneously. Moreover, for a given polymer membrane, both diffusivity and solubility depend strongly on the process parameters such as pressure difference, feed composition, and temperature, as well as the intrinsic nature of the polymer. Each selectivity-contribution (diffusivity or solubility) to total selectivity can be varied by such factors, but research has mainly focused on improving selectivity by modifying the chain structure of polymers.

3.2.5.1 Diffusivity-Selective Polymer Membranes

Molecular sieves are materials that can separate gas molecules by the size and shape on a molecular level. Most of molecular sieves are inorganic materials having well-defined pore sizes and shapes. For example, zeolite molecular sieves [27, 28], metal organic framework (MOF) [29], reduced graphene oxide (rGO) [30], and covalent organic framework (COF) [31] are well-known materials for catalysts or inorganic membranes. Carbon molecular sieves [32–36], known as amorphous carbons, in contrast to carbons, diamonds, or graphene, can be prepared by pyrolyzing polyimides that have a *d*-spacing around 3.65 Å, a value between the kinetic diameters of oxygen (3.46 Å) and nitrogen (3.64 Å). High gas selectivity (i.e. $\alpha(O_2/N_2) > 10$) of these molecular sieving materials is due to high diffusivity–selectivity caused by the differences of molecular sizes and shapes. Other two-dimensional nanosheets can be used as fillers for polymer composite membranes [37] for gas separation. For example, $\alpha(CO_2/N_2) > 91$ [38] and $\alpha(H_2/CH_4) > 275$ [39] can be obtained using rGO and boron nitride (BN) as fillers in polymer membranes.

Diffusivity (or mobility)–selectivity reflects the ability of the polymer matrix to be selective for the shape and size of penetrant molecules. Diffusivity–selectivity is

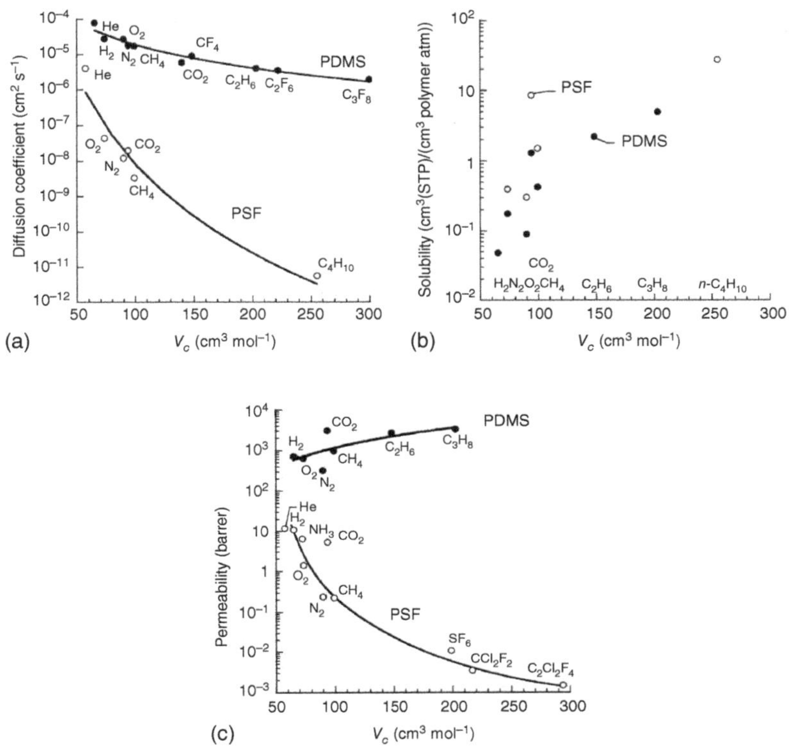

Figure 3.7 Correlation of (a) diffusion, (b) solubility, and (c) permeability with penetrant critical volume in polysulfone and PDMS [26].

governed by relative penetrant mobilities and by structural factors such as polymer chain stiffness and intersegmental polymer packing. However, in contrast to inorganic or carbon molecular sieves with a controlled molecular size, polymer membranes are seldom considered as molecular sieves. If an organic molecular sieve with high selectivity can be prepared, it will have vast advantages over zeolites or carbon molecular sieves because of its easy processability, thin-film formability, cost, and other reasons (Figure 3.7; Table 3.2).

Although many highly selective glassy polymer membranes have been developed and reported in the literature, newly synthesized polymers showing remarkable permselectivity are not yet practical for gas separation applications. Their complicated synthetic routes and poor solubilities in common organic solvents do not meet the commercial criteria for economic membrane modules and systems.

3.2.5.2 Solubility-Selective Membrane

Solubility coefficient demonstrates how many gas molecules are sorbed in polymer membranes. It depends on the condensability as well as the physical interactions between the penetrants and the polymer membrane. Solubility is determined by the concentration of the sorbed gas per unit volume of a polymer. Generally, the

Table 3.2 Permeabilities of oxygen and carbon dioxide through selected polymers.

Polymer	Permeability (Barrers) 25–35 °C	
	O_2	CO_2
Poly(vinyl alcohol) (extrapolated values)	7×10^{-7}	1.2×10^{-6}
Polyacrylonitrile	0.0002	0.0018
Vectra liquid crystalline polymer	0.00047	0.0070
Poly(vinylidene chloride)	0.0053	0.029
Polychlorotrifluoroethylene	0.012	—
Poly(ethylene terephthalate)	0.035	0.17
Nylon 6	0.038	0.16
Poly(vinyl chloride)	0.045	0.16
Polyethylene (density = 0.964)	0.40	1.8
Poly(vinyl acetate)	0.50	—
Cellulose acetate	0.80	2.4
Butadiene-acrylonitrile (39 wt% An)	0.96	7.4
Polysulfone	1.1	6.3
Bisphenol A polycarbonate	1.3	8.0
Poly(ethyl methacrylate)	1.3	7.0
Polypropylene	2.2	9.2
Polystyrene	2.6	12.4
Polyethylene (density = 0.914)	2.9	12.6
Poly(2,6-dimethyl-1,4-phenylene oxide)	16	70
Polybutadiene	19	138
Natural rubber	23	130
Poly(4-methyl pentene-1)	32	93
Polydimethylsiloxane	610	3250
Poly(4-methyl-2-pentyne)	2700	10,700
Polytrimethylsilylpropyne	10,000	33,100

concentration as a function of pressure at constant temperature shows a sorption isotherm with a characteristic shape that is concave to the pressure axis. The solubility–selectivity term in Eq. (3.8) is thermodynamic in nature and is governed by the relative polymer–penetrant interactions and the relative condensability of the penetrants. Solubility–selectivity terms contribute significantly to separations of condensable vapors and polar molecules.

In the absence of specific interactions of gases with the polymer, solubility can be determined by the critical temperature (T_c) of the gas. Although the unrelaxed volume in between the polymer chain segments, together with polymer–penetrant interactions, determine the concentration value, the effectiveness of these factors is reduced by the condensability of vapor, which is related to their critical temperatures. Thus, the solubility–selectivity of gases that do not interact with the polymer is primarily determined by the ratio of their critical temperatures. On the other hand, the critical volume (V_c) or kinetic diameter of the gas is more significant in the size-exclusive mechanism (diffusivity–selectivity). It is important to know the differences in polymer–penetrant interactions for a more exact evaluation of the contribution of solubility–selectivity to the permselectivity.

Figure 3.6 shows the correlation between diffusion, solubility, and permeability with penetrant V_c in polysulfone and poly(dimethyl siloxane) (PDMS). Polysulfone shows a clear distinction in diffusion and permeability with the V_c while PDMS exhibits less dependence with the gas kinetic diameter.

Typically, the permselectivity of rubbery polymers is governed by solubility–selectivity. In this respect, there are no large differences in permanent gas pairs nor in glassy polymers while, for vapor/gas or CO_2/permanent gas pairs, the solubility–selectivity is large enough to lead to a high permselectivity. Table 3.3 summarizes some physical properties of significant gases in membrane gas separation.

Table 3.3 Physical properties of simple gases [40].

Gas	T_b [a)] (K)	T_c [b)] (K)	ε/k [c)] (K)	σ [d)] (nm)	d_k [e)] (nm)
He	4.3	5.3	10.2	0.255	0.260
H_2	20	33	60	0.283	0.289
N_2	77	126	71	0.380	0.364
CO	82	133	92	0.369	0.376
O_2	90	155	107	0.347	0.346
CH_4	112	191	149	0.376	0.380
C_2H_4	175	283	225	0.416	0.390
C_2H_6	185	305	216	0.444	—
CO_2	195	304	195	0.394	0.330
H_2S	212	373	301	0.362	0.360
C_3H_8	231	370	237	0.512	0.430
NH_3	240	406	558	0.290	—
SO_2	263	431	335	0.411	—
n-C_4H_{10}	272	425	331	0.469	0.430
H_2O	373	647	809	0.264	0.265

a) Boiling temperature.
b) Critical temperature.
c) Lennard–Jones temperature.
d) Potential length constant or collision diameter of the molecule.
e) Kinetic diameter (Breck [41]).
Source: Van Krevelen [40] and Breck and Sieves [41].

Vapor/gas separation membranes or CO_2-selective membranes, such as silicone- and poly(ethylene oxide)-crosslinked rubbery polymer membranes, are typical examples representing high solubility–selectivity contributions to permselectivity. Lin and Freeman [42] suggested a material selection guideline for CO_2 selective membranes. To achieve high CO_2 permeability and high CO_2/light gas (i.e. CH_4, N_2, and H_2) selectivity, many polar groups such as ether oxygens (–O–), nitriles (–CN), carbonyls (–C=O–), acetates (–COO–), and amides (–NHCO–) were introduced into the polymers. These polar groups contribute to high CO_2 solubility and CO_2/light gas solubility–selectivity, but polar groups in a polymer decrease CO_2 diffusion coefficients due to the strong affinity between CO_2 and the polar groups. It is well-known that poly(ethylene oxide) (PEO) with ether oxygen units appears to provide excellent CO_2 separation and permeation properties. However, pure PEO materials tend to crystallize easily, making it difficult to improve CO_2 permeation and separation properties. To solve this problem, crosslinked poly(ethylene glycol) arylates (XLPEO) containing branches with methoxy end groups that do not crystallize at typical operation temperatures were prepared. At 35 °C, the CO_2 permeability and CO_2/H_2 gas selectivity reached up to 570 Barrers (1 Barrer = 10^{-10} cm^3 ((STP) cm (cm^2 s cmHg))$^{-1}$ and 12, respectively. Furthermore, the CO_2/H_2 pure gas selectivity of XLPEO reached a value of 40 at −20 °C (i.e. CO_2 permeability = 52 Barrers). This is due to the increased CO_2 solubility and CO_2/H_2 solubility–selectivity. It indicates that solubility–selectivity can be tuned by increasing the condensability of gas at lower temperatures, leading to high selectivity. However, economic process conditions (e.g. pressure and temperature) should be considered in practical applications.

3.3 Limitations of Gas Separations Using Polymer Membranes

> "Tradeoff relations of gas permeability and selectivity – there exists an empirical rule that more permeable polymers are statistically less selective and vice versa."

An ideal gas separation membrane should have high permeability and high selectivity. The gas permeation properties of polymer membranes have been extensively studied and a wide variety of polymers have been synthesized to be more permeable and selective. Nevertheless, there are still strong tradeoff relations of gas permeability and selectivity in polymer membranes that will not easily be broken in the future. Moreover, this tradeoff behavior of polymer membranes is not yet fully understood theoretically. However, it is believed that this tradeoff behavior for specific gas pairs is unique and related to parameters of the gas molecules.

Based on an enormous survey of approximately numerous references, including journals, books, and patents, Robeson [43, 44] proposed first in 1991 the tradeoff relations of polymer membranes for various gas pairs (i.e. O_2/N_2, H_2/CH_4,

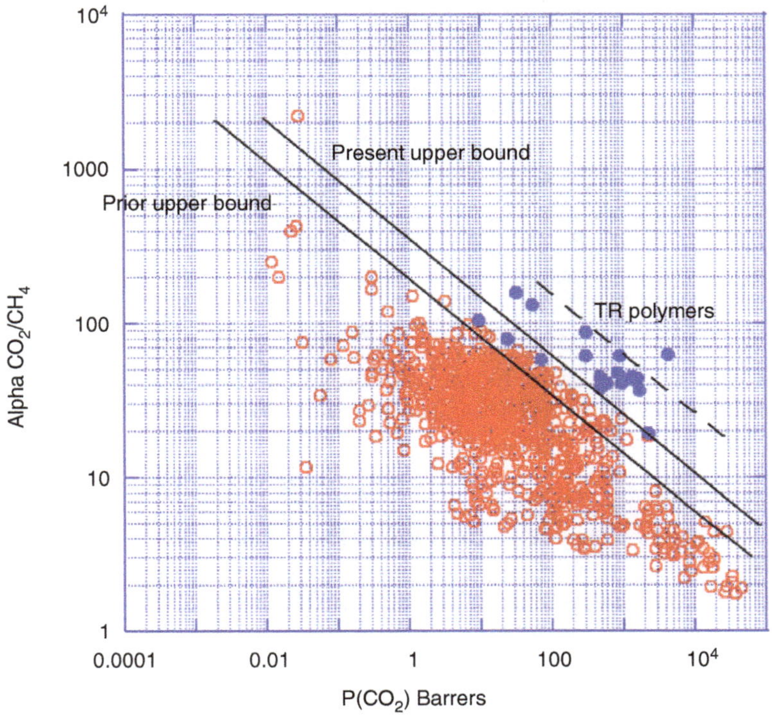

Figure 3.8 Upper bound line of CO_2 permeability and CO_2/CH_4 permselectivity [44].

CO_2/CH_4, H_2/N_2, He/CH_4, He/N_2, He/H_2, He/O_2, and H_2/O_2) and updated later in 2008 [44]. The transport parameters, selectivity α_{AB} and permeability P_A, make a cloud on this diagram located below an upper bound line. This line represents the best combination of α_{AB} and P_A values (Figure 3.8).

The following empirical equation was proposed to explain the upper bound lines for given gas pairs.

$$\log \alpha_{AB} = \log \beta_{AB} - \lambda_{AB} \log P_A \tag{3.15}$$

According to Robeson, the transport parameters, β_{AB} and λ_{AB}, for various gas pairs correlate with the Lennard–Jones kinetic diameters of gases.

$$\lambda_{AB} = d_B - d_A \tag{3.16}$$

d_A and d_B are the kinetic diameters (Å) of the gas molecules A and B, respectively.

Freeman [45] expounded upon these tradeoff relations. Using well-known correlations such as the compensation effect and dependence of the activation energy of diffusion on penetrant molecular size, Eq. (3.17), identical to Eq. (3.15), was obtained. Eq. (3.17) was derived from (a) the Arrhenius equation for the diffusion of small molecules (Eq. 3.18), (b) the correlation between the front factor and activation energy (Eq. 3.19), and (c) the correlation of the effect of penetrant size on activation (Eq. 3.20).

$$\ln \alpha_{A/B} = -\left[\left(\frac{d_B}{d_A}\right)^2 - 1\right] \ln P_A + \left\{\ln\left(\frac{S_A}{S_B}\right) - \left[\left(\frac{d_B}{d_A}\right)^2 - 1\right] \times \left(b - f\left(\frac{1-a}{Rt}\right) - \ln S_A\right)\right\} \tag{3.17}$$

$$D_A = D_{0A} \exp\left(-\frac{E_{DA}}{RT}\right) \tag{3.18}$$

$$\ln D_{0A} = a\frac{E_{DA}}{RT} - b \tag{3.19}$$

In these equations, a and b are independent of gas type. a is a universal value (0.64) independent of polymer type [7] and b has a value of 9.2 for rubbery polymers and 11.5 for glassy polymers [41].

$$E_{DA} = cd_A^2 - f \tag{3.20}$$

c and f are constants which depend on the polymer and d_A is the penetrant diameter (Å).

Equation (3.17) has the same mathematical form as the empirical relation by Robeson if β_{AB} and λ_{AB} are identified as follows:

$$\lambda_{A/B} = \left(\frac{d_B}{d_A}\right)^2 - 1 = \left[\frac{d_B + d_A}{d_A^2}\right](d_B - d_A) = k(d_B - d_A) \tag{3.21}$$

$$\beta_{A/B} = \frac{S_A}{S_B} S_A^{\lambda_{A/B}} \exp\left\{-\lambda_{A/B}\left[b - f\left(\frac{1-a}{RT}\right)\right]\right\} \tag{3.22}$$

If k in Eq. (3.21) is constant, Eq. (3.18) correlates well with Eq. (3.16). The value of k is approximately 0.8 and shows little variation. Figure 3.9 displays the linear relationship between these parameters for a given set of gas pairs. Table 3.4 lists the upper

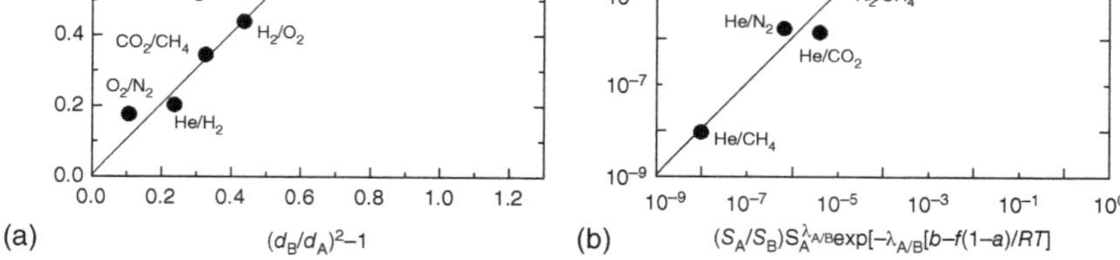

Figure 3.9 Comparison of slopes of ln $\alpha_{A/B}$ versus ln P_A plots, (a) $\lambda_{A/B}$ and (b) $\beta_{A/B}$ reported by Robeson with the theoretical prediction (solid line) [45].

Table 3.4 Overview of "upper bound" parameter of k and n.

Gas pair	Upper bound (1991 [43])		Upper bound revisited (2008 [44])		Upper bound redefined (2015 [46])		Upper bound redefined (2019 [47])	
	k (Barrer)	n	k (Barrer)	n	k (Barrer)	n	k (Barrer)	n
O_2/N_2	3.89×10^5	−5.80	1.40×10^6	−5.67	1.67×10^7	−5.7	—	—
CO_2/CH_4	1.07×10^6	−2.63	5.37×10^6	−2.64	—		2.26×10^7	−2.40
H_2/N_2	5.29×10^4	−1.53	9.77×10^4	−1.48	1.10×10^6	−1.46	—	—
H_2/CH_4	1.85×10^4	−1.21	2.72×10^4	−1.11	1.95×10^5	−1.1	—	—
CO_2/N_2	—	—	3.10×10^7	−2.89	—	—	7.56×10^8	−3.41

bound parameters. These parameters have been modified as the new microporous polymers were evolved to show high permeability of specific gas pairs.

The tradeoff relations of gas separation properties of polymer membranes were repeatedly considered using a free volume model and linear correlations [48]. Here, the equation identical to Eq. (3.10) was obtained from direct consequences of the free volume model for glassy polymers.

$$\log \alpha_{AB} = -x - (y - 1) \log P_A \tag{3.23}$$

These parameters determine the average lines of experimental correlations of permeability and selectivity and not the upper bound lines as suggested by Robeson and Freeman. They lead to the same recommendations of sorption selectivity enhancement for moving above the upper bound lines of Robeson's correlation. Tradeoff relations of gas permeability and selectivity were well-correlated with data obtained at room temperature.

3.4 Polymer Membrane Materials

3.4.1 Cellulose Acetate

Cellulose acetate (CA, see structure in Figure 3.10) asymmetric membranes have been developed by Loeb and Sourirajan as a reverse osmosis membrane for desalination [49]. To compete with the cryogenic process, permeation rate of helium, neon, oxygen, argon, and nitrogen gases through asymmetric cellulose acetate membranes have been measured and reported [50]. As the membrane thickness was very thin (<0.5 μm), the flux was very high and therefore its permeation rate was comparable to fluoropolymer (Teflon FEP). Before cellulose acetate was commercialized for this application, however, Monsanto (now Air Products) introduced its Prism polysulfone membrane system in 1979 [9]. First commercial CA membrane system was introduced in 1982 for hydrogen recovery and the first commercial system for CO_2 removal from natural gas [9, 51]. CA has been employed in the world's

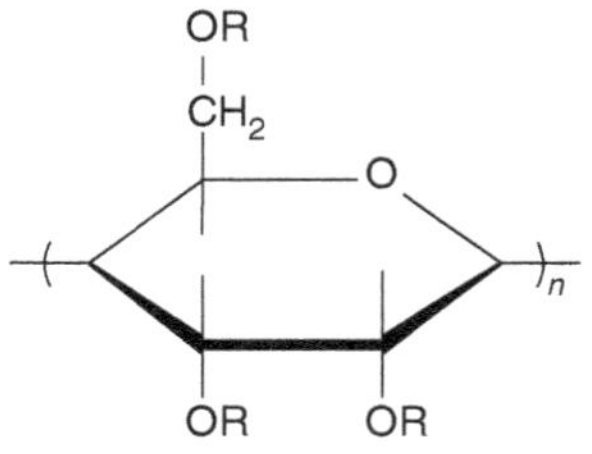

Figure 3.10 Chemical structure of cellulose derivative membrane. When R is hydrogen, it is called cellulose. The number of R = acetyl group ($-CH_3C=O$) is called the degree of substitution or DS. Cellulose derivative with DS = 2 is called cellulose diacetate and that with DS = 3 is called cellulose triacetate. Cellulose diacetate is generally called CA.

largest-capacity spiral wound membrane CO_2 removal system and the world largest capacity membrane system for hydrogen recovery. Cellulose derivative membranes (such as CA and cellulose triacetate or CTA) cover up to 80% of the membrane markets in natural gas processing. Two major CA membrane manufacturers supplying cellulose acetate-based modules are Cynara® (part of Cameron) and UOP Separex (part of Honeywell). The Cyanara® systems are based on asymmetric hollow fibers from Toyobo in Japan while Separax® uses a spiral wound module. For large-scale natural gas processing, CA is the only economically competitive with solvent absorption when the natural gas contains over 10–20 mol% CO_2, depending on the wellhead pressure.

There are two critical factors influencing the performance of cellulose acetate membranes. First is the degree of acetylation in the glucoside repeating unit. As acetyl groups are replacing three hydroxyl groups in the repeating units, the number of substitutions is called the degree of acetylation. For instance, if three hydroxyl groups are replaced with acetyl groups, the material is called cellulose triacetate and vice versa. Since the acetyl groups are larger than hydroxyl groups, cellulose triacetate membrane shows larger CO_2 permeability compared with cellulose diacetate membrane while the selectivity of CO_2/CH_4 remains the same. This is because the large acetyl groups in the polymer chain enhances the free volume and lessen the packing density, resulting in the flexibility and the mobility of the polymers, while more hydroxyl groups in cellulose derivatives strengthen the hydrogen bonding and increase the chain packing density [52].

Second critical factor is the fabrication of the membrane. Since cellulose acetate membrane is fabricated from solvent exchange, coagulation medium is usually water. When drying or evaporating the membrane, removal of water from the wet membrane leads to capillary forces collapsing the microporous structure, creating a denser active layer [53]. Solvent exchange approach has been proposed to prevent the pore collapsing. Water is a high surface tension liquid with surface tension of ($72.53\,mN\,m^{-1}$). If large surface tension liquid such as high hydrocarbons like nonane or decane with surface tension of 22.9 and $23.9\,mN\,m^{-1}$, respectively, is used, CO_2 permeance in GPU (gas permeation unit) will increase while the selectivity remains the same [54]. The evaporation temperature is critical according to Minhas et al. [55] showing that 80 °C provides the best membrane morphology for selectivity but generally a low permeability depending on solvent combinations.

Although cellulose acetate dominates the natural gas processing market, there are several issues that need considerable attention for long-term market position. One of the significant concerns is the plasticization, either by CO_2 or by heavy hydrocarbon components in the feed gas. Plasticization increases the flexibility of the polymer chains and dramatically decreases the separation performance, reducing the mechanical strength, speeding-up aging, and resulting in catastrophic membrane failure [56]. Plasticization pressure depends on the membrane thickness: 10 atm for a dense membrane of 76 μm thickness, while <5 atm for asymmetric membranes. There are considerable amount of research to decrease the plasticization such as composites with inorganics [57] and polymers [58]. Cellulose acetate is cheap and fabrication methods are well known.

3.4.2 Polysulfone

Since the development of the first asymmetric cellulose acetate membrane by Loeb and Sourirajan, most polymers are difficult to fabricate into useful asymmetric membranes in film or hollow fiber form, for gas separation. In fact, asymmetric cellulose acetate membranes possess pores with only 5–10 Å in diameter that permit the passage of most permanent gases across the membrane. Most glassy polymers with relatively high permeabilities are easy to spin as asymmetric hollow fibers of relatively low surface porosity. Such fibers are still too porous to be useful for gas separations (see Figure 3.11a). Spinning the hollow fibers with low surface porosities results in low gas permeation since the effective layer thickness is increased significantly. When the morphology is controlled to have a thin surface layer, small

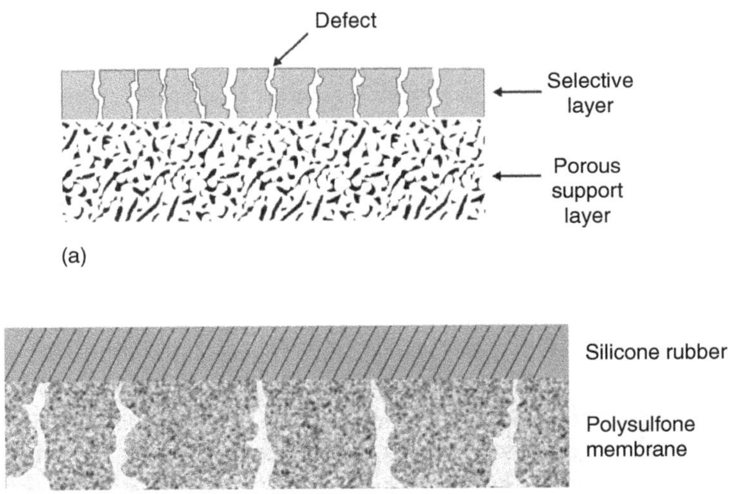

(b)

Figure 3.11 Schematic of (a) Loeb–Sourirajan membrane with defects. (b) Defects in surface of asymmetric polysulfone membrane in hollow fiber or flat sheet form can be cured with silicone rubber, resulting in high permeance and selectivity suggested by Henis et al. [9, 59–61].

unavoidable surface defects as well as pores are present in the membrane surface, resulting in gas leaks. Henis et al. [9, 59–61] noticed that micropores formed in glassy polysulfone hollow fibers can be plugged with rubbery PDMS solution through after-treatment to ensure the high flux and selectivity of small gases such as hydrogen and helium or even oxygen. They called this as a composite or resistance model membrane and analyzed using the electrical analog (see Figure 3.11b). Silicone rubber-coated polysulfone hollow fiber composite membrane shows unchanged hydrogen flux and separation factor over a wide range of surface porosity of membranes, compared with the conventional porous membrane. This is important in practice that after spinning hollow fibers, mere coating of silicone rubber in solvent (even in a module form) can plug the surface pores to ensure the flux and separation factors of useful gas molecules from a mixture. This type of membrane is widely used in industry for hydrogen separation from hydrocarbon and oxygen separation from air for nitrogen blanketing.

3.4.3 Polyimides

Aromatic polyimides (PIs) exhibited high permselectivity for various gas pairs (e.g. CO_2/CH_4 and O_2/N_2), high chemical and thermal stability as well as mechanical strength. Aromatic PIs are produced by a condensation between aromatic diamines and aromatic dianhydrides. Table 3.4 lists typical commercial polyimides used in membrane technology. Aromatic PIs have received much attention in gas separation [11, 62]. Du Pont (USA) and Ube Industry (Japan) have pioneered the commercial application of polyimides in separation processes. Du Pont began industrial application of aromatic PI membranes in 1962 using Kapton film with separation of helium from natural gas which was then applied to separate CO_2 from various gas streams. Chemical structures of PI are linked with their intrinsic gas transport properties. Chemical structure–property relationships have been reported in several articles [63, 64]. Manipulating the chemical structure of polyimides changes the transport properties of the resulting membranes. Modification of the existing polyimides can produce the better membranes.

In general, chain rigidity of polyimides governs the permselectivity while inter-chain spacing, and chain mobility determines the permeability. The main factors influencing the gas transport properties are (i) configuration of spatial linkage, (ii) type of bridging groups, and (iii) bulkiness and polarity of pendant groups in the polymers. For spatial linkage, meta-linked polyimides display better chain packing and limited rotational freedom compared with their para-linked isomers. Therefore, the meta-isomers exhibit lower fractional free volume (FFV) and gas permeability but higher gas pair selectivity. Bridging groups in the polymer chain influences gas separation performance as well. Bridging groups with low energy barriers (e.g. O and S) facilitate chain motions and results in higher gas permeability while the incorporation of bulky linkages such as CH_2 and O=S=O lowers gas permeability. Higher electronegativity leads to stronger chain–chain interactions and enhanced selectivity. Bulky side chains increase interchain spacing and reduces the packing density of the polymer chains. Gas permeability increases with the bulky pendant groups (e.g. CH_3, CF_3 and C_2H_5) due to the increase in FFV. The polarity of the pendant groups influences the chain–chain interactions and thus chain rigidity and chain packing density. For example, polarity of C=O groups in polymer is the reason for higher permselectivity and lower permeability.

Chain rigidity can be increased by using the polymers of intrinsic microporosity (PIMs) which was first reported by McKeown and Budd [65]. Polyimides can copolymerize with PIM moieties to produce so called PIM-PI membranes. These copolymers showed much higher permeability and reasonable selectivity. Copolymerization or blends of polyimides with poly(ethylene oxide) (PEO) produces high CO_2 permeability and selectivities possibly due to the improved CO_2 solubility of PEO segments. Poly(imide siloxane) copolymers showed better CO_2 permeability with a sacrifice of CO_2/CH_4 selectivity. Poly(imide pyrrolone) showed high CO_2/CH_4 selectivity that is close to or above the upper bound (Table 3.5).

Polyetherimide was introduced by General Electric in 1982 (Ultem). Combining imide and ether groups provides easy processability and thermal resistance and melting characteristics. Polyamide-imide (PAI) is prepared by the reaction between diamines and trimellitic anhydride derivative. An example of a PAI is Torlon from Solvay. PAI shows the advantages of amide and imide bonds in the polymer chain, resulting in an excellent resistance to chemicals and heat as well as mechanical load. P84 is a co-PI of BTDA with 80% methylphenylene diamine and 20% methylene diamine (BTDA-MPDA/MDA). Hollow fiber (HF) PI membranes are specifically prepared first by Ube Industries in Japan using BTDA/ODA for separation of hydrogen. Later, Air Products and others used HF PI for separation of air as well as dehydration.

Crosslinking of polyimides with diamines, diaols, UV, heat, and other sources have been reported [66]. Crosslinking of PI offer several advantages such as the suppression of plasticization in gas separation and pervaporation processes [67]. Un-crosslinked PI membranes usually show instability in many organic solvents due to the presence of imide groups, while crosslinked PI membranes seems to be stable in many organic solvents depending on the degree of crosslinking. Crosslinked PI membranes are important for separation of small molecules such as dyes and small organics using solvent resistance nanofiltration membranes.

3.4.4 Siloxane Polymers

Siloxane-based elastomers are high-performance materials with excellent oxidative and thermal stability, biocompatibility, and high gas and vapor permeability [68]. Siloxane-based polymers are typically prepared by a ring-opening polymerization of a trimer or tetramer [69], as shown in Figure 3.11.

PDMS is the most important siloxane polymer [70], for gas and vapor separation. PDMS is known as the most permeable rubbery polymer. Due to chain flexibility, rotational mobility, and large free volume, the low T_g (−123 °C) of PDMS explains its high gas permeability (Figure 3.12).

Table 3.5 Commercially available polyimides in membrane technology and their chemical structure.

Polyimide	Chemical structure
Kapton® (DuPont Electronics)	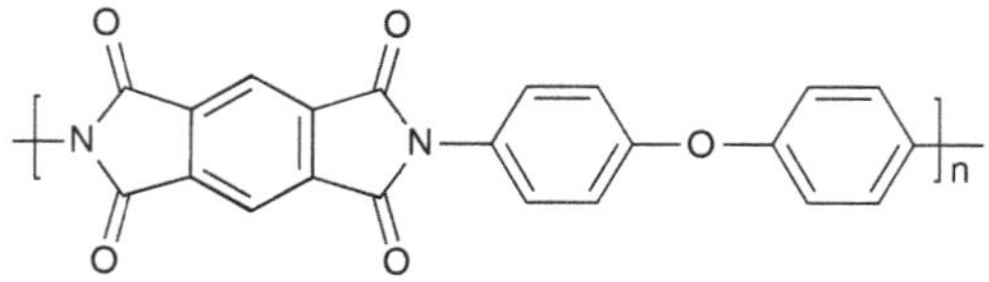
Ultem® Polyetherimide (Plastics International)	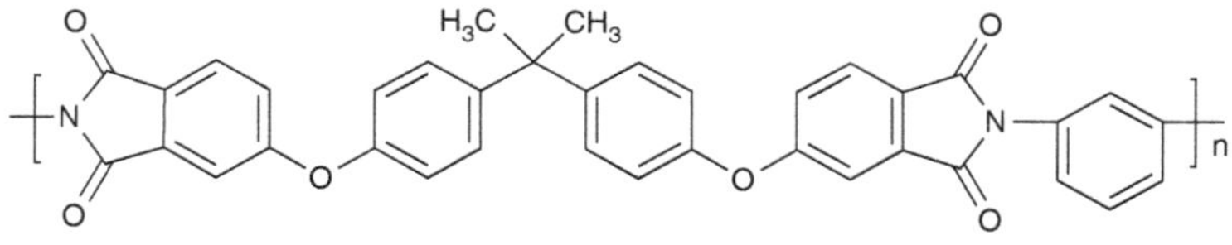
Matrimid® (Hunstman, Switzerland)	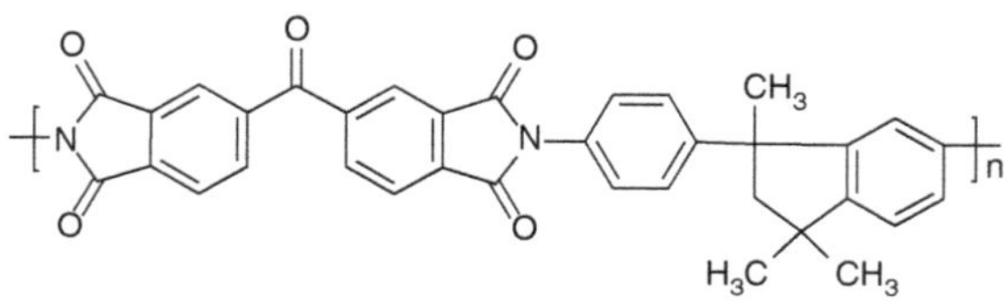
P84® (Evonik)	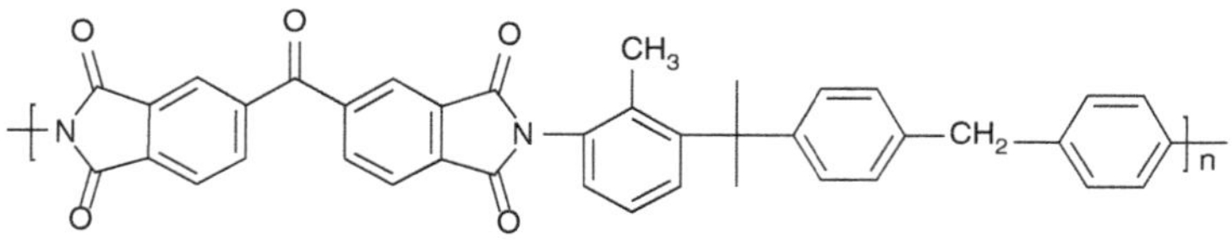
Torlon® (Solvay Advanced Polymer)	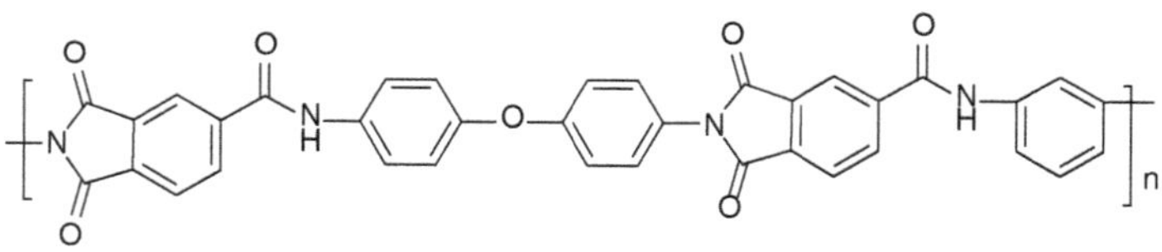
6FDA-durene	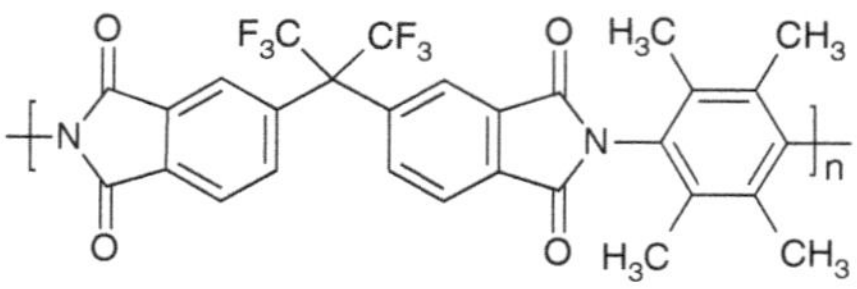
Upilex®-RN (Ube Industries)	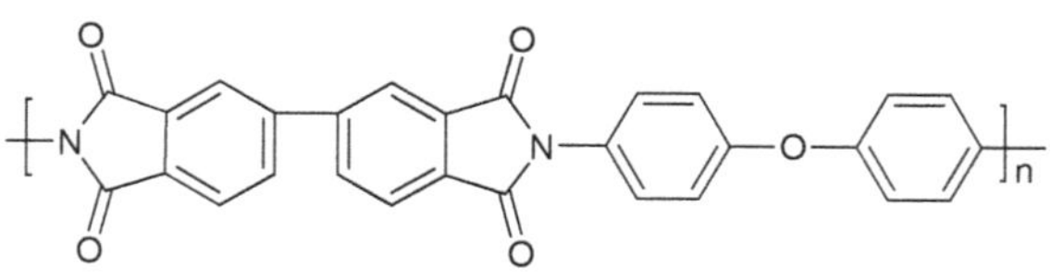
Aurum® (Mitsui Chemicals)	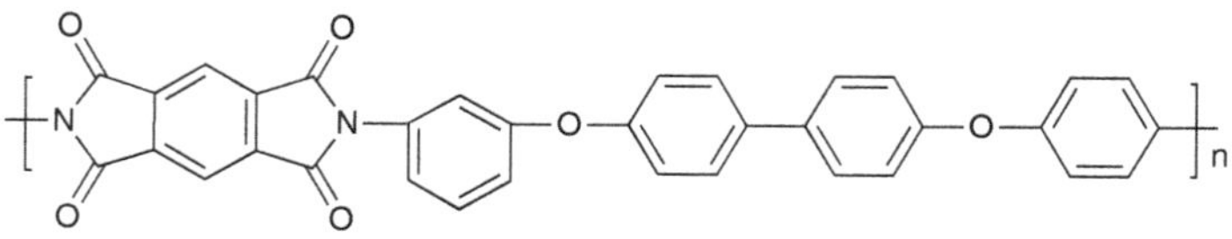

Figure 3.12 D3 or D4 monomer and poly(dialkyl siloxe) where R = CH_3, C_2H_5 etc.

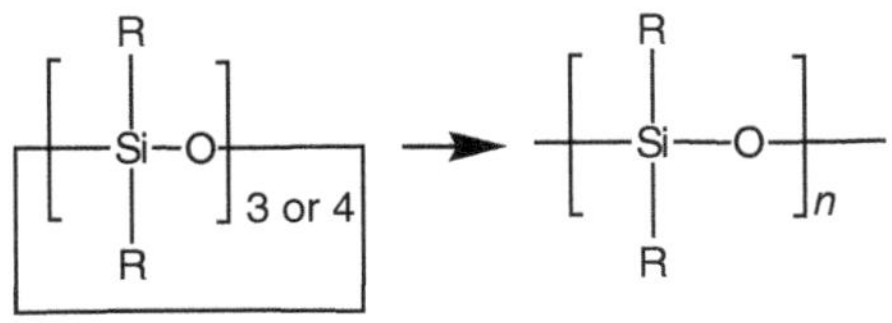

Although siloxane polymers show much higher gas permeability than many other rubbery polymers, the gas selectivity is lower than that of glassy polymers such as polysulfones and polyimides. For instance, oxygen/nitrogen selectivity of PDMS is approximately 2.1. There have been many efforts to improve selectivity without decreasing permeability by varying the basic siloxane structure. Modified siloxane polymers were investigated [68] but these structural changes did not greatly improve gas separation performance. Therefore, for permanent gas separation, such as O_2/N_2 separation, siloxane polymers have been largely used as a protective coating to avoid the significant decrease of separation performance caused by the probable formation of cracks or defects of asymmetric glassy polymer membranes in bulk production of hollow fiber or flat sheet membranes.

The most promising application of siloxane polymer membranes is the separation of nitrogen from air and organic vapors from permanent gases, such as the recovery of high-value organic vapors, ethylene, propylene, gasoline, and vinyl chloride from industrial waste gas streams [71]. So far, the permeation properties of PDMS membranes have been reported for various pure gases and organic vapors or mixed gases [72–82]. Although the poly(1-trimethylsilyl-1-propyne) (PTMSP) membrane, a stiff chain and ultra-high free volume glassy polymer, shows the highest hydrocarbon permeability and excellent selectivity, PDMS is still the most practical polymer membrane. This is because PTMSP exhibits poor chemical resistance towards aliphatic and aromatic hydrocarbons [83].

Usually, PDMS membranes easily sorb condensable gases and vapors and are swollen by them. Preferable swelling by these penetrants leads to a decrease of selectivity of a vapor or gas over a smaller penetrant in a mixed gas separation. However, a study by Pinnau et al. [84] showed that vapor-induced swelling of PDMS membrane results in improvements in both permeability and selectivity, particularly for hydrocarbon/methane and hydrocarbon/hydrogen separation. They explained that vapor-induced swelling of PDMS results in increased permeability and leads to higher diffusion coefficients of all penetrants. In addition, the relative permeability increase becomes greater as the size of the components in the mixture increase. This is a good example of improving vapor/gas selectivity by increasing vapor/gas diffusivity selectivity. Accordingly, the vapor/gas selectivity increases by increasing vapor concentration in the feed. Pinnau et al. [84] found that a decrease in temperature results in a significant increase in hydrocarbon/methane and hydrocarbon/hydrogen selectivity (Figure 3.13).

3.4.5 Substituted Polyacetylenes

Polyacetylene-based polymers have been evaluated for gas separation applications because these amorphous glassy polymers are characterized by T_g above 200 °C,

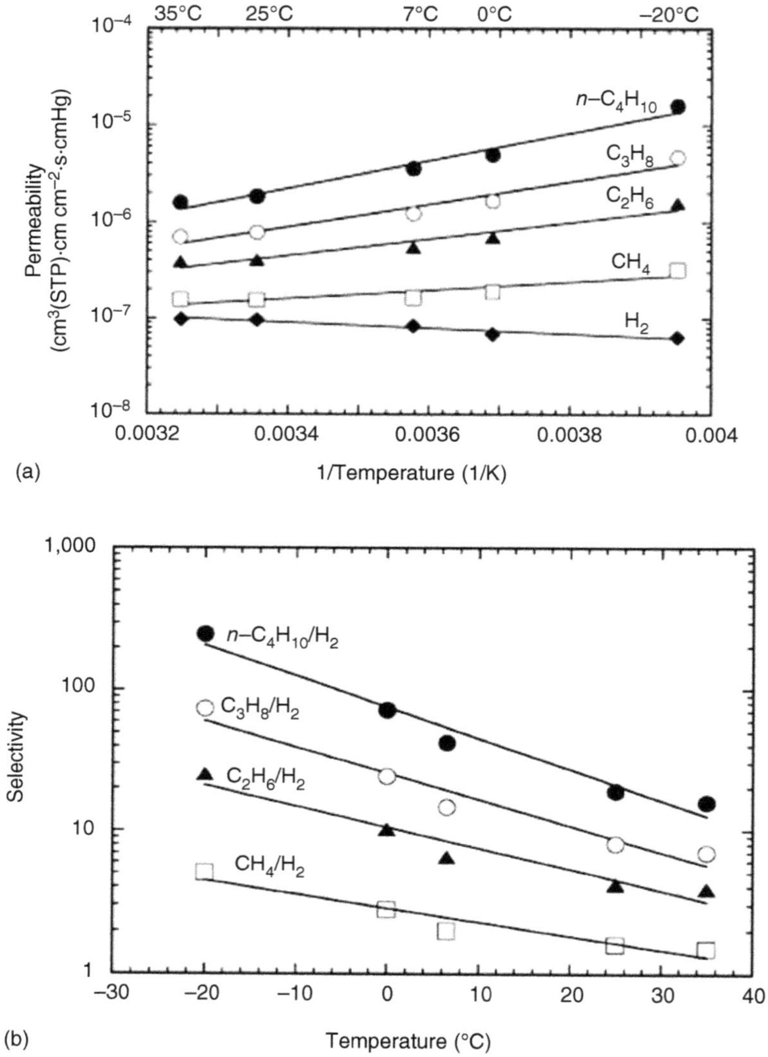

Figure 3.13 For a PDMS membrane. (a) Mixed-gas hydrogen, methane, ethane, propane, and n-butane permeability versus inverse feed temperature and (b) *n*-butane/hydrogen, propane/hydrogen, ethane/hydrogen, and methane/hydrogen selectivity versus temperature. Source: Pinnau and He [84].

low densities, and high gas permeabilites. PTMSP, a substituted polyacetylene containing bulky substituents, has been studied for membrane use because of its unique structure and excellent gas and vapor permeation properties. The first report of the synthesis of PTMPS appeared in 1983 by Masuda et al. [85]. The chemical structures of several polyacetylenes are shown in Figure 3.14. The synthesis of high molecular weight PTMSP was accelerated by the discovery of WCl_6 analogous catalysts ($NbCl_5$ and $TaCl_5$). PTMSP is a glassy polymer with high gas permeability

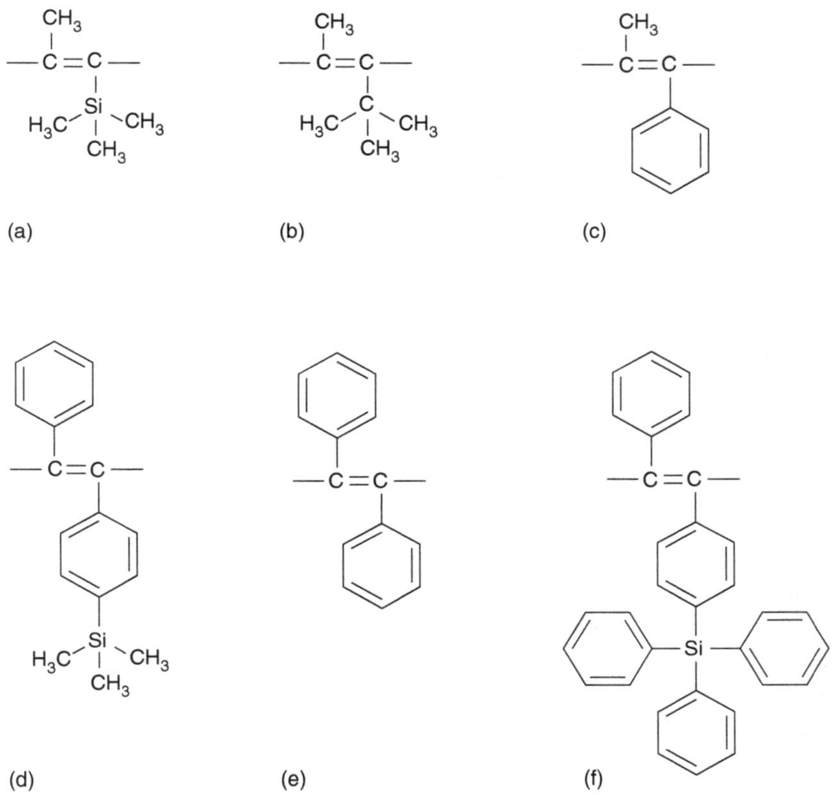

Figure 3.14 Chemical structures of disubstituted polyacetylenes: (a) PTMSP, (b) PMP, (c) PPP, (d) PTMSDPA, (e) PDPA, and (f) PTPSDPA.

and low selectivity. It is soluble in nonpolar solvents such as toluene, cyclohexane, and carbon tetrachloride, thus permitting an easy film formation by a solution casting method.

Nagai et al. [86] reported a detailed review on the synthesis and characterization of substituted polyacetylenes and its use in membrane applications. Discovery of the PTMSP membrane with high gas permeability prompted membrane scientists to explore the relationship between gas transport behavior and PTMSP microstructure. High gas or vapor permeability in the PTMSP membrane is due to the ultra-high free volume of this polymer. Free volume of this polymer can be two or three orders of magnitude higher than free volumes of conventional glassy polymers such as polyimides, poly(phenylene oxide)s, and polysulfones. The fractional free volume (FFV) of PTMSP is about 0.34. Typically, it is believed that the large amount of free volume (~30%) in freshly prepared PTMSP is interconnected, thus permitting the rapid diffusion of gas or vapor [87]. After the synthesis of PTMSP, other substituted polyacetylenes such as poly(4-methyl-2-pentene) (PMP) and poly(*tert*-butylacetylene) (PTBA) were synthesized. These polymers, analogous to

PTMSP structures, exhibit similar physical properties (densities, free volumes, and glass transition temperature) and are classified as "superglassy polymers".

Despite excellent gas permeation properties of PTMSP, potential application of PTMSP for gas separation is limited because of its rapid loss of free volume by physical aging or relaxation and its lack of chemical stability in heat, radiation, or ultraviolet light in the presence of oxygen. The accessible free volume of PTMSP seems to be very dependent upon the method of preparation or sample history. Therefore, these polymers require improving their long-term stability [83] using further crosslinking or mixing with inorganic nanoparticles.

3.4.6 Amorphous Fluoropolymers

Fluoropolymer has been synthesized from fluorine-containing monomers such as tetrafluoroethylene, trifluoroethylene, and vinylidene fluoride. Fluorinated polymers have many potential characteristics compared to common hydrocarbon polymers. Polytetrafluoroethylene (PTFE), known as a commercial trade name of Teflon®, possesses extraordinary chemical, thermal, mechanical, and electric properties. Whereas most PTFE- or PVDF-based fluoropolymers are either crystalline or semicrystalline, perfluoro-alicyclic Teflon® AF polymers (Figure 3.15) have been reported to exhibit similar desirable electric, chemical, thermal, and mechanical properties, compared to semicrystalline fluoropolymers [88]. Moreover, these amorphous fluoropolymers have unique physical properties such as high T_gs and high free volume elements, similar to the most famous high free volume glassy polymer, PTMSP. These fluoropolymers can be dissolved in a few fluorinated solvents due to low cohesive energy density (CED), which contributes to low activation energies for diffusion of small molecules as well as solubility of gases [89].

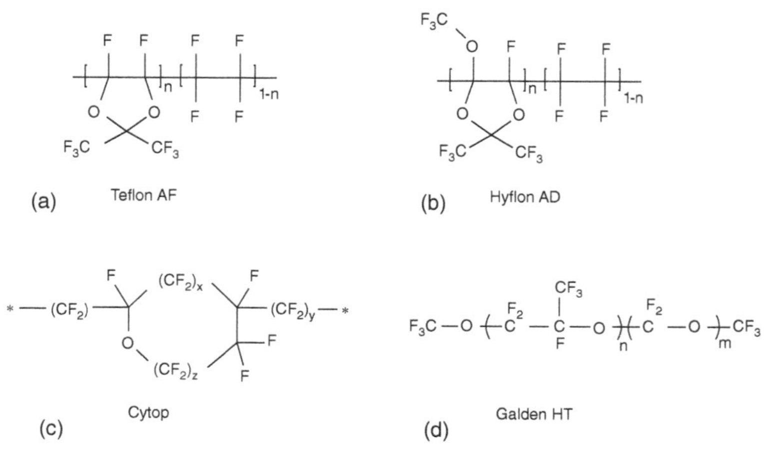

Figure 3.15 Chemical structures of Teflon (a) AF-2400, (b) AF-1600, (c) Cytop, and (d) Galden HT. Structures of major perfluoropolymers and fluorinated solvent.

To prepare Teflon® AF series, reactions of hexafluoroacetone (HFA) and ethylene oxide (EO) give rise to defluorinated 2,2-bis(trifluoromethyl)-4,5-difluoro-1,3-dioxole (BDD) monomer, followed by the copolymerization with tetrafluoroethylene (TFE). The physical properties of these amorphous copolymers are dependent upon the compositions in the copolymers. Currently, there are two commercial grades of Teflon® AF polymers produced by DuPont, AF-2400, and AF-1600, which contains the composition of copolymers for which $n = 0.87$ and 0.65, respectively, where n is the molar ratio of the BDD monomers [90].

Hyflon® AD is an amorphous perfluorinated copolymer of TFE and 2,2,4-trifluoro-5-trifluorometoxy-1,3-dioxole (TTD) produced by Solvay. Compared to Teflon® AF series, the reduced intermolecular distance due to the structural change result in moderate fractional free volume of 23%. Thus, the gas permeabilities are lower than those of Teflon® AF membranes, allowing the increase of permselectivities. While this fluoropolymer has the advantage of solubility in fluorinated liquid, such as Galden® HT 55 (Solvay) and HFE 7100, the permeation properties are affected by the kind of residual solvents as well.

Because the syntheses of these fluorinated polymers should treat very toxic fluorine chemicals, they are mainly produced by major chemical companies such as Solvay Specialty Polymers. Like PTMSP, these polymers are permeable to permanent gases. As summarized in Table 3.6, carbon dioxide and nitrogen permeability of a Teflon AF-2400 membrane is 2200 and 480 Barrers, respectively. Jansen and Drioli investigated the transport properties of Hyflon® AD membranes [95–97]. Because the molar weight of the solvents used for film casting of Hyflon® have relatively high, the residual solvents affected the membrane performances. Treatment of the membranes above their T_g allowed generation of the reproducible permeability and permselectivity without plasticization by the residual solvents. Due to the smaller and packed alicyclic structure of the fluorinated monomer, the density of Hyflon® AD membranes are higher than those of Teflon® AF membranes and thus the fractional free volume relatively diminishes. Therefore, the gas permeabilities

Table 3.6 Physical properties and gas transport properties of amorphous perfluorinated polymers.

	Physical properties			Gas permeability (Barrers)					Permselectivity				
	T_g (°C)	Density (g cm^{-3})	FFV	CO_2	H_2	O_2	N_2	CH_4	O_2/N_2	CO_2/N_2	CO_2/CH_4	H_2/CO_2	References
Teflon AF 2400	240	1.74	0.33	2200	2090	960	480	390	2.0	4.6	5.7	0.95	[91]
Teflon AF 1600	160	1.8	0.31	520	550	270	110	80	2.4	4.7	6.5	1.05	[92]
Hyflon AD 80	134	1.92	0.23	150	210	67	24	12	2.8	6.3	13	1.4	[93]
Hyflon AD 60	134	1.92	0.23	130	180	57	20	10	2.9	6.5	13	1.38	[93]
Cytop	108	2.03	0.21	35	59	16	5.0	2.0	3.2	7	18	1.69	[93]
PVDF	30	2.15	—	1.2	2.4	1.4	0.7	1.3	2.0	1.7	0.9	2.0	[94]

of Hyflon® AD membranes do not reach those of Teflon® AF membranes while the permselectivities of permanent gas mixtures are higher.

PTMSP and AF-2400 are both ultra-high free volume glassy polymers, but their gas permeation properties are somewhat different. Permeabilities of small gas molecules in AF-2400 membranes are higher than those of large, condensable gases while PTMSP membranes exhibit the opposite permeation behaviors. The main contribution to selectivity of AF-2400 is due to diffusivity-selectivity while the selectivity of PTMSP is controlled by large differences in solubility. Also, AF-2400, like PTMSP, has negative activation energies of permeation for small gases, but exhibits weak temperature dependence of gas permeability as compared to PTMSP. Because the vapor/gas selectivity of AF-2400 membrane is sensitive to vapor activity, vapor/permanent gas selectivity varies depending upon the feed pressure. Therefore, AF-2400 membranes are not suitable for vapor/gas separations, such as the separation of air from organic vapors. On the other hand, AF-2400 membrane can be used as an effective coating material for composite and asymmetric polymer membranes such as siloxane polymers because of its high gas permeability.

3.4.7 Polybenzimidazole

Polymer membranes are severely limited in their uses in extremely high temperature and corrosive environments. At temperatures above 100 °C at oxidative atmospheres, most of organic polymers are thermally decomposed or deformed. For that reason, microporous inorganic membranes such as zeolite, silica, modified alumina, and carbon molecular sieve have been regarded as potential membrane materials able to survive extreme operating conditions such as high temperature, high pressure, and oxidative streams. These microporous inorganic membranes exhibit high selectivity and high permeability as well as thermal, chemical, and mechanical stability. They can be used in specific applications in which polymer membranes fail to display the desirable separation performance.

However, the cost of inorganic membranes or their modules is still too expensive to be competitive against the current polymer membranes or modules.

Pesiri et al. [98] reported the availability of polybenzimidazole (PBI, see Figure 3.16) for gas separations at high temperatures. PBI was chosen because of its high T_g above 427 °C and its onset of thermal decomposition near 600 °C. Figure 3.16 shows the chemical structure of poly[2,2′-(m-phenylene)-5,5′-bibenzimidazole] (PBI). It can be synthesized from condensation polymerization of 3,3′,4,4′-tetraaminobiphenyl (TAB) and diphenyl isophthalate (DPIP) in poly(phosphoric acid) or in a hot molten nonsolvent such as sulfolane or diphenyl sulfone [99]. Gas permeation tests were carried out at 300–350 °C, near the temperature of reformate streams used in hydrogen separations. Poor gas separation performance of PBI at room temperature is due to its rigid-rod structure. However, higher separation performance can be achieved by using operational temperatures above 100 °C. For instance, hydrogen permeability of the PBI membrane at room temperature is low (0.09 Barrer) but increases up to 18 Barrer at 340 °C. Carbon dioxide permeability also increases from 0.01 Barrer at 25 °C to 4 Barrer at 340 °C. In contrast, H_2/CO_2

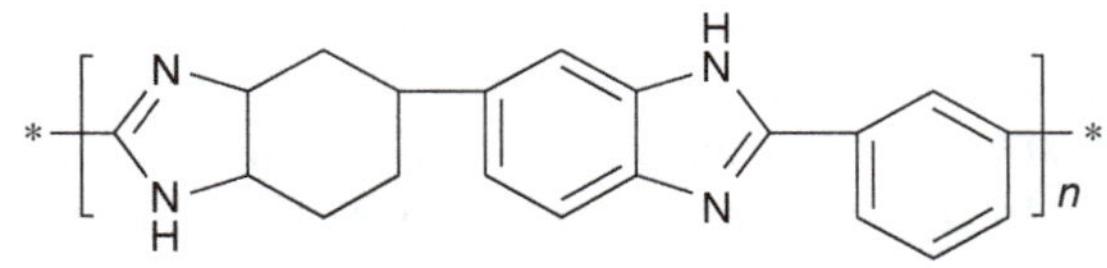

Figure 3.16 Chemical structure of poly [2,2′-(m-phenylene)-5, 5′-bibenzimidazole] (PBI).

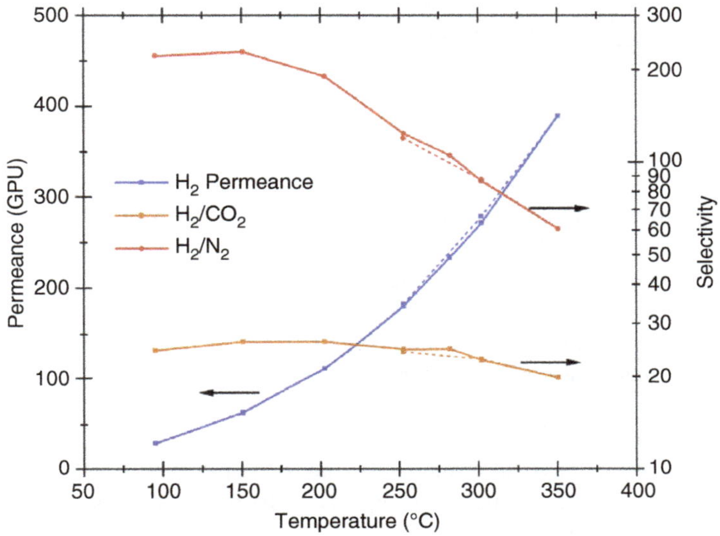

Figure 3.17 H_2 permeance and H_2 selectivity over CO_2 and N_2 as a function of temperature measured for the PBI hollow fiber membrane [100]. Dotted and solid lines represent data collected during increasing and decreasing temperature cycles, respectively.

selectivity decreased from 9 to 4.5. Interestingly, H_2/CO_2 selectivity at temperature ranges between 200 and 270 °C increased to about 20 (Figure 3.17). Thin layer of PBI was coated on stainless steel and evaluated for syngas separations at elevated temperatures for H_2 production [101]. The PBI composite membrane demonstrated exceptional long-term thermos-chemical stability in the syngas environment in the presence of H_2S. The H_2 permeance and H_2/CO_2 selectivity for the PBI composite membrane in simulated dry syngas was recorded at 7 GPU and 47, respectively. They recently developed macrovoid-free PBI hollow fiber membranes for elevated H_2/CO_2 separations [100]. Dope composition containing an acetonitrile diluent enables asymmetric PBI hollow fibers using a water coagulant. H_2 permeances of 400 GPU and H_2/CO_2 selectivity >20 was achieved. It is interesting that hydrogen permeability through membranes of course is ubiquitous in nature with membrane transport of gases and liquids as an inherent element of life itself.

Although this study provides insight into polymer membrane materials for separating valuable gases from high-temperature gas streams, more information is required to understand gas transport and the mutual interactions that affect gas selectivity at temperatures above 100 °C. There are very few reports so far, and the single or mixed gas permeation data (including diffusivity and/or solubility) is not sufficient for analyzing gas transport behavior at high-temperature ranges.

Although H_2/CO_2 selectivity of PBI is potentially useful for commercial hydrogen production applications, the H_2 permeability of PBI is only about 100 Barrers at 300 °C. If PBI membrane could be considered for practical applications, such as hydrogen purification or as a water gas shift membrane reactor, its properties must be tailored to achieve high H_2 flux (productivity) in a practical module and system.

3.4.8 Nanocomposites and Mixed Matrix Membranes

Over the last several decades, the introduction of inorganic nanoparticles into a continuous polymer matrix has been tried to improve mechanical and thermal stability of conventional polymer membranes and to improve gas separation performance. These membrane materials are called "mixed matrix membranes" (MMMs) or "hybrids or composite membranes". Inorganic particles mixed with organic polymer matrix can be classified into two categories: porous and nonporous inorganic particles. Colloidal silica (from in situ sol–gel reaction), zeolite, metal organic frameworks (MOF), and boron nitride (BN) belong to porous inorganic particles, while fumed silica, clay, graphene, graphene oxide (GO), and reduced graphene oxide (r-GO) belong to nonporous inorganic particles.

Although there are many publications dealing with mixed matrix membrane materials in the literature during the last two decades, the original concept dates to the past decades. Acting as physical crosslinkers, inorganic fillers such as fumed silica or carbon black were already used for improving the mechanical properties of natural rubbers or synthetic rubbery polymers. However, with the advent of nanotechnology, the combination of polymer and nano-sized inorganic components becomes useful beyond the mixing to improve the physical and chemical properties. Composite technology is promising for tailoring polymer microstructures that might affect gas transport behavior in the membrane-based gas separation.

Typically, the addition of inorganic particles may improve the gas selectivity by increasing the diffusional pathway for gases and can also strengthen the anti-plasticization effect by restricting the thermal motion of polymer chains. Unexpected gas transport behavior of polymer nanocomposite membranes was reported [102]. Polymer nanocomposites consisting of PMP and hydrophobic surface-treated nano-sized fumed silica were prepared in the wide range of fumed silica content. The fumed silica is so nonporous that gas permeability was expected to decrease with the increase of silica content. Classically, the incorporation of impermeable filler particles into a polymer leads to a systematic reduction in molecular transport [103, 104]. The original idea of this work might be to prevent the severe decline of gas permeabilities caused by the progress of physical aging in ultrahigh free volume glassy polymers such as PTMSP and PMP, and simultaneously to improve the gas selectivity by increasing the diffusion pathway because nonporous inorganic fillers can act as both a physical hardener and barrier. In contrast to this initial expectation, it was reported that the gradual addition of nanostructured fumed silica to several glassy high free volume polymers results in an increase in permeability [102, 105] as shown in Figure 3.18. The unexpected improvement of gas permeability is supposed to be due to the disruption of polymer chain packing by the silica particles, leading to an increase in the free volume available for molecular transport.

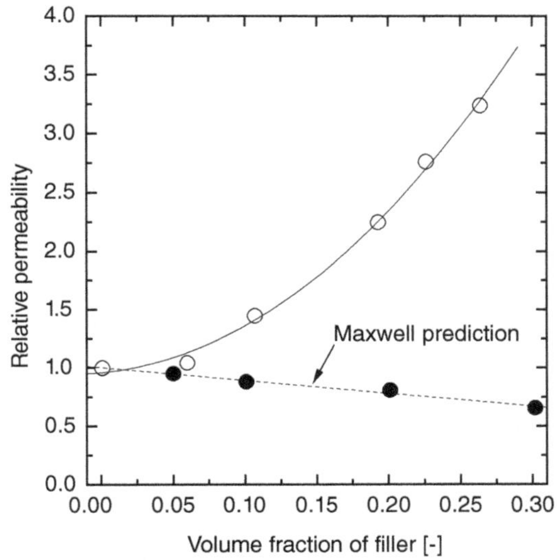

Figure 3.18 The ratio of penetrant permeability in the nanocomposite to that in the pure polymer as a function of filler volume fraction. Source: Merkel et al. [102].

When the high free volume polymer PMP is combined with inorganic nanoparticles, an increase in vapor/gas selectivity was also reported [106]. However, when the inorganic particles are added in PTMSP, different results on selectivity have been obtained [107]. The incorporation of fumed silica increases the free volume of PTMSP without creating non-selective defects. However, it can also create free volume elements large enough to allow non-selective Knudsen transport [105]. To fully understand these phenomena, more systematic studies of important factors (e.g. the dispersion state of inorganic nanoparticles into the polymer matrix, the effect of different kinds of polymers, particle sizes, particle shapes, and kinds of particles) are needed.

MMM incorporating polymers and inorganic particles may be promising materials for improving the weaknesses of organic polymer membranes. Composite membranes or MMM can be prepared from inorganic particles coupled with an organic polymer matrix by covalent bonding or ionic bonding or van der Walls interaction. These couplings between different materials can prevent undesirable reductions of selectivity due to large interfaces between the inorganic particle and the polymer. The incorporation of inorganic particles or molecular frameworks into polymer matrices can significantly reduce the loss of selectivity at elevated temperatures by retarding thermal motion of the polymer chain. Park and Lee [108] reported the gas permeation properties of imide-siloxane copolymer/silica hybrid membranes. They used silane-coupling agents to prepare imide-siloxane copolymer/silica hybrid membranes. Even at high silica content (up to 50 wt%), the hybrid membranes did not experience macroscopic phase separation due to the existence of coupling agents. This might be due to the improvement of compatibility between organic polymer matrix and inorganic particles.

Two-dimensional (2D) nanosheets have shown great potential for separation applications because of their exceptional molecular transport properties. Nanosheet materials such as GO, MOF, and covalent organic framework (COF) show unique and fast molecular transport through nanopores and/or nanochannels [109, 110]. However, dimensional instability of nanosheets in harsh environments diminishes the membrane performance and hinders their long-term operation in various applications such as gas separation, water desalination, and ion separation. Nanosheet membranes have been modified by crosslinking or functionalization that improved the stability of the membranes, their separation functionality, and the scalability of membrane formation while retaining the excellent molecular transport properties of the membranes. These improvements have enhanced the potential of nanosheet membranes in practical applications such as separation processes. Recently developed 2D nanosheet materials, such as MOFs [29, 111, 112], COFs [31], and GO [30] have been considered as attractive building blocks to construct high-performance membranes (Figure 3.19) because of their thin atomic thickness, robustness, and potential for functionalization. The precise and fast molecular transport properties of 2D nanosheet membranes come from their intrinsic nanopores on nanosheets as well as their nanochannels between stacked nanosheets.

Graphene-based membranes are the most widely reported nanosheet membranes and include porous graphene, GO, and rGO [30]. Among them, GO membranes have demonstrated unique molecular transport properties caused by hydroxyl groups on GO, which provide nanochannels suitable for transport of ions and small molecules. GO membranes are usually prepared by coating methods such as spin-coating or filtration, which form ultrathin GO layers on a porous supporting layer with excellent mechanical stability. GO-based membranes are promising membrane materials that can be prepared with large effective areas. The remarkable molecular transport and separation properties of GO membranes can improve the efficiency of an entire process by increasing productivity and product purity of the separation process. Recently developed nanosheets and their composite membranes have enhanced stability and more possibilities for practical applications. The modification of 2D materials have improved the overall materials properties for membranes, solving the concerns of previously developed 2D membranes. The modifications, such as chemical and physical crosslinking, functionalization, and inclusion in polymer nanocomposites, tightly bind the 2D nanosheets, which regulates the space between nanosheets and its effects on molecular transport properties and increases dimensional stability. A variety of crosslinkers, such as linear and aromatic monomers with amide bonds, or polymers to physically entwine nanosheets with polymer chains, has been incorporated into 2D materials to form ultrathin nanocomposite membranes. After development of graphene materials, numerous graphene-like 2D materials have been investigated and widely studied about their physiochemical properties, fabrication approaches, and applications [118–124]. Synthetic methodologies in solution-based methods for constructing layered structures and exfoliating 2D materials in relatively mild conditions were developed resulting in uniform-sized, high-quality, and single- or few-layered nanosheets. 2D materials have shown great potential in applications such as electronic and optical devices, sensors, catalysis, and energy storage [118–120], but still presented some issues as a membrane for practical separation processes [121–124].

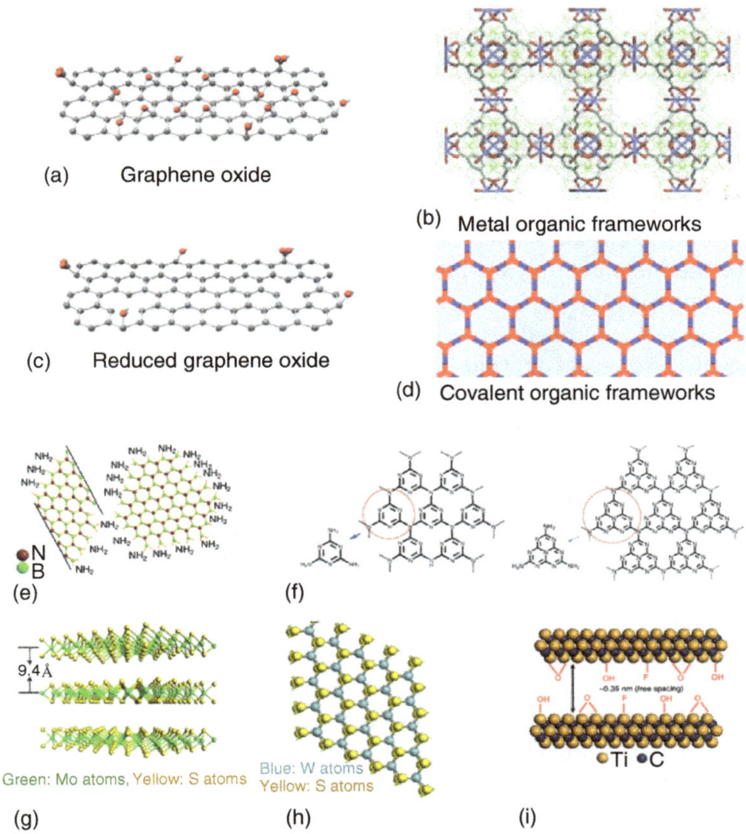

Figure 3.19 Examples of 2D materials. (a) Graphene oxide [30], (b) metal organic frameworks [29], (c) reduced graphene oxide [30], and (d) covalent organic frameworks [31]. Examples of other 2D materials. (e) Boron nitride (BN) [113], (f) graphite carbon nitride ($g-C_3N_4$) [114], (g) molybdenum disulfide (MoS_2) [115], (h) tungsten disulfide (WS_2) [116], and (i) MXene [117]. Source: Copyright 2015, Elsevier. Copyright 1999, The American Association for the Advancement of Science. Copyright 2013, American Chemical Society. Copyright 2015, 2018, Nature Publishing Group. Copyright 2012, John Wiley & Sons. Copyright 2013, Springer.

Molecular transport through nanosheet membranes was first theoretically investigated by quantum mechanical calculations that showed that small pores generated on graphene or GO enable selective gas transport for specific gases [125]. Stacked graphene or GO also form nanochannels between nanosheets, called the interlayer spacing, and the width of the nanochannels is determined by functional groups of graphene or GO that influence gas permeation and separation properties. Kim et al. demonstrated selective gas transport of small molecules through a few, layered GO membranes with thickness <10 nm [126]. The membranes had the ultrafast gas transport property of H_2 and selectivity over other gases. Interestingly, the gas transport properties varied depending on the experimental conditions such as temperature and humidity. GO membranes are likely to swell in high-humidity environments that increase the interlayer spacing of GO nanosheets, but water molecules accumulated in nanochannels hinder gas transport, except for CO_2 transport. Usually, gas transport properties of conventional polymer membranes are changed by temperature and humidity mainly due to changes in gas activation behavior [11], but for GO membranes, the gas transport properties change because the nanochannel structure becomes unstable.

One approach to improve the mechanical properties of a membrane and form an ultrathin layer is to crosslink the GO nanosheets with borate. Wang et al. prepared a GO-borate membrane with 8 nm thickness by vacuum-assisted filtration and thermal treatment [127]. Depending on the thermal treatment temperature, the interlayer spacing was controlled between 0.68 and 0.79 nm, the change of interlayer spacing between wet and dry states was small, and the membrane did not lose its separation performance in a humidified condition with an industrial gas stream. The gas permeation properties of the membrane had a CO_2 permeance of 650 GPU ($1\ \text{GPU} = 10^{-6}\ \text{cm}^3\ ((\text{STP})\ \text{cm}^2/\text{s/cmHg})^{-1} = 3.3 \times 10^{-10}\ \text{mol}\ (\text{m}^2/\text{s/Pa})^{-1}$) and selectivity of 75 over methane (CO_2/CH_4), which exceeded other GO membranes and was even comparable to zeolites or ultrathin polymers. Yan et al. introduced a small amount of GO to a chitosan membrane and followed the crosslinking with borate with a thermal treatment to improve its mechanical and barrier properties such that only 1.0 wt% each of boron and GO in chitosan increased the tensile strength by 160% and surpassed the original oxygen permeability by 10% [128], mainly because the GO nanosheets and crosslinked network increased the tortuosity of the pathway for gas transport and decreased the gas diffusion. Yang et al. prepared thiourea-crosslinked GO (TU-GO) membranes through a facile hydrothermal self-assembly synthesis by immersing a ceramic support tube in a GO and thiourea solution at 80 °C in an alkaline condition (pH = 8) that caused covalent crosslinking [129]. Thickness of the TU-GO layer was regulated from 70 nm to several microns by controlling reaction time and solution concentration. This membrane presented a unique H_2 permeance of $>68 \times 10^{-8}\ \text{mol}\ (\text{m}^2/\text{s/Pa})^{-1}$ (2060 GPU) and an ideal selectivity of over 150 for any other gas molecules.

GO nanosheets are used as a building block or a scaffold of ultrathin membranes, although they are often used to suppress gas transport and improve barrier properties. Because membrane thickness is important to increase gas permeance, membrane fabrication with an ultrathin membrane layer has been investigated but has often led to defects on the membrane that eliminated its separation properties. Kim et al. introduced GO nanosheets into polybenzoxazole-co-imide, called a thermally rearranged polymer, with a thickness of 35 nm [130]. A polymer solution with GO nanosheets was spin-coated on an alumina disc as a substrate and GO nanosheets uniformly coated the substrate without defects. The membrane demonstrated an improved CO_2 permeance of 1784 GPU while maintaining a gas selectivity that was 482 times higher than the freestanding membrane film.

MOFs have widely been investigated as gas separation membranes due to their unique crystalline and pore structures. 2D MOFs provide well-ordered nanopores as well as nanochannels that are advantageous for membrane applications, compared to graphene-based nanosheets. MOF membranes usually have been introduced as a thin film composite on a substrate or a MOF-polymer composite, or a MMM, but micron-sized 3D crystals often agglomerated and restricted the integration of MOFs in the polymer phase. 2D nanosheet fillers of MOFs are advantageous in incorporating fillers into polymers or building ultrathin 2D membranes for membrane processes. Rodenas et al. introduced 1,4-benzenedicarboxylate (CuBDC) MOF to synthesize MOF nanosheets from bulk CuBDC MOF by the diffusion-mediated modulation of the MOF growth kinetics [131]. This unique MOF crystal morphology was evolved based on the synthesis temperature from ultrathin nanosheets (<10 nm) at a low temperature of 298 K to thicker platelets at a high temperature of 333 K. The same methodology can be adapted to other layered MOFs by using alternative metal nodes such as Co^{2+} and Zn^{2+} to tune the porosity and functionality of MOFs. These 2D MOF nanosheets were incorporated into a commercial polyimide called Matrimid® 5218 to fabricate MMMs and MOF nanosheets that were uniformly distributed through the entire membrane, whereas the regular MOF crystals usually formed fractures or defects between crystals and the polymer phases. Pustovarenko et al. prepared nonlayered MOF nanosheets through a surfactant-assisted method and fabricated MMMs with Matrimid® 5218 [132]. This membrane presented an improvement of CO_2 permeability from pristine Matrimid® 5218 by almost 200% and CO_2/CH_4 selectivity after adding 16 wt% MOF nanosheets. Peng et al. prepared poly(Zn_2(benzimidazole)$_4$) ($Zn_2(bim)_4$) nanosheet membranes coated on α-Al_2O_3 discs as molecular sieve nanosheets [133]. $Zn_2(bim)_4$ nanosheets were prepared from bulk $Zn_2(bim)_4$ by ball milling and were coated by vacuum filtration. The resulting membrane demonstrated H_2 permeance of up to 2700 GPU with H_2/CO_2 selectivity of 291. Wang et al. introduced $Ni_8(5\text{-bbdc})_6(\mu\text{-OH})_4)$ structure, called Mesh Adjustable Molecular Sieve (MAMS-1) due to their excellent hydrothermal stability and 2D layered structure with interplanar distances of 2.9 and 5.0 Å [134]. This MAMS-1 membranes presented a sharp cut-off of the permeance between He and H_2 gases and larger gases. Hu et al. incorporated GO as a seeding layer during ZIF-8 synthesis and formed ZIF-8/GO nanosheet membranes on alumina as well as on nylon substrates [135]. Incorporating GO into ZIF-8 greatly reduced the membrane thickness to ~100 nm and improved gas permeance. This membrane had an H_2 permeance of over 5.46×10^{-8} mol $(m^2/s/Pa)^{-1}$ (165 GPU). Similarly, g-C_3N_4 was also used as a template layer of ZIF-8, exhibiting H_2 permeance of 203 GPU [136]. Shan et al. prepared ultrathin layer of BILP-101x on γ-Al_2O_3 support that showed high temperature stability and exhibited H_2 permeance of up to 24.3 GPU with H_2/CO_2 selectivity of 39.7 [137].

Other 2D materials, such as COF, MXene, MoS_2, WS_2, BN, or g-C_3N_4 also presented performance in gas separation application. Ultrathin MoS_2 membranes prepared by Wang et al. showed very high H_2 permeance up to 27,440 GPU based on Knudsen diffusion [138]. Ding et al. presented freestanding 2D MXene membranes with thickness of 2 μm that showed excellent H_2 permeability and selectivity (H_2/CO_2 selectivity of 160) with stable long-term performance for >700 hours of continuous operation [117]. Shen et al. showed another MXene membrane fabricated on AAO substrate by vacuum filtration that exhibited defect-free layer with controllable thickness of 5–80 nm [139], demonstrating precise H_2 transport with

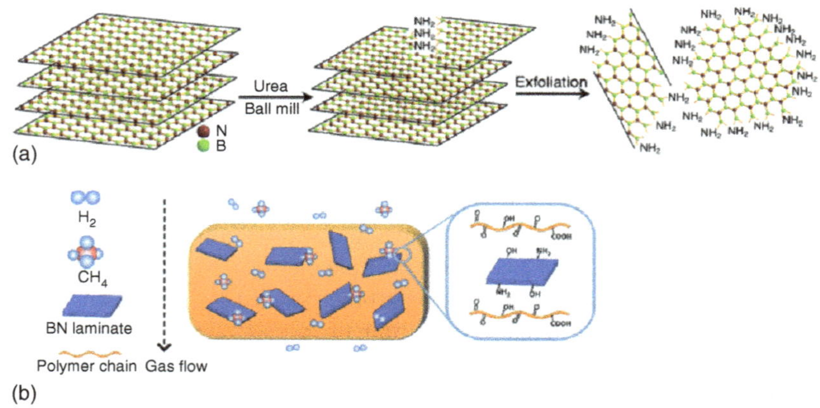

Figure 3.20 Schematic illustration of the exfoliation and dispersion process. (a) Schematic illustration of the exfoliation [113] and (b) schematic illustration of the structure of functionalized BN-polymer membrane and selective gas permeation through the membrane [141]. Copyright 2015, Nature Publishing Group. Copyright 2018, John Wiley & Sons.

H_2 permeance of 1584 GPU and H_2/CO_2 selectivity of 27. However, after crosslinking with borate and polyethylenimine followed by 75 °C treatment (MBP-75), CO_2 permeance was remarkably enhanced. Incorporation of nanosheets such as WS_2 or BN into polymer phase to form MMMs improved gas permeability and/or selectivity. 10% introduction of WS_2 into fluorinated poly(phenylene oxide) (FPPO) increased CO_2 permeability, CO_2/N_2 selectivity, and CO_2/CH_4 selectivity by 216, 148, and 137%, respectively, because WS_2 provided transport channels for gas molecules [140]. Similarly, introduction of functionalized BN nanosheets into thermally rearranged polymers by 1 wt% increased H_2/CH_4 selectivity from 24.1 to 275 [141] (see Figure 3.20).

3.4.9 Other Promising Polymers

3.4.9.1 Pebax

Poly(amide-6-b-ethylene oxide) (PEBAX®) can increase solubility selectivity. PEBAX® is a block copolymer consisting of a polyamide (PA) block as a hard segment and a polyether (PE) block as a soft segment (Figure 3.21). A crystalline amide block in PEBAX® functions as an impermeable phase, whereas an ether block acts as a permeable phase because of its high chain flexibility.

Rezac et al. [142] reported the sorption and diffusion characteristics of water and methanol in a series of PEBAX® copolymers. They revealed that those polymers were used to separate methanol selectively from air, rather than separating methanol from water. Bondar et al. [143] found that in applications such as the removal of CO_2 from mixtures of hydrogen in syngas, PEBAX® shows high

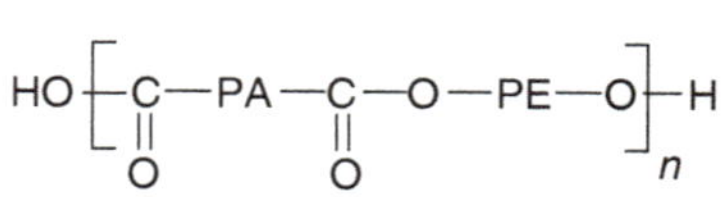

Figure 3.21 Structure of poly(amide-*b*-ethylene oxide) (PEBAX®).

Table 3.7 Structure–property relationships of poly(amide-*b*-ethylene oxide) membranes for CO_2 separation.

Copolymer	Soft segment	Hard segment	Temp.	P_{CO_2} (Barrer)	α_{CO_2/N_2}	α_{CO_2/H_2}	References
PEBAX 2533	Tetramethylene oxide 80 wt%	Nylon-12	35	180	18	3.5	[143]
PEBAX 4033	Tetramethylene oxide 53 wt%	Nylon-12	35	95	16	3	[143]
PEBAX 4011	Ethylene oxide 57 wt%	Nylon-6	35	57	42	6.9	[143]
PEBAX 1074	Ethylene oxide 55 wt%	Nylon-6	35	110	46	7.9	[143]
PEBAX 3533	Tetramethylene oxide 70 wt%	Nylon-12	25	132	61	6.1	[144]
PEBAX 1657	Ethylene oxide 60 wt%	Nylon-6	25	122	71	—	[144]
PEBAX 1657 /TEOS 10 wt%	Ethylene oxide 60 wt%	Nylon-6	25	154	72	—	[145]
PEBAX 1657 /TEOS 19 wt%	Ethylene oxide 60 wt%	Nylon-6	25	205	75	—	[145]
PEBAX 1657 /TEOS 27 wt%	Ethylene oxide 60 wt%	Nylon-6	25	277	79	—	[145]

selectivity in polar or quadrupolar/nonpolar systems (e.g. CO_2/H_2 or CO_2/N_2). The sorption and permeation results indicate strong interactions between the polar gas CO_2 and the PEO blocks in the copolymer. Kim et al. [144, 145] found that strong affinities of polar species to the PE block is attributable to the high permeability and permselectivity of polarizable gases through the PEBAX® copolymer. This leads to high CO_2/N_2 selectivity up to 61 and SO_2/N_2 selectivity up to 500. The permeability of CO_2 in PEBAX® ranged between 110 and 180 Barrer depending on the soft PEO or tetramethylene oxide contents (Table 3.7). The CO_2 permeability in PEBAX® decreased with an increase of hard PA content from 20% to 46% in PEBAX® as shown in Figure 3.22. In the presence of uniform and rigid chain packing, a crystalline polymer lacks in both the sorption sites as well as the mobility of the chains to allow transport of gas molecules.

3.4.9.2 Polymers with Intrinsic Microporosity

As many conventional glassy polymers have been known to be dense without micropores except free volume elements resulting from chain rotation in the matrix, polymers with intrinsic microporosity (PIMs) are a kind of microporous polymers containing microporous cavities due to the mixed conformation of highly stiff ladder-like domain and flexible benzodioxane structure as described in Figure 3.23. Their rigid and contorted molecular structures interfere with chain packing, and

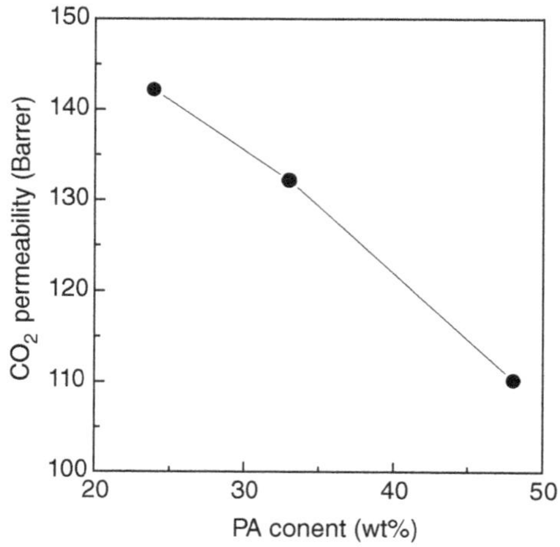

Figure 3.22 Change of CO_2 permeability with the amount of crystalline region in PEBAX® membranes. Source: Kim et al. [144]/Elsevier.

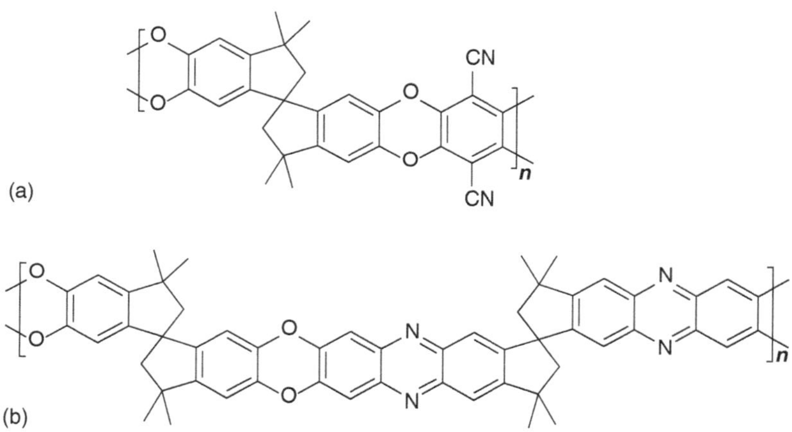

Figure 3.23 Typical polymers with intrinsic microporosity: (a) PIM-1 and (b) PIM-7.

can retain an efficient size of free volume elements for transporting small gas molecules, and thus achieve high diffusion and permeation coefficients [146].

Since the first synthesis of PIM was reported by Budd and McKeown in 2004, many studies have been performed regarding their microporous characters, sorption, hydrogen storage, and transport properties [146–148]. Up to now, more than 700 scientific papers have been published focusing mainly on the syntheses of new PIMs, the specific surface areas by structural modifications and the sorption capabilities of hydrogen as a promising candidate for hydrogen storage materials [148, 149]. Molecular concept of PIMs is presented in Figure 3.24. PIMs are consisted of structural units and linking groups that produce a rigid and contorted polymer that cannot fill space efficiently. Therefore, PIMs have a large number of free volume elements by which they have high surface areas (500–1000 $m^2\ g^{-1}$) with microcavity diameters ranging between 0.6 and 0.8 nm as measured by positron annihilation

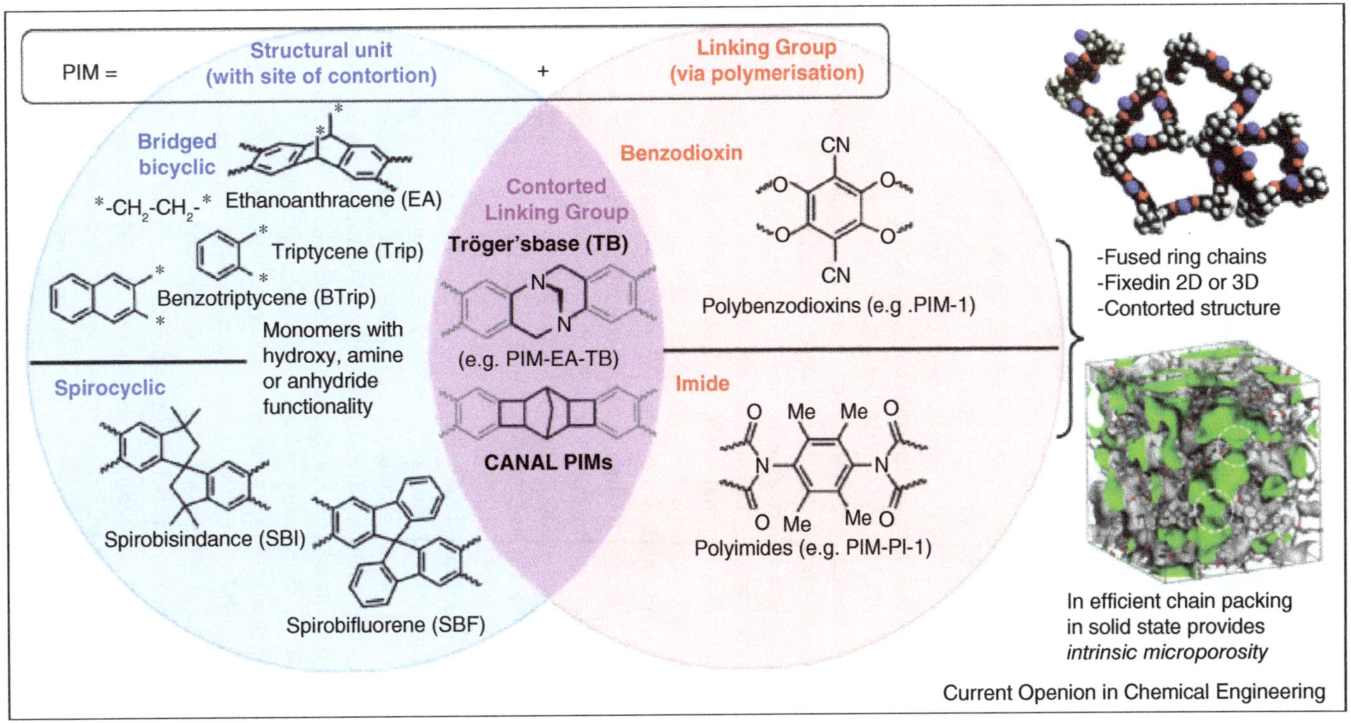

Figure 3.24 Molecular concept of PIMs in which structural units are fused together via linking groups to produce a rigid and contorted macromolecule which cannot fill space efficiently [149].

life time spectroscopy (PALS) [150]. PIMs have represented high gas permeation properties; O_2 permeabilities and O_2/N_2 selectivities of PIM-1 and PIM-7 are 370 Barrer, 4.0 and 190 Barrer, 4.5, respectively [151]. Gas permeabilities of these high free volume polymers can vary depending on the solvent during the synthesis and casting methods of the membrane, physical aging, and deformation of the free volume elements. In several references, CO_2 permeabilities of PIM-1 are ranged from 2300 to 6500 Barrers. Typically, PIM-1 polymer is dissolved in $CHCl_3$ to cast the membrane and evaporation takes several days to eliminate the residual solvent. During this slow evaporation of PIM-1 membrane, high free volume elements are formed and act as a permeable pathway to transport small gas molecules. After storing the film for a long time at ambient condition, these free volume elements undergo physical aging, indicating that the free volume elements are transiently destructed, and become smaller after time passes. These transient changes in free volume elements in PIM-1 results in a reducing in the original transport properties.

Typically, spiro-centered PIM-1 synthesized using spirobisindane (SBI) units present high BET surface area in the range of 720–820 $m^2\ g^{-1}$ and outstanding hydrogen gas permeability ($P(H_2)$ = 5010 Barrer), but these are still located in the vicinity of the upper bound. Therefore, microporosity and gas transport behaviors of PIMs have also been controlled by incorporation of cardo or spiro units. Adamantane and fluorene moieties were used to fine-tune their microporosity. However, the reported literature mainly investigated the effect of introducing these units into the PIM main chains. Shrimant et al. designed a new adamantine-containing bis(catechol) monomer, namely, 4,4′-((1r,3r)-adamantane-2,2-diyl)bis(benzene-1, 2diol) (THADM), and performed copolymerization reactions with typical spiro-centered PIM-1 to tailor the physicochemical properties as well as gas transport characteristics by changing the ratio of adamantanes to SBI units [152]. Note that controlled gas separation properties of the resulting PIM copolymers were located close to the 2008 Robeson upper bound, particularly for gas pairs such as H_2/N_2, CO_2/N_2, O_2/N_2, and CO_2/CH_4.

Several structural modifications have been reported to prevent efficient space packing of the stiff polymer chains and consequently show intrinsic microporosity. An exchange of a biscatechol monomer into a bis(phenazine)-based tetrahalide monomer provides the potential diversity of polymers suitable for making PIMs [153]. Tetrahalide monomers containing the spirocenter allows the use of a greater number of biscatechol monomers, which would otherwise lead to linear polymers that pack efficiently in space to give a dense, non-microporous material. Catechol-containing PIM-7 to PIM-10 represent diverse microporous sorption behaviors; surface area of PIM-10 is 713 $m^2\ g^{-1}$. Cardo-PIM-1 based on monomers containing fluorene moiety, a polycyclic aromatic hydrocarbon, showed CO_2 permeability of 430 Barrer and CO_2/N_2 selectivity of 33, respectively.

To regulate the free volume in PIMs, Du and Guiver introduce several bulky and rotational groups, and copolymerize those functional monomers. (i) Trifluoromethyl phenylsulfone groups are relocated within the interchain free volume of the rigid and contorted ladder polymer, acting to reduce gas permeability and increase selectivity [154]. Therefore, gas permeabilities for O_2 and CO_2 decreased with increasing content of the trifluoromethylphenylsulfone compared with the dinitrile monomer within the copolymer, while the selectivities of gases against nitrogen increased. (ii) 5,5′,6,6′-Tetrahydroxy-3,3,3′,3′-tetramethylspirobisindane (TTSBI) with tetrafluoroterephthalonitrile (TFTPN) and disulfone-based monomers: the incorporation of disulfonyl side groups into PIM structures increases selectivity [155]. The gas permeabilities and selectivities of DS-PIMs can be tuned by the size and nature of the pendant side chains and the copolymer ratio. (iii) A series of copolymers from tetrahydroxy dinaphthyl, tetrahydroxy spirobisindane, and tetrafluoro terephthalonitrile monomers: DNPIM-33 exhibits CO_2 permeability of 4646 Barrer and CO_2/N_2 permselectivity of 19.2, exceeding the polymeric upper bound [156]. This work demonstrates that significant improvements in properties may be obtained through development of copolymers with intrinsic microporosity (CoPIMs) that extends the spectrum of high-molecular-weight ladder structures of poly(dibenzodioxane)s. While bulky side chains in common glassy polymers enlarge the interchain distances and increase fractional free volume, these functional groups in PIM do not contribute to increment of torsional energies for the ladder-like main chain, but act as barriers for gas transport although the permselectivities are slightly improved.

McKeown and Budd present criteria for polymers with intrinsic microporosity by using more than 100 monomers [157]. From BET sorption measurement, positron annihilation lifetime spectroscopy and atomistic computer simulation, they emphasized the importance of microporous materials in several applications.

Investigation of shape-persistent iptycene units in PIM polymers is interesting ways to tailor polymer microporosity [47, 158–162]. Triptycene-containing PIMs (TPIMs) exhibited a higher energy barrier to deformation than spiro-centered PIM-1, resulting in enhanced gas separation performance. They presented superior gas selectivity (α_{H_2/N_2} ~ 50) compared to other ladder-type PIMs (α_{H_2/N_2} of PIM-1 ~6.1 and PIM-7 ~20) while maintaining a high gas permeability (P_{H_2} ~ 1549 Barrer). In particular, TPIMs showed improved H_2/gas selectivity compared to other triptycene-containing PIs since the entirely fused-ring backbone of TPIM contained more rigid than rotatable triptycene units in the PIs [158]. McKeown et al. advanced the chemistry of triptycene-based PIMs by grafting substituents such as methyl, trifluoromethyl, tetramethylnaphthalene, and hexamethylindane groups into the TPIM chains (PIM-DM-BTrip, PIM-TFM-BTrip, and PIM-HMI-BTrip, respectively). As a result, they obtained unexpected BET surface areas from 848 to 1034 $m^2\ g^{-1}$ with dramatic improvement in CO_2 separation performances.

Fluorene-containing spriobifluorene (SBF) units have been incorporated into PIM-1 (SBF-PIMs) that showed enhanced BET surface areas of 803 $m^2\ g^{-1}$ with notable gas permeability ($P(H_2)$ = ~6320 Barrer) [163]. Compared to SBI units, spiro-centers in the SBF units have additional fused benzene rings, which can offer more restricted movement. They helped to enhance gas diffusion selectivities to surpass the 2008 upper bounds for O_2/N_2, CO_2/N_2, CO_2/CH_4, and H_2/N_2. As in TPIM polymers, various kinds of substituents were introduced into pendants at SBF spiro-centers, PIM-SBF-1-7 [164], and PIM-TMN-SBI [159]. All the resulting polymers presented an improvement in gas transport characteristics, especially for tetramethyl substituted PIM-SBF-2, which showed a significant BET surface area ($938\,m^2\,g^{-1}$) and hydrogen permeability of 9160 Barrer.

Instead of SBI modification using iptycene, adamantane, and fluorene units, there have been growing studies on replacing spiro-centers with three-dimensional and bicyclic EA units [165–167] or TB units [168]. Co-polymerizations were conducted by changing the concentration of monomers bearing EA units, and they showed higher gas permeability than PIM-1, resulting from a ~10% increase in BET surface area. TB units also have been replaced at the dioxane site of PIM containing SBI or EA units (PIM-SBI-TB or PIM-EA-TB, respectively). In comparison to relatively flexible dioxane linkages, the TB unit has a contorted heterocycle feature with V-shape bridged bicyclic structures, leading to improved polymer rigidity (Figure 3.25). The resulting PIM-EA-TB presented a high BET surface area of $1028\,m^2\,g^{-1}$ and better gas performances than either PIM-1 or PIM-SBI-TB, especially for the upper bounds of O_2/N_2, H_2/N_2, and H_2/CH_4.

Further optimizations in TB chemistry have been conducted together with iptycene units [159, 170, 171]. In comparison to PIM-EA-TB, PIM-Trip-TB showed more desirable gas transport behaviors especially in O_2/N_2, CO_2/N_2, and CO_2/CH_4 applications. The fixed propeller-like structure of the triptycene units provided

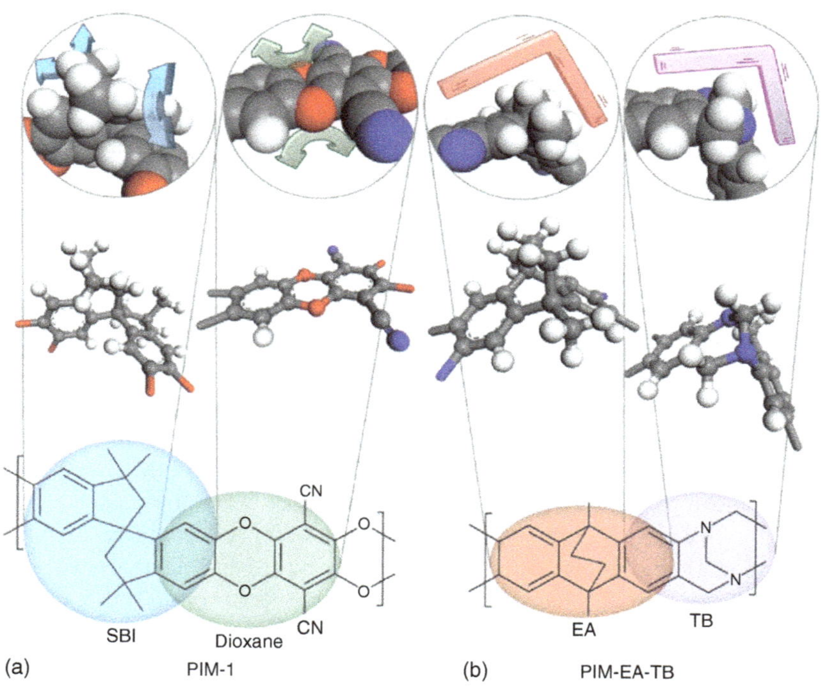

Figure 3.25 Chains of microporous polymers that have a contorted ladder structure resist rotation, which allows them to maintain a more uniform pore structure. (a) The PIM-1 polymer contains SBI and dioxane linkages, but both can bend and flex to a considerable extent. (b) The PIM containing EA and TB units has a more rigid architecture. Source: Reprinted with permission from Guiver and Lee [169]. Copyright (2013) AAAS.

additional interchain distances, generating intrinsic microporosity within polymers. In this regard, TB-containing PIMs with extended triptycene units such as benzotriptycene (PIM-BTrip-TB) or TMN (PIM-TMN-Trip-TB) have attracted interest from membrane communities [47]. Due to their greater internal molecular free volumes from the additional fused rings at the top of triptycene units, PIM-BTrip-TB, for instance, showed higher gas permeability than PIM-EA-TB and PIM-Trip-TB with a CO_2 permeability of 13,200 Barrer.

Apart from EA and TB units, another three-dimensional bicyclic structure was adopted to increase the stiffness of the existing SBI-type PIM-1 polymer. A new PIM polymer with locking polymer rigidity, PIM-C1, was deliberately prepared by introducing an eight-membered ring locking in the flexible spiro-carbon pivot point [172]. The internal strain energy of newly locked SBI units (20 kcal/(mol rad^2)$^{-1}$) was much higher than that of existing SBI or SBF (8.6 and 10 kcal (mol rad^2)$^{-1}$, respectively), demonstrating that the vibration of planes around the spiro-carbon center should be largely suppressed. PIM-C1 exhibited a large BET surface area of 818 m^2 g^{-1} and faster gas transport properties ($P(H_2) = 9870$ Barrer) than SBI-containing PIM-1 and PIM-SBF.

Polyimides constructed of the ladder-like structure are prepared by the reaction of benzodioxane-containing dianhydride and bulky diamines to acquire both high free volume of PIM and several superior stabilities of polyimides [173, 174]. PIM-polyimides represent good solubilities for most polar aprotic solvents as well as extraordinary thermal stabilities, possessing the advantages of the two polymers. The gas permeabilities of PIM-like polyimides are the highest among the every reported polyimide membranes at close regions to upper bounds; PIM-PI-8ii have 490 Barrer of O_2 permeability and 3.5 of O_2/N_2 selectivity [174]. Note that the CH_4 permeability is higher than the N_2 permeability, different from the phenomena that permeation of N_2 is 1–2 times faster than that of CH_4 in usual glassy polymer, revealing that the high free volume characteristics governing PIM are dominant for the gas transport.

Table 3.8 summarizes the physical properties, gas transport properties of representative gases, and selectivity of PIM and related polymer membranes reported so far.

3.4.9.3 Thermally Rearranged (TR) Polymer Membranes

Thermally rearranged (TR) polymer is a novel microporous material prepared by chain rearrangement of polyimide with ortho-functional groups in solid state [176, 177]. As shown in Figure 3.26, conversion of aromatic polyimides containing functional groups into highly rigid heteroaromatic polybenzazoles or ladder-like polymers (e.g. polybenzoxazole (PBO), polybenzothiazole (PBT or PBZ), polybenzimidazole (PBI), and polypyrrolone (PPL)) at elevated temperature result in the evolution of microcavities with an intermediate cavity size, a narrow cavity size distribution and a shape reminiscent of bottlenecks connecting adjacent chambers, such as those found elegantly in nature in the form of ion channels and aquaporins [176]. The free volume elements in TR polymers are three-dimensional networks of intermolecular microcavities, which are accessible for transport of small gas molecules. This peculiar free volume structure accounts for both the outstanding

Table 3.8 Physical properties and permeation properties of polymers with intrinsic microporosity (PIMs).

Polymer	Physical properties			Gas permeability (Barrers)					Permselectivity				References
	Density($g\,cm^{-3}$)	FFV	Surface area	CO_2	H_2	O_2	N_2	CH_4	O_2/N_2	CO_2/N_2	CO_2/CH_4	CO_2/H_2	
PIM-1	1.09	0.26	860	2300	1300	370	92	125	4.0	25	18	1.8	[65, 151]
	—	—	—	3496	2332	786	238	360	3.3	15	10	1.5	[150]
	—	—	—	4030	1900	990	270	350	3.6	15	12	2.1	[175]
PIM-7	—	—	750	1100	860	190	42	62	4.5	26	18	1.3	[151]
Cardo-PIM-1	—	—	621	430	320	59	13	22	4.5	33	20	1.3	[153]
TFMPS-PIM1	1.214	0.22	—	731	—	156	33	—	4.7	22	—	—	[154]
TFMPS-PIM2	1.196	0.22	—	1476	—	308	75	—	4.1	20	—	—	
TFMPS-PIM3	1.156	0.24	—	2841	—	561	158	—	3.6	18	—	—	
TFMPS-PIM4	1.089	0.26	—	3616	—	737	217	—	3.4	17	—	—	
DS-PIM1-33	1.187	0.19	—	1408	—	322	88	—	3.7	16	—	—	[155]
DS-PIM2-33	1.198	0.19	—	1077	—	216	52	—	4.2	21	—	—	
DS-PIM3-33	1.162	0.21	—	2154	—	369	93	—	4.0	23	—	—	
DN-PIM-33	—	—	—	4646	2347	907	242	—	3.7	19	—	2.0	[156]
PIM-PI-1	1.15	0.232	680	1100	530	150	47	77	3.2	23	14	2.1	[173, 174]
PIM-PI-2	—	—	500	210	220	39	9	9	4.3	23	23	1.0	
PIM-PI-3	1.26	0.226	471	520	360	85	23	27	3.7	23	19	1.4	
PIM-PI-4	1.26	0.228	486	420	300	64	16	20	4.0	26	21	1.4	
PIM-PI-7	1.19	0.223	485	510	350	77	19	27	4.1	27	19	1.5	
PIM-PI-8i	—	—	683	3700	1600	545	160	260	3.4	23	14	2.3	
PIM-PI-8ii	1.14	0.231	—	2270	1020	320	100	170	3.2	23	13	2.2	

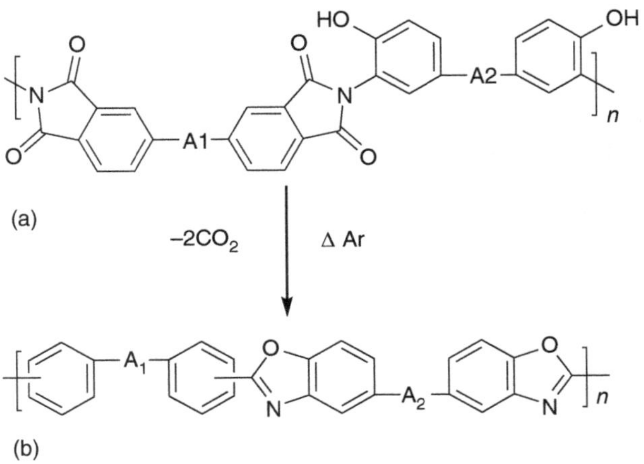

Figure 3.26 Mechanism of thermal rearrangement: (a) hydroxyl-containing polyimide [177] and (b) thermally rearranged polybenzoxazole (TR-PBO).

permeability of TR polymers with fast diffusion of gases and their still high permselectivity in the separation of small gas molecules. The fractional free volume in TR polymer matrix is up to 25% [177, 178].

The gas permeabilities of almost all TR polymer membranes are enhanced by at least two orders of magnitude over those of the pristine orthro-functional polymers and typical glassy polymers, confirming the presence of larger interconnected free volume elements. The gas permeabilities of TR polymer membranes are lower than those of PTMSP membrane, but gas selectivities of CO_2 separations (e.g. CO_2/CH_4, CO_2/N_2) are two to three times higher than for PTMSP. The CO_2 permeabilities reach up to 4000–5500 Barrer while the CO_2/CH_4 selectivities lied between 20 and 40. Moreover, the additional advantage of TR-polymers is that it is convenient to control the permeabilities and permselectivities by thermal conversion ratio determined by thermal treatment times and temperatures [177].

The concept of TR polymers involves the preparation of rigid and rod-like aromatic polymers bearing hetero-cyclic benzoxazole, benzimidazole, and benzithiazole rings formed by thermal treatments with aromatic polyimide [176], polyetherimide[179–181], or polyamide [182] membranes. The rigid nature of the phenylene-heterocyclic ring units can inhibit rotations of the polymer backbone, resulting in microporous characteristics [169]. However, due to their complex and high-temperature fabrication procedure [183], a new post-fabrication methodology with ready-made membranes made of solution-processible polyimide or polyamide precursors, known as *thermal rearrangement*, was devised. Initial explorations of TR polymers have focused on extensive studies with commercially available *ortho*-hydroxyl diamines, i.e. TR-able monomers to establish structure–property–performance relationships [181, 184, 185]. Copolymerization is a way to control the gas transport performances of copolymer membranes as well as to confirm the effect of thermal rearrangement. Poly(benzoxazole-co-imide) membranes were obtained from the thermal rearrangement of poly(hydroxyl imide-co-imide) [186].

Depending on the chemical structures in the TR-able monomers, their unique pore size distributions, which are hourglass-shaped bimodal distributions, can viably be tailored. Various factors influence the microporosity and gas separation performances in TR polymers, i.e. precursor preparation history [187], thermal treatment conditions (temperature and dwell time) [177], copolymerization ratio [188–190], and integration of functional materials [39, 130, 191–193]. Accordingly, attempts to fabricate industrially interesting hollow fibrous platforms or electrospun mat geometries have been reported [194–197].

However, most of the previously mentioned works involved incorporation into non-TR-able diamines (diamines without hydroxyl groups) or dianhydrides, which offset benefits from the functional groups due to a lower TR conversion ratio. Therefore, current studies are more devoted to synthesizing new TR-able diamines with these bulky units. TR polymers with bulky adamantyl groups were proposed from nucleophilic adamantane-containing hydroxyl diamines [198]. The precursors showed good thermal stability capable of TR processing, and all adamantane-based TR polymers exhibited increases in *FFV* and gas permeability at the thermal rearrangement temperature. AD (0.5), which is a copolymer containing 50 mol% of adamantane moiety, presented a ~15-fold increase in CO_2 permeability after being thermally treated at 450 °C. It surpassed the 2008 upper bound of CO_2/CH_4 without considerably sacrificing gas selectivity.

Yeong et al. prepared cardo TR-PBOs by introducing new hydroxyl diamines containing fluorene units, and they investigated the effects of fluorene ratio in the polymer main-chain on gas transport properties [199]. With 10 mol% of fluorenes, *FFV*s were improved by 31% from 0.15 to 0.20, with a three-fold increase in CO_2 permeability up to ~1539 Barrer compared to the non-fluorene-containing TR polymers.

To maximize polymer rigidity, incorporation of inflexible and stiff segments into TR polymers has been investigated with spiro-centered and fluorene-based SBI or SBF units (Figure 3.27) [200–202]. A series of spiro-centered HPIs was successfully synthesized from an SBI-based diamine with four kinds of dianhydrides, and their TR analogs were obtained. SpiroTR-PBO-6F, a spiro-based TR polymer prepared from 6FDA, exhibited the highest *FFV* of 0.27 and a CO_2 permeability of 675 Barrer, and it showed improved performance in comparison to a prototype TR polymer made of APAF-6FDA (*FFV* = 0.22 and $P(CO_2)$ = 398 Barrer). Moreover, Ma et al. expanded the concept of spiro-centered TR polymers by introducing SBI units into dianhydride moieties (SPDA) as well as SBF unites into *o*-hydroxyl diamine moieties (HSBF) [201]. The resulting SPDA–SBF–PBO exhibited an improved CO_2 permeability of 1280 Barrer, which is higher than those of half-substituted spiro-centered sprioTR-PBO-6F *or* 6FDA-SBF-PBO.

For rigid and microporous TR polymers with bicyclic moieties, Meckler reported TR-PBOs featuring TB monomer subunits [203]. TB units play a crucial role as exceptionally rigid sites of contortion along the polymer backbones as well as sites for slow segmental chain motion for enhanced diffusion selectivity. For O_2/N_2 separation, the resulting TB-based TR polymers presented much higher

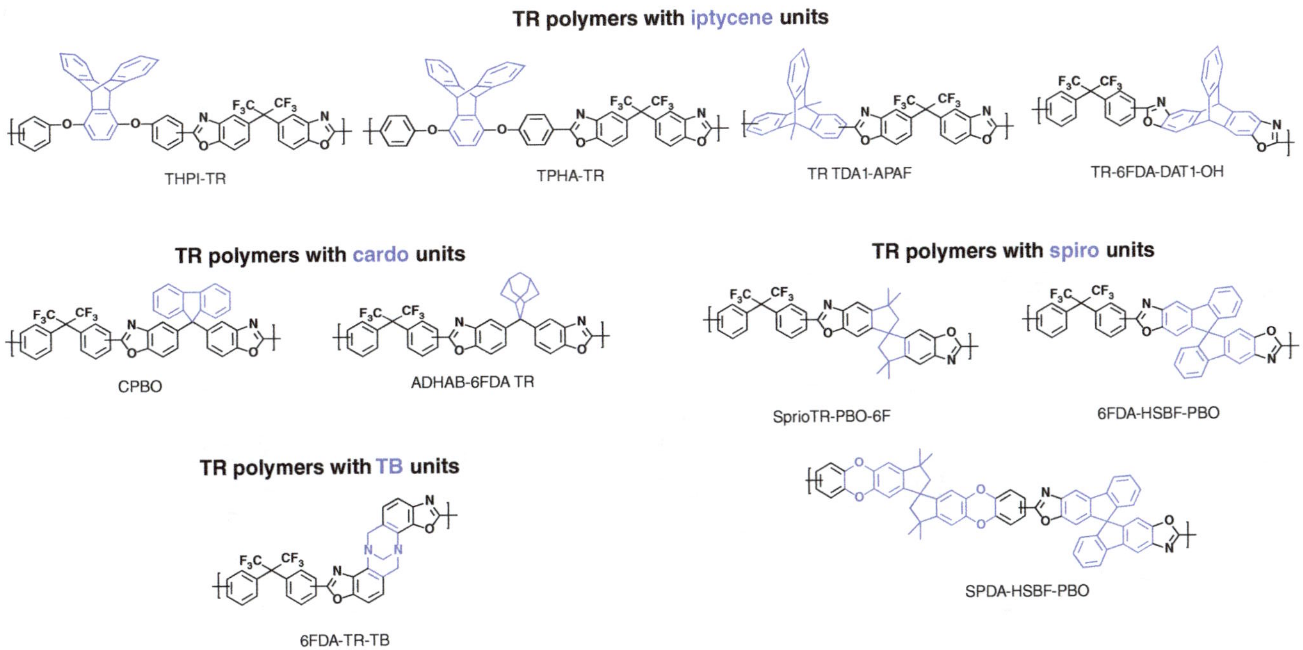

Figure 3.27 TR polymers containing shape-persistent units.

O_2 diffusivity than N_2, resulting in improved O_2/N_2 separation performances with an O_2 permeability of 108 Barrer and an O_2/N_2 selectivity of 9.

Cross-linked TR polymers (XTRs) were fabricated *via* imide-to-benzoxazole thermal rearrangement procedures with cross-linkable polyimides. Cross-linkable polyimides were prepared using poly-condensation reactions of either thermal cross-linking [204] or chemical cross-linking [205] of HPI precursors bearing crosslink-able diamines with pendant carboxyl groups. XTRs retained the same bimodal pore size distributions as uncross-linked TR polymers. However, the population of larger pores, which regulates gas permeable characteristics, increased with degree of cross-linking. During the TR procedure, rigid inter-chain bridges induced by cross-linked phenylene segments in polyimide precursors did not allow polymer relaxations to fill the micropores. These bridges promote enhanced gas permeabilities for the resulting XTR polymers and provide higher resistance to plasticization responses up to 40 bar for a CO_2/CH_4 binary mixture.

To boost the synergistic effects of cross-linking structure and ideal TR conversion rates, entirely inter-connected TR network polymers were proposed by designing crosslink-able polyimide precursors as bis(*o*-hydroxyl)maleamic acid (BHMAA)) monomer building blocks [206]. During membrane formation *via* solvent evaporation of the BHMAA solutions, each electron withdrawing carbonyl group in the BHMAA block was reacted with adjacent double bonds and was then cross-linked and changed into bismaleimides (BMIs). Further thermal treatments afford BMI with TR polymers. The resulting bismaleimide-based TR networks (TR-BMI) exhibited a tremendous increase in BET surface area from ~10.8 (BMI) to ~1130 $m^2\ g^{-1}$ (TR-BMI) with a sharp distribution of differential pore volumes located around 7.3 Å. The finely tuned microporous frameworks in TR-BMI networks contributed to enhanced gas transport behavior (e.g. CO_2 and O_2 permeabilities of ~5440 and ~1171 Barrer, respectively), which is far above the upper bounds, particularly for industrially important gas pairs (CO_2/N_2, CO_2/CH_4, and O_2/N_2).

Modification of TR polymer membranes and PIM membranes via fluorine gas molecules block the part of microcavities in the membrane pores so that the selectivity of small gas molecules such as hydrogen and helium, over methane or carbon dioxide increases up to about 800 times [207]. This concept can be applied and extended to other separation methods such as organic nanofiltration and so on.

3.4.9.4 High-performance Polyimides

To increase the microporosity in polymeric membrane, two strategies are suggested (Figure 3.28): (i) introducing intrinsically microporous units in the polymer and (ii) maximizing polymer chain rigidity to hinder efficient chain configurations. Commercially available monomers with bulky fluorinated groups or non-coplanar structures are the most viable and traditional approach to induce intrinsically microporous characteristics. Non-coplanar monomers have kink moieties, which result in twisted and crank-like structures of the resulting polymer. Many researchers have extensively reported on the effects of changes in kink structures of commercial monomers on microporosity and gas transport behaviors [208–212]. A series of polyimides prepared from various aromatic dianhydrides displayed

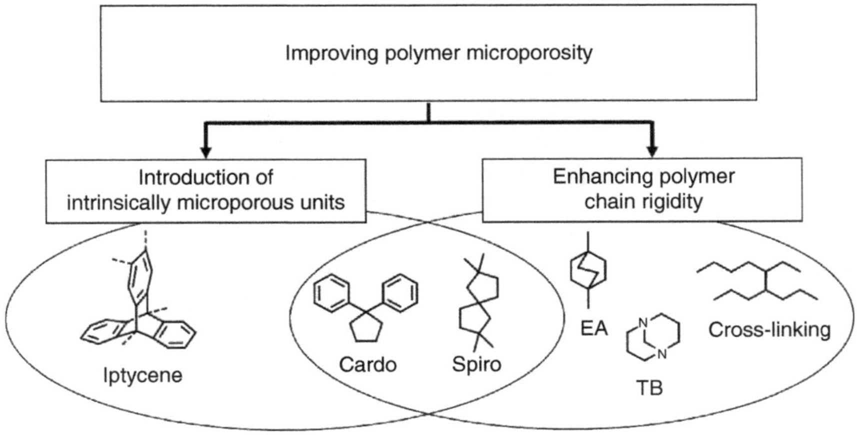

Figure 3.28 Strategies for improving microporosity in polymeric membrane materials by incorporating bulky and rigid groups in the polymer backbone or by crosslinking.

different *FFV*s, and their gas transport properties changed with chemical structure. *FFV*s and gas permeabilities of polyimides of bulk 2,3-bis(3,4-dicarboxyphenyl)hexafluoropropane dianhydrides (6FDA) were higher than those of 3,3′4,4′-biphenyltetracarboxylic dianhydrides (BPDA) or 4,4′-oxidiphthalic anhydrides (ODPA), indicating that kinked and bulky hexafluoropropane groups in 6FDA moieties can efficiently inhibit polymer chain packing. As a result, ~35-fold increases in H_2 permeability values were observed for 6FDA moieties in comparison to BPDA counterpart.

Many researchers have endeavored to design new synthetic monomers bearing bulky substituents for hindering efficient polymer chain packing. Shape-persistent iptycenes [213–224] such as triptycene, benzotriptycene, or pentiptycene have been intensively investigated for gas separation membranes by virtue of bulky three-dimensional [2,2,2]-ring configurations with 120° contortion sites and unique internal free volume elements. Iptycene units in polyimides serve to expand inter-chain distances and suppress chain motions. Thus, the resulting polymers showed larger *FFV*s and enhanced gas permeability. For instance, triptycene-based polyimides with dibenzodioxine linkages presented a high BET surface area of 612–862 $m^2\ g^{-1}$ and a superior CO_2 permeability of ~2839 Barrer [213].

Other types of alicyclic synthetic monomers, cardo monomers and spiro monomers, were also considered to enhance microporosity. 'Cardo' means "loop" or "hinge" in Latin, meaning that cardo polymers contain bulky cyclic units. On the other hand, spiro monomers have a double-strand structure linked together at tetrahedral bonding atoms (or spiro-centers). Recently reported cardo monomers or spiro monomers are mainly designed with adamantane [225–228] units or fluorene units [229, 230]. The presence of such groups on the sidechain in polymer backbones induces microporosity; however, those on the main chain even regulate polymer rigidity. Maya et al. introduced three kinds of adamantyl ester pendant groups into the side-chain of polyimides, resulting in a ~19% increase in *FFV* with

~threefold improvement in CO_2 permeability [225]. Plaza-Lozano et al. synthesized adamantane containing diamines and incorporated them into the main-chain of polyamides as well as polyimides. Polymers with 6FDA exhibited enhanced CO_2 permeability while retaining CO_2/CH_4 separation performance [226].

Kazama et al. intensively studied the effect of fluorene moieties on the change in T_g, *FFV*, and gas transport characteristics with diverse polymers including polyamides, polycarbonates, polysulfones, and polyimides [229]. For all cases, integration of fluorene moieties induced restricted molecular movements in the main chain with the increase in T_g from 94 (polysulfones) to 135 °C (polycarbonates), resulting in increase in *FFV*s and gas permeabilities [231, 232]. For example, bis(phenyl)fluorene-based polyimides showed a CO_2 permeability of ~130 Barrer, which was 4.8 times higher than that of polyimides with bis(phenyl)isopropane groups [229].

SBI and SBF units in polymer backbones offer highly microporous characteristics and outstanding gas transport behavior in polyimides [233, 234]. For example, SBF containing polyimides displayed a remarkable BET surface area of over 600 $m^2\ g^{-1}$ with a CO_2 permeability of ~6674 Barrer [233].

As depicted in Figure 3.29, the degree of torsional freedom of the spiro-center is almost twice that of triptycene units. New types of three-dimensional and bicyclic units including ethanoanthracene (EA) and Troger's Base (TB) have been incorporated [235–239]. EA-derived polyimides (PIM-PI-EAs) showed enhanced gas separation performances especially in O_2/N_2, H_2/N_2, CO_2/CH_4, and CO_2/N_2 applications [235], compared to the spiro-centered analog PIM-PI-SBI [234].

Zhuang et al. first explored the concept of "polyimides incorporating TB units (PI-TBs)" for gas separation membranes. They succeeded in designing the polymers by *in situ* formation of the TB unit during polymerization. This is different from the case of EA units, that employ synthetic monomers with pre-formed EA units [235]. PI-TBs exhibited high *FFV*s of above 20% and presented better gas permeability [236]. Furthermore, tertiary amines in the TB units offer favorable sorption

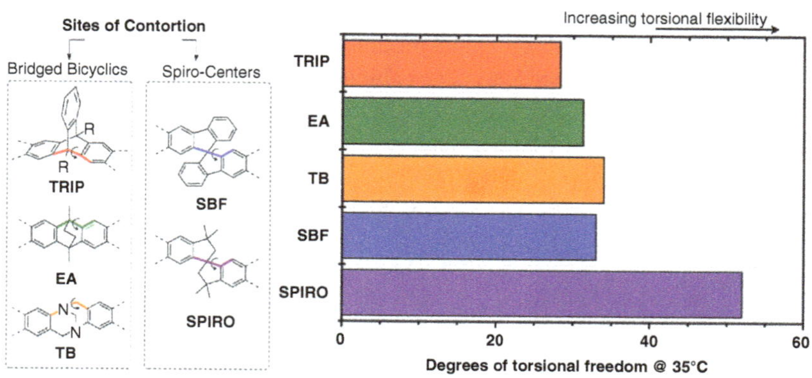

Figure 3.29 Molecular simulation of dihedral angle distributions for bridged-bicyclic and spiro-based sites of contortion. Source: Reprinted with permission from Swaidan et al. [214]. Copyright (2014) American Chemical Society.

sites with CO_2 molecules. Fluorene moieties were introduced into PI-TBs *via* copolymerization to fine-tune their microporosity. The obtained copolymers presented improved gas transport properties in comparison to fluorene-based non-TB polyimides [237].

Crosslinking or network formation is another way to restrict polymer chain movement and to improve polymer rigidity. Crosslinked polyimides by incorporation of diamino benzoic acid (DABA) crosslink-able monomers into the main-chains have been introduced from 1990s, and in fact, the polyimides containing 10 and 40 mol% of DABA exhibited 1.5 and 2.9-folds increase in CO_2 permeability, respectively, after crosslinking [240, 241]. Dose et al. further demonstrated changes in *d*-spacings and gas permeability of the crosslinked polyimides with different degrees of crosslinking [242]. Moreover, other crosslinking agents including thermally decomposable cyclodextrins have been also considered to tailor polymer rigidity as well as gas transport characteristics [243].

3.5 Membrane Gas Separation Applications

The membrane gas separation market is growing and changing. Nitrogen separation from air is the largest market where hydrogen separation from ammonia gas purge stream is the second largest market using gas separation membranes. Growing market for gas separation membrane process includes the biogas separation from landfill gas, carbon capture from enhanced oil recovery, and flue gases due to the global climate change and the sustainability of the environment.

3.5.1 Air Separation

The largest market for gas separation membrane in current usage is the production of nitrogen from air. Gas separation membranes for removing nitrogen from air have been available since the mid-1980s. Although a small number of developed membranes have been introduced using poly(4-methyl-1-pentene) (TPX) that showed the O_2/N_2 selectivity of around 4. Polyimides and polyamides exhibited the selectivity of around 6-7, providing favorable economics, particularly for a small nitrogen plant. For small nitrogen production plant, membrane systems are low-cost and thus are favorable. Table 3.9 lists the permeabilities and selectivities of polymers of interest in air separation membrane systems. There is a sharp trade-off between permeability and selectivity. This trade-off was illustrated in the Robeson's 2008 trade-off plot shown in Figure 3.30. The first membranes were spiral wound, but today, hollow fiber membranes have been used. For the main gases involved in air separation, the following applies: water-vapor permeates fastest, followed by CO_2, O_2, and finally N_2. This order of gas permeation is utilized in nitrogen separation since oxygen can be favorably removed from compressed air. Currently, polymer membrane systems are commonly used if a residual oxygen content of 1% is acceptable. If greater degrees of purity are needed, pressure swing adsorption (PSA) systems are more cost-effective.

Table 3.9 Permeabilities and ideal selectivities of selected polymers in air separation.

Polymer	P_{O_2} (Barrer)	P_{N_2} (Barrer)	α_{O_2/N_2}
Poly(1-trimethylsilyl-1-propyne) (PTMSP)	7600	5400	1.4
PIM-1	990	270	3.6
Teflon AF2400	960	480	2.0
Teflon AF1600	270	110	2.4
Silicon rubber	600	280	2.2
Poly(4-methyl-1-pentent) (TPX)	30	7.1	4.2
Poly(phenylene oxide) (PPO)	16.8	3.8	4.4
6FDA-6FpDA (polyimide)	16.3	3.5	4.7
Matrimid	1.7	0.25	6.6
Polysulfone	1.1	0.18	6.2

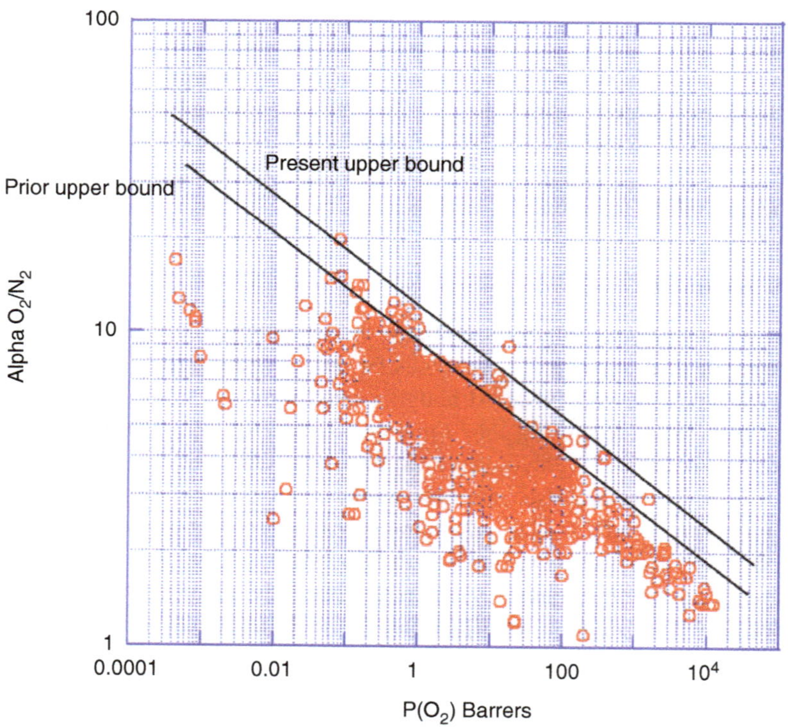

Figure 3.30 Upper bound curve of oxygen permeability and selectivities of oxygen and nitrogen. Source: Robeson [44]/with permission from Elsevier.

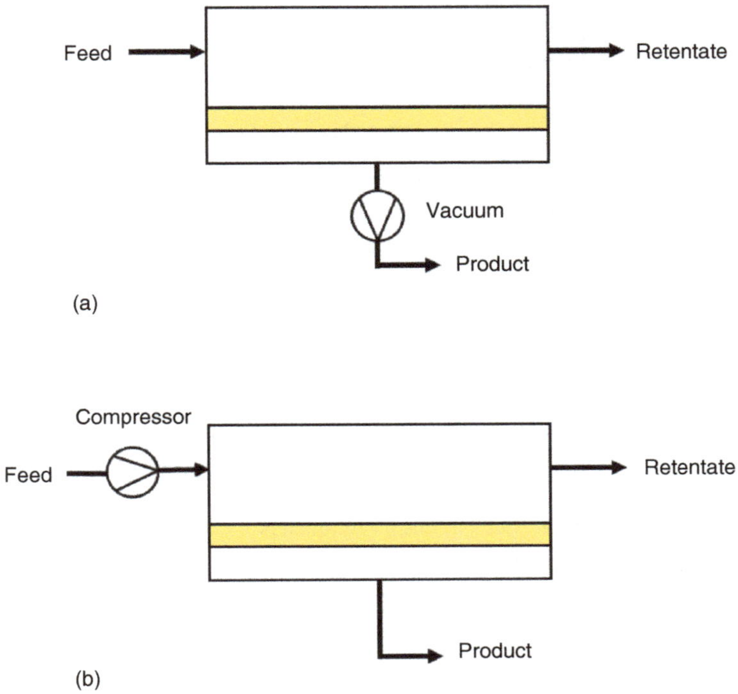

Figure 3.31 Schematic representation of (a) vacuum and (b) pressure mode for air separations.

Air separation membrane units can be operated in either pressure or vacuum mode as shown in Figure 3.31. In the pressure mode, feed air is typically pressurized to several bars, while permeate is maintained at close to atmospheric pressure (~1 bar). The differential pressures are higher than those of the vacuum mode, leading to reduced membrane area requirements. In the vacuum mode, feed air is pressurized slightly above atmospheric pressure and a vacuum is maintained on the permeate side of the membrane. The retentate is vented at atmospheric pressure. The vacuum mode is typically more energy-efficient than the pressure mode because a vacuum is applied only to the permeate stream.

Depending on the processes and applications, single-step, two-step, and three-step processes are commonly applied in the industry to produce nitrogen from air (Figure 3.32). More complex process designs are required to produce nitrogen gas of >99% purity. Figure 3.33 introduces the nitrogen recovery as a function of product nitrogen concentration for membranes with selectivities between 2 and 10.

One of the big markets for nitrogen production using membrane systems is on-board inert gas system (OBIGS) installed in the airplane. Airplane wing usually are filled with the fuel liquids while the electrical lines are passing through these tanks. If there is an electrical spark in the tank, there will be an explosion which is a disaster. In fact, in 1996, airplane crossing the Atlantic Ocean exploded and all the passengers have passed away. Several years later, the accident was blamed

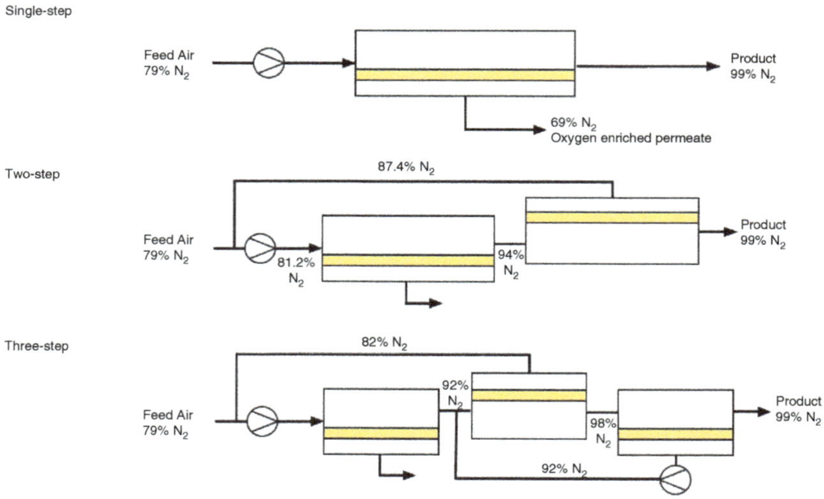

Figure 3.32 Single-, two-, and three-step designs for nitrogen production from air.

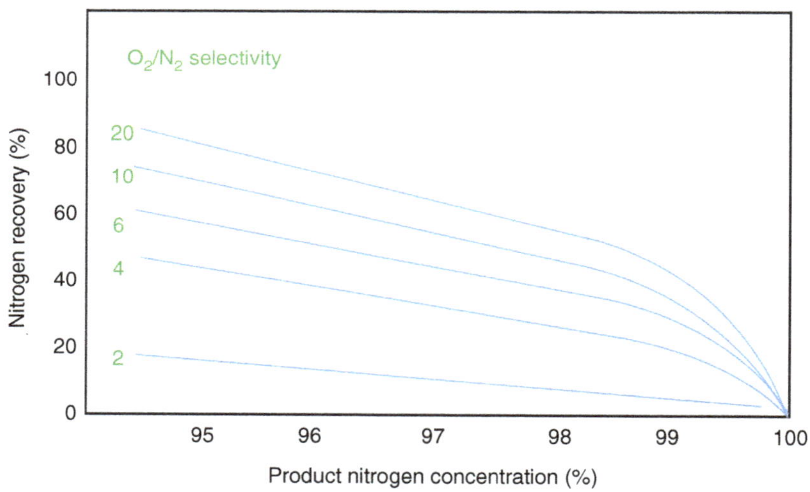

Figure 3.33 Nitrogen recovery as a function of product nitrogen concentration for membranes with selectivities between 2 and 20.

for the electrical spark that caused the explosion because the fuel vapor formed in the empty spaces of the fuel tank is much more explosive upon exposure to an electrical spark. Since this accident, Federal Aviation Agency in US requires the OBIGS system for all the airplanes flying to and from the US. Air Products Co., LTD is leading this industry while they use PSF and PI hollow fibers as base materials for this system. Hollow fiber membranes used for this application is first introduced by Henis and Tripodi [9, 60] in 1978 at Monsanto (now Air Products). They invented Loeb-Sourirajan type asymmetric membranes typically made with

PSF hollow fibers. PSF hollow fibers usually have defects that reduces the selectivity of oxygen/nitrogen. PDMS is usually much more permeable to most of the gases than other polymers. Permeable PDMS solution coated on top of PSF hollow fibers produces a thin layer of PDMS layer that typically plugged these defects in PSF hollow fiber surface and eliminate the convective flow through these defects (see Figure 3.10). The silicone rubber layer also protects the membrane during handling. Silicone rubber coating is usually conducted inside of a module before shipping to protect any explosion due to the solvents in PDMS solution. The development of silicone rubber-coated PSF membrane was the first successful gas separation membranes that were commercialized for hydrogen/nitrogen separation by Monsanto.

Oxygen-rich air is used in the chemical industry, in refineries, and in various fermentation and biological digestion processes, but low-cost production is required for these applications. Figure 3.34 compares the range of current membrane nitrogen production systems with the cryogenic and pressure swing adsorption systems. Cryogenic process has long been used to produce low-cost oxygen that is diluted with atmospheric air. The quantity of pure oxygen that should be mixed with air to produce the desired oxygen enrichment determines the cost. When comparing with membrane process and cryogenic process to produce oxygen, membrane process is generally not competitive. However, small oxygen-enriching devices developed for students working in polluted rooms were developed in early 2000s in Asia. This is a high-end market for oxygen generating membrane systems. These devices produced 25–28% oxygen and were filled with silicone rubber or PTMSP.

A continuous development of better oxygen-enriched air separation membranes is an area of interest because of the potential application in combustion processes. When fuels such as methane and oil are burned with air, oxygen helps combusting the fuels and generates the heat and energy. If oxygen concentration is elevated from 21% in the air to 35% as an oxygen-enriched air, fuel efficiency will boost up

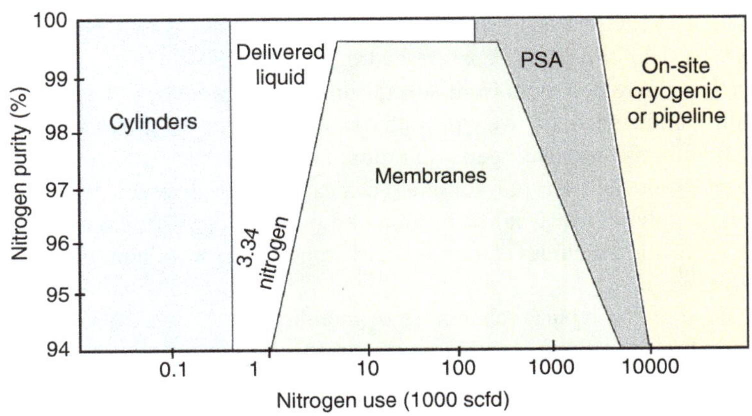

Figure 3.34 Approximate competitive range of current membrane nitrogen production systems.

and energy lost with the hot exhaust gas due to the nitrogen decreased. Carbon dioxide separation from the exhaust gas would be easier. Useful energy that can be extracted from the same amount of fuel increases significantly even if air is enriched to 25–35%.

For oxygen-enriched air separation, the vacuum mode is more energy-efficient and therefore more suitable. However, the vacuum mode requires a larger membrane area than the pressure mode because of the limited differential pressure. The pressure mode is more suitable for nitrogen production because the retentate can be obtained at conditions closer to feed conditions. In some cases, vacuum and pressure modes can be used together for a synergetic effect. In this mixed mode of operation, the feed air is pressurized, and a partial vacuum is maintained on the permeate side to increase both the feed-to-permeate pressure ratio and the feed-to-retentate pressure ratio. The oxygen-enriched permeate is then mixed with an ambient air stream to obtain the desired air flow and oxygen concentration. Significant applications of air separation are the production of oxygen and water-depleted air for oil and gas drilling, controlled atmosphere, maritime transport, inerting, beverage dispensing, tire inflation, and laboratory gas supply.

3.5.2 Hydrogen Separation

Hydrogen is now considered as an alternative fuel source as the global climate issue has been an urgent matter for the governments in the world. Currently, hydrogen is mostly produced from petrochemical plants and thus is called as a "grey hydrogen". If the hydrogen is produced from water without evolving carbon dioxide emission, it is called "green hydrogen". Here, we will discuss about how we can produce the grey hydrogen.

The first large-scale commercial membrane plant was applied in the separation of hydrogen from nitrogen from ammonia purge gas by Monsanto in 1980 [59], followed by a number of similar applications such as H_2/CH_4 separation in refinery off-gases and H_2/CO in oxo chemical synthesis plant [244]. Hydrogen is a small gas molecule which is highly permeable compared with other gases. Glassy polymers have long been used to efficiently separate hydrogen. Table 3.10 lists the typical hydrogen-selective glassy polymers from several membrane produces. Because of high permeability and selectivity, we can understand why the first application of gas separation membrane was hydrogen separation. Early developed gas separation membranes were produced from polysulfone (Permea) and cellulose acetate (UOP Separex), but now polyimides (Ube) or brominated polyimides (Permea) are also used. Another application includes extraction of hydrogen from petroleum naphtha cracking streams.

Figure 3.35 illustrates a typical schematics of membrane separation process to recover hydrogen from ammonia plant purge gas stream. During the ammonia production stage from hydrogen and nitrogen, argon enters the reactor as an impurity with the nitrogen stream, and methane enters the reactor as an impurity with the hydrogen. Two-step membrane design is used to reduce the cost of recompressing the hydrogen permeate stream to the very high pressures of ammonia reactors.

Table 3.10 Selected hydrogen separation membranes.

Membrane (Producer)	Hydrogen permeance (GPU)	Selectivity		
		H_2/CO	H_2/CH_4	H_2/N_2
Polysulfone (Permea)	100	40	80	80
Cellulose acetate (Separex)	200	30–40	60–80	60–80
Polyimide (Ube)	80–200	50	100–200	100–200
Polyaramide (Medal)	100	100	>200	>200

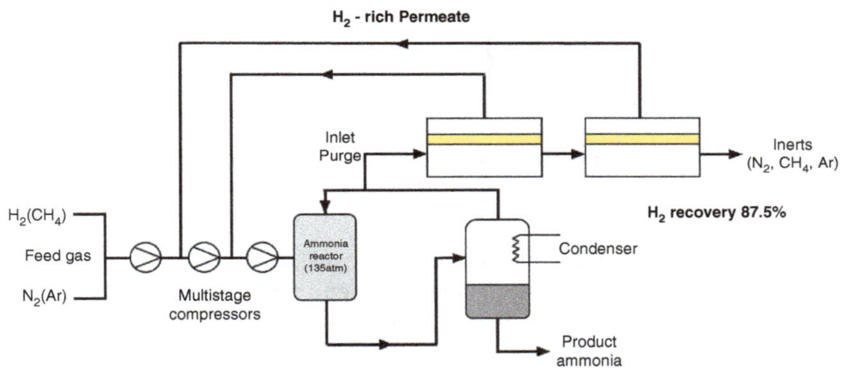

Figure 3.35 Simplified flow schematic of the PRISM membrane system to recover hydrogen from an ammonia reactor purge system. A two-step membrane system is used to reduce permeate compression costs.

The major processes for hydrogen upgrading in refineries are the PSA, cryogenic separation, and membrane separation. The increased need for hydrogen in refineries is due to increased environmental regulation and heavier crude. Product purity can be maintained for small changes in feed composition by adjusting the feed-to-permeate pressure ratio. But the relatively strong impact of product purity on recovery for these systems indicates that hydrogen recovery can be significantly affected. In refinery industry applications, however, the major impurity is methane. Figure 3.36 shows the hydrogen recovery from a hydrotreater. Spillman [245] compared the economics for a membrane-based system with a cryogenic separation. The investment costs were comparable for both processes (US \$1.35 million, for 450,000 m^3/day separation) and the hydrogen recovery rates were similar.

Soon, the hydrogen economy will expand, and hydrogen production will need to be increased from what it is today. Green hydrogen can be produced from water by multiple sources. These include renewable sources such as solar, wind, hydropower, and biomass, and non-renewable sources such as fossil fuels and nuclear energy. Water can be electrolyzed to produce hydrogen, and this will be discussed in Chapter 7 in more detail. Hydrogen-based power generation systems, such as fuel cells (see Chapter 4) have the potential to eliminate emission of all

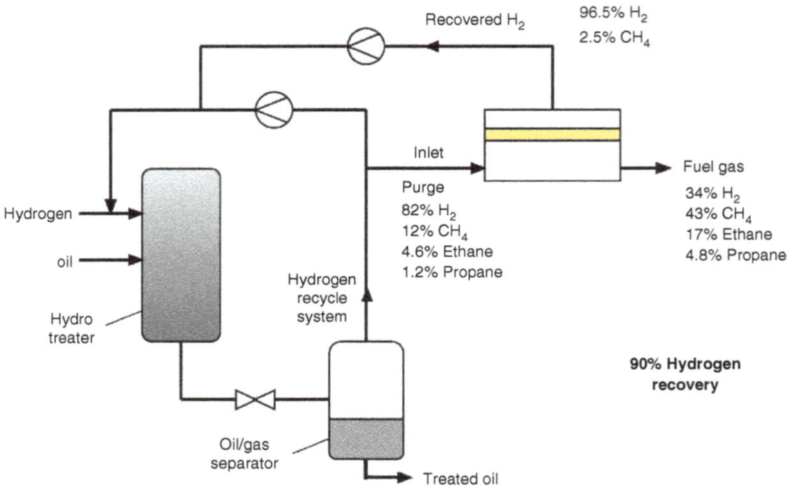

Figure 3.36 Hydrogen recovery from a hydrotreater used to lower the molecular weight of a refinery oil stream. Air Products polysulfone membranes (PRISM) are used.

pollutants, including nitrogen oxides, carbon monoxide, volatile organic chemicals, and particulates. Fuel cells are also potentially useful for clean power generation at small, distributed facilities as well as for emergency backup and portable power generation. The availability of inexpensive yet highly pure hydrogen is a prerequisite for the transition to fuel-cell-based power generation. It would be ideal to produce pure H_2 from electrolysis of H_2O using solar energy generated by solar cells. Water electrolysis using proton exchange membranes and anion exchange membranes has been actively investigated now. However, for now, the inevitable choice is to obtain H_2 from fossil and biomass fuels. Hydrogen can be recovered from several kinds of gas streams. Many of them include low levels of hydrogen and undesirable impurities. One significant separation is the separation of CO_2/H_2. H_2 is produced *via* steam reforming of hydrocarbons to make CO and H_2 and then a water–gas-shift is used to make CO_2 and H_2. Syngas (or synthesis gas) can be prepared from methane as an intermediate step in the production of basic chemicals, such as methanol, ammonia, and oxo-alcohols. Carbon dioxide is produced in the syngas reaction and must be removed before further catalytic conversion of CO and H_2 can proceed toward the desired chemical. Depending upon the kinds of basic chemicals to be produced, the H_2/CO syngas ratio may need to be adjusted. Water–gas-shift reaction is an old and established technology that may find new applications in fuel cells and coal processing systems. Water–gas-shift is a reversible, exothermic reaction that is thermodynamically unfavorable at elevated temperatures. Membrane technology may potentially provide opportunities for new applications of water–gas-shift. By combining the water–gas-shift reaction with selective removal of one of the reaction products, a single reactor can operate simultaneously at high temperature and high conversion, without the requirement of excess steam. Such a membrane reactor needs a membrane that is highly permselective for either H_2 or CO_2.

3.5.3 Hydrocarbon/Hydrocarbon Separation

Olefins, especially ethylene and propylene, have been one of the most important raw chemicals because of the wide application of their downstream products, such as polyethylene (PE), polypropylene (PP), poly(vinyl chloride) (PVC), and ethanol, since the chemical and petrochemical industry began to develop in 1920s [246]. The global production of ethylene and propylene is 201 and 130 million metric ton (MMt) per year in 2020. It is forecast that this will increase to almost 300 and 160 MMt per year in 2025. Olefins are produced mainly from steam cracking of naphtha that contains about 5 wt% of paraffin. Ethane cracking is useful for shale gas development with low-cost processing, but the product contains more than 40 wt% paraffin. The downstream industry needs high purity olefins, e.g. ethylene purity of >99.9 wt% for PE production and propylene purity of >99.5 wt% for PP production. Therefore, highly efficient separation of olefin/paraffin is required in chemical industry.

The existing technology for olefin/paraffin separation is cryogenic fractional distillation that required low temperature (~90 °C), high pressure (usually 2 MPa), and more than 100 stages to produce polymer-grade olefins because of the low relative volatility for these olefin/paraffin systems. Currently, purification of olefin from paraffin requires roughly 0.3% of global energy consumption. One of the reasons for high energy consumption to separate olefin/paraffin is their similar boiling points and evaporation properties. Alternative process such as absorption and adsorption have been developed. However, subsequent desorption is necessary to obtain purified olefins, which demands an inevitable energy penalty and process complexity.

Membrane-based separation technology is a mild technology that does not require high energy and only require a high chemical potential difference between thermodynamic, kinetic, and molecular properties of olefin and paraffin molecules. Savings of energy consumption can be about 80%, or up to 0.3–1.5 GJ of energy per metric ton of olefin. If the selectivity is >10, a membrane process can be used along with distillation to achieve energy saving. When the selectivity is >35, membrane-based separation technology can be applied to olefin/paraffin separations. Membrane selectivity is more important than permeability if a minimum permeance of 32 gas permeation unit (GPU) is achieved. Membrane stability over time along with mixed gas transport data are important aspects to consider.

Unfortunately, current polymer membranes show poor selectivities in olefin/paraffin separations [247–253]. High free volume polymers such as PIM-1 and analogs and TR membranes show high permeability (~100 Barrer) but low selectivities (<10) for ethylene/ethane. Some rigid and high free volume moiety incorporating polymer membranes show high permeability up to 1000 Barrer propylene permeability but selectivity of <10 for propylene/propane pairs. The effect of plasticization and temperature on the selectivity must be further improved. Figure 3.37 shows the schematic of channer-based and carrier-based membranes toward olefin/paraffin separation.

High flux zeolitic imidazole framework (ZIF) membranes for propylene/propane separation has been achieved by post synthetic linker exchange [254]. Thickness of the ZIF-8 layer was effectively reduced, resulting in an enhancement in permeance

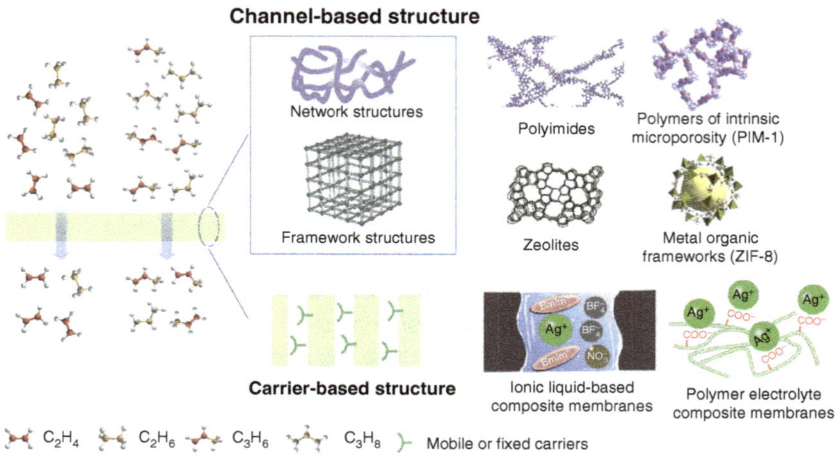

Figure 3.37 Schematic of channel-based and carrier-based membranes toward olefin/paraffin separations. Source: Adapted from [246].

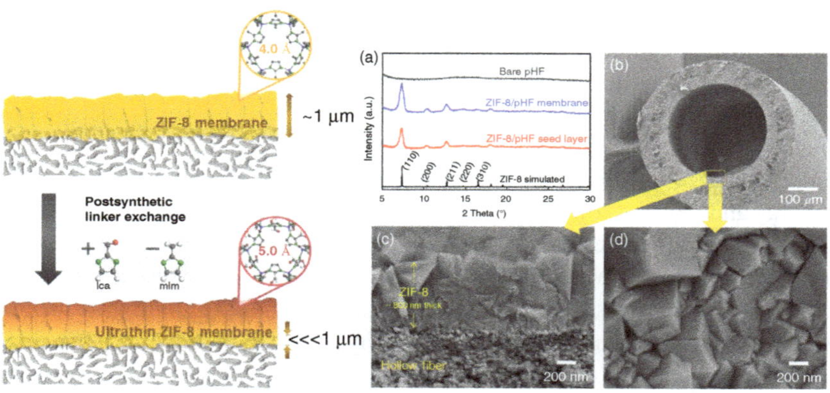

Figure 3.38 (a) Ultrathin ZIF-8 layer with a thickness less than 1 μm can be fabricated by post-synthetic linker exchange. (b) ZIF-8 thin layer can be grown inside of polymer hollow fibers. Source: (a) Adapted from [254]. (b) Adapted from [255].

of propylene (800 GPU) and separation factor of 40 (Figure 3.38a). These thin film of ZIF-8 membranes were coated on the bore-side of Matrimid polymeric hollow fibers and showed propylene permeance of 55 GPU and separation factor of 46 [255] (Figure 3.38b). ZIF-8 membranes had an effective aperture size of ~4 A, that can be promising for separation of propylene/propane. Well-intergrown polycrystalline membranes of ZIF-67 were heterogeneously grown on ZIF-8 seed layers. A tertiary growth of ZIF-8 layers was applied to heterogeneously grown ZIF-67 membranes, the membranes exhibited propylene/propane separation factor of ~200 possibly due to enhanced grain boundary structure [256].

Many efforts have been focused on improving separation using MMMs by introduction of zeolites [257–259], organic [260, 261] and MOFs [262–270], carbon molecular sieves (CMS) [271] in the polymer matrix. Extensive report on MOF-based membrane separation has been reported [272]. Permeability and selectivity of olefins/paraffins are sometimes promising but in general still not reaching the required selectivity to compete with the existing technology. Other works include CMS [273–277] that are brittle, and it is difficult to scale-up production. CMS showed aging phenomena due to its highly porous structures. Propylene permeance of CMS hollow fiber membranes pyrolyzed from polyimides was about 50 GPU with selectivity of about 30 [276], After aging, for two weeks, propylene permeance dropped to <20 GPU while selectivity increased to 17.

For olefin/paraffin separations, facilitated-transport polymer membranes were proposed (see Figure 3.39). Olefin compounds having a π-bond can penetrate through the membranes with transition metals such as cuprous (Cu^{++}) and silver (Ag^+) ions by a "facilitated transport mechanism". Here, π-bond in olefin acts as a nucleophile that attracts an electrophile (Ag^+, for example). An active transport of specific molecules through a membrane is achieved by utilizing a carrier species. The carrier shuttles a specific penetrant across membrane boundaries. The penetrant is quickly transported from the side with higher penetrant

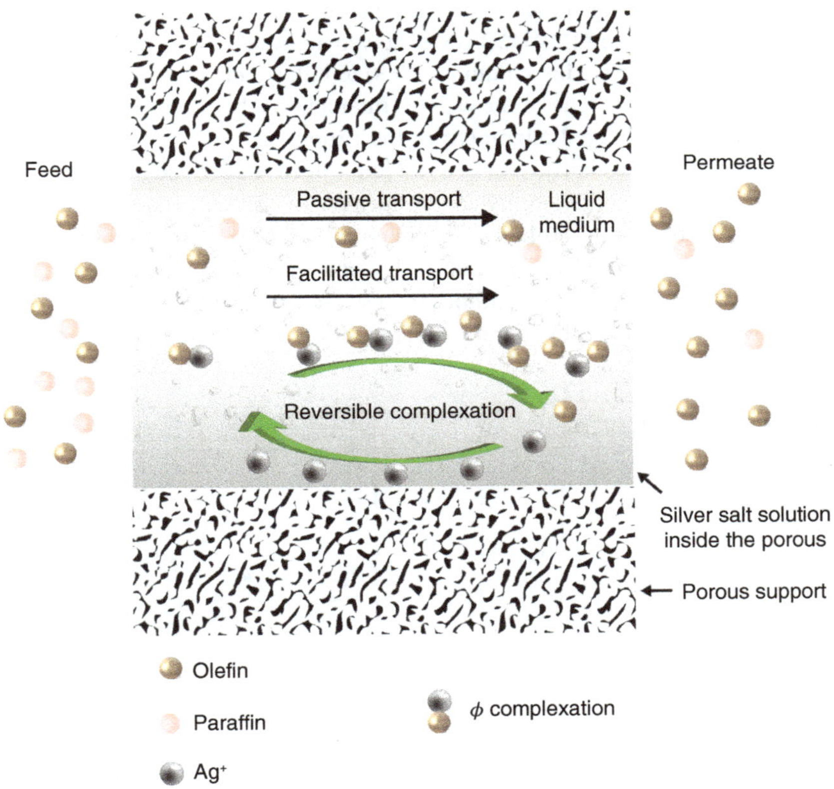

Figure 3.39 Supported liquid membrane concept for facilitated transport of olefin in olefin/paraffin mixture. Source: Adapted from [278].

concentration to the side with lower penetrant concentration. When a feed mixture contains only one species with which the carrier can react, only that species is transported through the membrane. The driving force in facilitated transport is a concentration gradient of penetrant-carrier complexes across the membrane. When silver nanoparticle is incorporated in a polymer matrix, propylene/propane selectivity of about 60 and mixed gas permeance of about 3.5 GPU was obtained [279]. Ionic liquid with silver ion in PVDF-HFP polymer showed 50/50 vol% mixed propylene/propane permeability of 6630 Barrer and selectivities of over 700, followed by stability testing for 10 days [280]. PIM-1 showed poor selectivity and high permeability of propylene/propane gases [281]. Hydrolyzed PIM-1 exhibited smaller fractional free volume than the pristine PIM-1, resulting in low propylene permeability of 421 Barrer and selectivity of ~9. When Ag^+ ion was incorporated, selectivity improves to 11.7, while Mg^{++} ion treated membranes showed improved selectivity of 15, but low permeability of 568 Barrer (Figure 3.40). For ethylene/ethane separation, composite membrane consisted of polystyrene support coated with thin ionic liquids (ILs, such as pyrrolidinium nitrate and 1-ethylimidazolium nitrate) showed selectivity of over 50 and permeability of about 240 Barrer [282], and the stability remained up to 6 months [283]. Recently, BN membranes with a distinct nanoconfinement were reported for efficient ethylene/ethane separation [284]. 2D BN endowed abundant nanochannels which was decorated with reactive ionic liquids (RILs) to tailor their sizes and to achieve nanoconfinement effect. The noncovalent interactions between RIL and BN nanosheets favor the ordered alignment of the cations and anions of RIL within BN nanochannels, while contributes to a fast and selective ethylene transport. The resultant membranes exhibited ethylene permeance of 138 GPU and electivity of 128 as well as stability for 180 hours. Combination of Ag^+/IL in graphene oxide nanosheets facilitated ethylene transport showing ethylene permeance of 72.5 GPU and selectivity of 215 [285]. Although these facilitated transport membranes (FTMs) show promising olefins/paraffins permeability/permeance and selectivity,

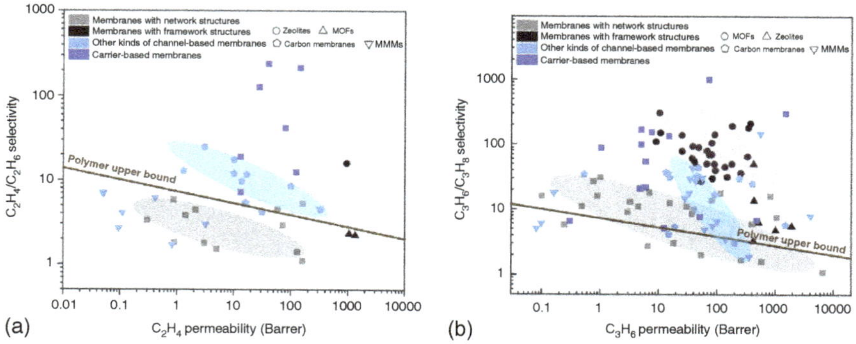

Figure 3.40 (a) Ethylene/ethane, (b) propylene/propane separation performance of polymer membranes, mixed matrix membranes, inorganic membranes, and carbon molecular sieve membranes reported up to 2020. The lines inside of the figures are the upper bound curve for polymer membranes. Adapted from [246].

they need higher long-term physical stability and chemical instability as there are concerns such as degradation of the carrier ion over time. For further readings on Ag^+ ion instability and long-term stability issues, readers are encouraged to read a recent review [278].

Even though a significant development has been realized, remaining challenges to overcome before membrane separation can be adopted for olefin/paraffin separation are the following.

High and improved selectivity is highly required. Except solubility enhanced membranes with carriers, the ethylene/ethane separation is about 10, which is far below the requirements of practical application. Combination of molecular sieving and facilitate transport is proposed. Incorporation of appropriate MOF material is desirable toward ethylene/ethane separation. ZIF-8 materials show high propylene/propane selectivity, however, currently no MOF materials for ethylene/ethane separation has been reported. COFs with reasonable permeability and selectivity can be developed. Large-scale fabrication for MOF membranes, especially in hollow fiber configuration is of great importance for industrial realization. Growth of thin MOF layer in bore side of polymer hollow fibers has been reported and is interesting rather than growth of thin MOF layer in the shell side of hollow fibers, because of the mechanical robustness.

3.5.4 Carbon Dioxide Separation

Carbon dioxide-selective separation is becoming an important issue in areas such as petrochemical engineering (e.g. CO_2 removal from natural gas), environment (e.g. CO_2 removal from flue gas), agriculture (e.g. control of CO_2 concentration), and other related industries. Membranes can selectively separate CO_2 from industrial processes.

Recently, global warming and severe climate changes provide further opportunities for large-scale applications, separation of flue gas or synthetic gas from coal-based power generation. Carbon dioxide, the most important green-house gas (GHG), is mostly emitted during production of electricity in power plants over 5000 sites worldwide [286–288]. Total CO_2 emission of flue gas composed of carbon dioxide, oxygen, and balanced nitrogen in a 600 MW coal-combustion power plant reaches to 3 million TCO_2 per year, corresponding to an enormous flow rate of 500 $Nm^3\ s^{-1}$. Driving force of the gas mixture for gas separation membrane is very low, which is usually 0.1–0.3 bar. Therefore, it is essentially required to develop highly permeable even at low pressure and highly selective membranes to treat the enormous amount of post-combustion flue gas from pulverized carbon (PC) power plants. Moreover, pre-combustion process, an alternative process substituting the present power plants due to the most efficient process ability and economic feasibility, needs high-performance membranes with the high thermal stability as well as mechanical stabilities at high pressure [286, 289]. Pre-combustion CO_2 capture process called as integrated gasification combined cycle (IGCC) is composed of coal gasification by steam reforming, desulfurization, water–gas-shift (WGS) reaction, and H_2/CO_2 separation [290]. For H_2/CO_2 separation, membrane materials have

Table 3.11 Comparison among three representative CO_2 separation processes.

	Natural gas sweetening	**Flue gas separation**	**Water gas shift recovery**
Objective	Enrichment of methane for fuel	Capture of CO_2 against global warming	Recovery of hydrogen with CO_2 capture
Major constituents	CO_2: 20–40% CH_4 : 60–80 %	CO_2: 8–15% N_2: 60–70% H_2O: 15–25%	H_2: 50–60% CO_2: 40–45%
Minor constituents	Hydrocarbons, H_2S	O_2, SO_x, NO_x	CO, H_2S, NH_3
Temperature	10–100 °C	50–200 °C	200–600 °C
Pressure	High (30–50 bar)	Low (0.9–1.5 bar)	High (20–60 bar)

been considerably limited only to metal or inorganic materials due to the high operation temperature over 300 °C. However, thermally stable rigid-rod polymers have also potentialities as candidates of the gas separation membrane if the materials possess high permeabilities as well as thermal and chemical stabilities [290, 291]. For the major CO_2 separation processes, conditions, and gas components of three representative processes are summarized in Table 3.11. Figure 3.41 displays the comparison between post-combustion, pre-combustion, and natural gas processing with industrial process to capture CO_2. Among them, post- and pre-combustion membrane processes are the main topic of discussion here.

3.5.4.1 Post-combustion Flue Gas CO_2 Capture

The world fossil fuel power plants emit about two billion tons of CO_2 per year which should be significantly reduced according to the Kyoto protocol. Carbon capture and storage (CCS) is one of the most promising ways to reduce CO_2 emissions. For this end, several technologies, such as physical absorption (e.g. Selexol, Rectisol), chemical absorption (e.g. MEA, MDEA), physical adsorption (e.g. molecular sieves, metal organic frameworks), and gas separation membranes, can be applied to capture CO_2 from flue gas in power plants. Among them, amine absorption has been widely investigated for CO_2 removal and considered to be the most mature technology. However, conventional amine absorption is an energy-intensive and high-cost process, which results in the large incremental costs of electricity generation. National Energy Technology Laboratory (NETL) estimated that amine unit will increase the cost of electricity production by 70% [293]. As an alternative, gas separation membranes and/or hybrid systems (e.g. membrane contactor and membrane-cryogenic process) for carbon capture [294], may have potentials to reduce the CO_2 capture cost in this application.

The Polaris® membranes developed by MTR has been demonstrated at pilot-scale for CO_2 capture from a natural gas combined cycle power plant [295–298]. A 20 ton/day skid was tested to validate the advanced modules (multi-tube and plate-and-frame) designed for low-pressure drop and small footprint, and the

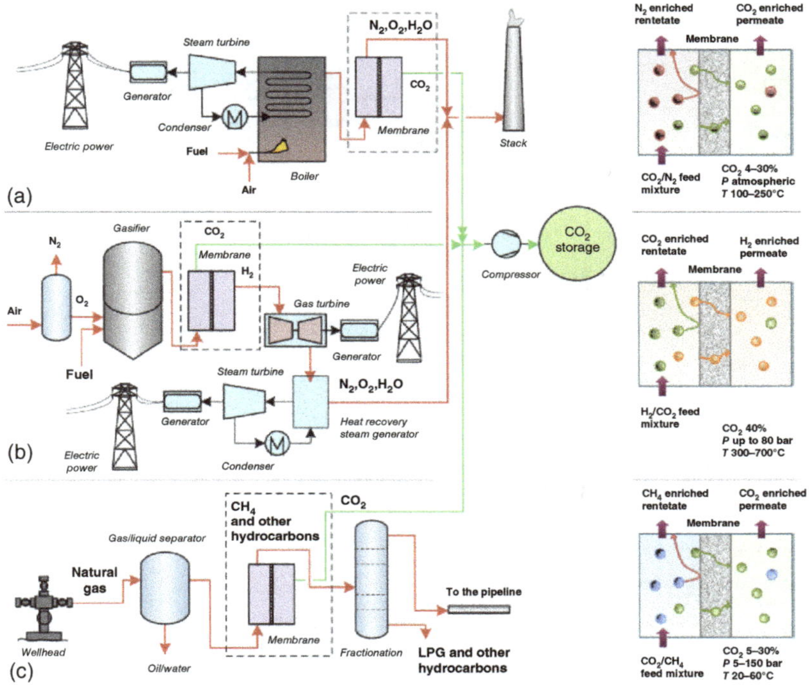

Figure 3.41 Potential application of membranes in CO_2 separation: (a) post-combustion, (b) pre-combustion, and (c) natural gas processing. The corresponding membrane gas separation and its process conditions are shown in the right picture [292].

system showed quite stable performance over ca. 1000 hours [298]. Moreover, MTR patented their process by feeding high CO_2 content air stream (air as sweep gas in the permeate side of the second stage membrane unit) into the boiler to increase the CO_2 concentration in the flue gas [299], which can greatly reduce the required membrane area and energy consumption for this application. Figure 3.42 shows the schematic of a membrane capture process using selective exhaust gas recycle that produces high-purity CO_2 ready for sequestration at an NGCC power plant. Figure 3.43 displays a simplified block diagram of an integrated gas turbine-membrane CO_2 capture system. The gas turbine is equipped with two separate air compressors. This is an oxygen-rich gas to be sent to the natural gas combustor and low oxygen, high carbon dioxide recycle gas to be used as the bypass coolant gas. Figure 3.44 is the picture comparing the size and footprint of the spiral-wound MTR Polaris CO_2 membrane capture system with two different solvent capture systems of similar capacity at NCCC where they have tested during 2015 in a pilot test unit installed in Alabama Power Plant Gaston Unit 5. The Polaris system achieved 90% CO_2 capture during testing and consistently captured CO_2 at a rate of 85% or higher throughout the 1000 hours testing.

The PolyActiveTM membranes developed by Helmholtz-Zentrum Geesthacht in Germany were tested for CO_2 capture from real flue gas using a pilot module

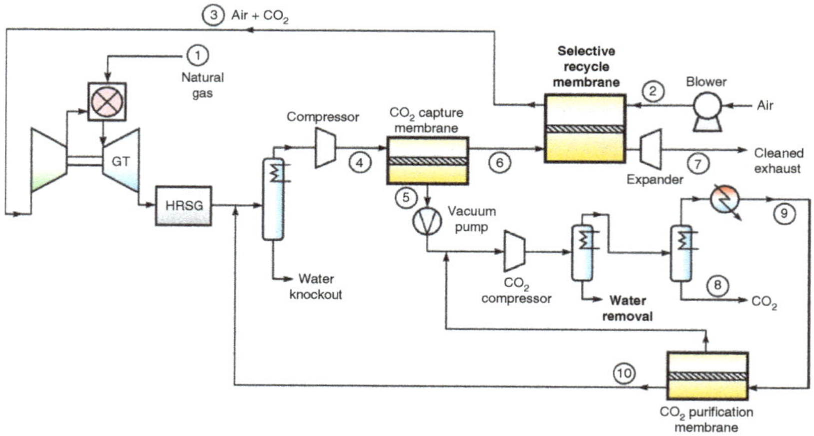

Figure 3.42 Schematic of a membrane capture process using selective exhaust gas recycle that produces high-purity CO_2 ready for sequestration at an NGCC power plant. Source: Adapted from [298].

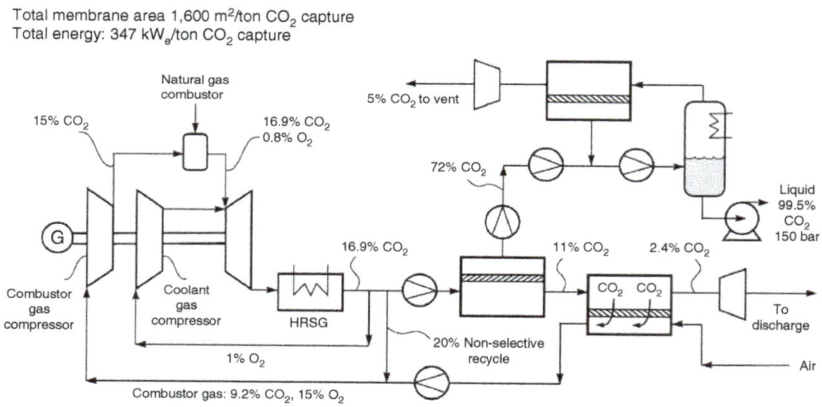

Figure 3.43 A simplified block diagram of an integrated gas turbine-membrane CO_2 capture system. The gas turbine is equipped with two separate air compressors. This is an oxygen-rich gas to be sent to the natural gas combustor and low oxygen, high carbon dioxide recycle gas to be used as the bypass coolant gas. Source: Adapted from [296].

with a membrane area of 12.5 m² [297]. The membrane system also showed stable performance over 740 hours continuously. They also reported that a CO_2 purity of 68.2 mol% in the permeate and a recovery of 42.7% could be achieved in a single-stage process. Figure 3.45 shows a schematic of the cross-sectional structure of PolyActive multilayer composite membranes and schematic representation of the 12.5 m² membrane module incorporating PolyActive membrane for CO_2 capture.

Ho et al. reported on FTM made of polyvinylamine (PVAm) as a fixed carrier [300–303] for CO_2 capture from flue gas. High molecular weight PVAm is prepared

Figure 3.44 A picture comparing the size and footprint of the MTR Polaris CO_2 capture system with two different solvent capture systems of similar capacity at NCCC. Source: Adapted from [299].

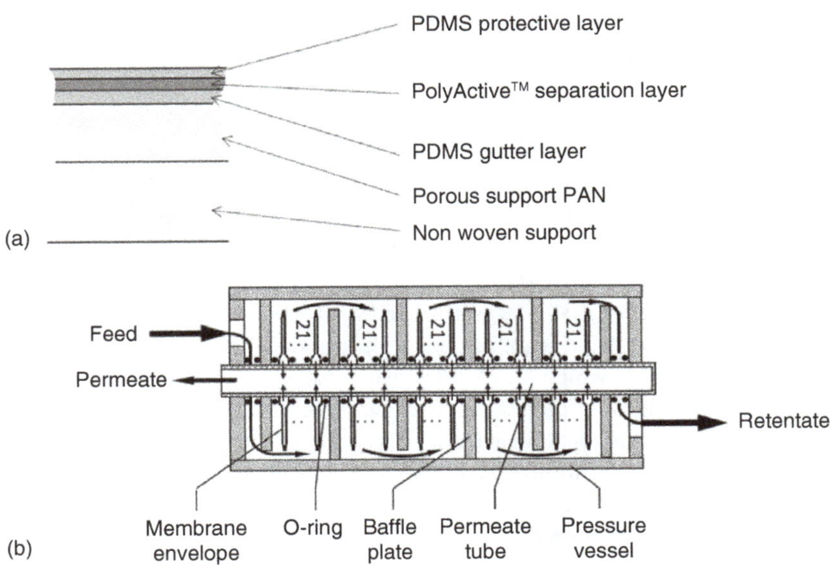

Figure 3.45 (a) Schematic of the cross-sectional structure of PolyActive multilayer composite membranes. (b) Schematic representation of the $12.5\,m^2$ membrane module incorporating PolyActive membrane for CO_2 capture. Source: Adapted from [297].

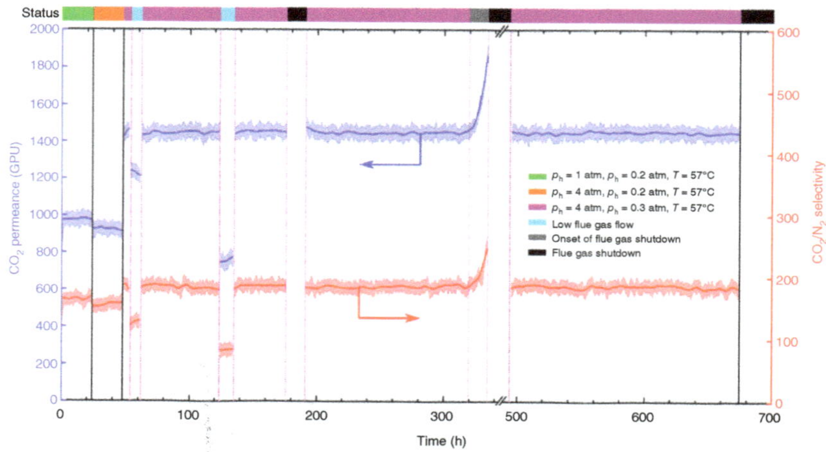

Figure 3.46 Stability plot of the SW module tested with actual flue gas at NCCC for 500 hours. CO_2 recovery and permeate CO_2 purity from the SW module were 47% and 95%, respectively. Source: Adapted from [303].

by the free radical polymerization of *N*-vinylformamide. Amino acid salts were selected as mobile carriers and blended with PVAm, resulting in CO_2 permeance of over 1000 GPU and CO_2/N_2 selectivity of over 170 at 57 °C and a partial pressure of 0.17 atm. The membrane and spiral wound module showed a CO_2 permeance of 1450 GPU and a CO_2/N_2 selectivity >160 at 67 °C [304]. Spiral wound membrane module with 1.4 m^2 membrane area was tested at the National Carbon Capture Center, Wilsonville, Alabama, USA, with actual flue gas slipstream containing 2 ppm SO_2, 1.5–4 ppm NO_2, and 7.6% O_2. As shown in Figure 3.46, despite of the multiple flue gas outages and supply upsets, the proto-type modules demonstrated a 500-h stable operation with transport performance in good agreement to the flat-sheet membranes.

Merkel et al. [295] initially reported that membrane with a CO_2/N_2 selectivity above 50 and a CO_2 permeance of 4000 GPU could offer a capture cost below \$15/ton CO_2, which is lower than US Department of Energy's (DOE) target goal (\$20/ton CO_2). They also reported that improving membrane permeance is more important than increasing selectivity (if selectivity > 30) to further reduce the cost of CO_2 capture from flue gas. They studied four cases CO_2 capture system process [296]: (i) 1.1 bar feed without enhanced gas (EGR), (ii) 1.1 bar with 20% EGR, (iii) 2 bar feed with 20% EGR and (iv) integrated design. In Case 4, the gas turbine is equipped with two separate air compressors which allow the oxygen-rich gas to be sent to the natural gas combustor and low oxygen. Compressing the feed flue gas to 2 bars will not reduce the size and cost of the downstream membrane and purification equipment, but using this case yields a capture cost of about \$64/ton CO_2. They indicated again that pressure ratio limitations render increases in membrane selectivity largely ineffective while increased membrane permeance will reduce the system capital costs and footprints. Integrated system design changes the way gas turbine is made but can save the CO_2 capture cost down to \$44/ton CO_2 using the 346 kWe/ton CO_2 range.

Simulation results [305] showed that membrane process using the high-performance FTM was feasible for CO_2 capture achieving >90% CO_2 recovery and high CO_2 purity above 90%, even from a flue gas with a low CO_2 concentration (~10%). Ramasubramanian et al. reported a CO_2 capture cost of \$25/ton CO_2 using an assumed membrane performance of CO_2 permeance of 3000 GPU (~8.2 $m^3(STP)/(m^2\ h\ bar)^{-1}$) and CO_2/N_2 selectivity of 140 [306]. More recently, membrane properties required for post-combustion carbon capture were systematically investigated [307, 308], and a permeance of at least 3 $m^3(STP)/(m^2\ h\ bar)^{-1}$ with high selectivity should be achieved to be competitive to MEA absorption system.

Reduction of CO_2 from industrial sectors such as steel production, cement factory contributes more than 10% of total CO_2 emissions. CO_2 removal from natural gas and biogas is compulsory to increase methane purity and avoid pipeline corrosion. CCS is the only solution to reduce CO_2 emissions from cement factory as 50% of CO_2 is the by-product of a cement production process. Therefore, CO_2 capture from those energy-intensive industries should be implemented. The first pilot-scale PVAm-based flat sheet and hollow fiber FTM membranes were investigated for CO_2 capture from a 17 vol.% CO_2 flue gas in cement factory. Plate and frame module showed low membrane efficiency and exhibited water condensation and corrosion issues, while hollow fiber FTM membrane modules with 20 m^2 membrane area [309] were evaluated for 6 month and showed stable permeate CO_2 purity of 65%. Techno-economic evaluation was reported to achieve 80% CO_2 recovery and >90% CO_2 purity. However, the designed two-stage membrane system was difficult to achieve specific CO_2 purity of >95% requirement (especially low O_2 limitation) for enhanced oil/gas recovery. The potential solutions are to introduce a third-stage membrane unit or a low-temperature liquefaction unit.

There is a report that PVAm/PVA blend FTM membranes can become a potential candidate for CO_2 capture from flue gas in steelmaking industry with 15–17.5 euro/tons CO_2 [310].

Therefore, it is strongly recommended to develop high-performance membranes with superior thermal, chemical, mechanical, and long-term stabilities for CO_2 separation. Recently developed polymer membranes are suggested to be the excellent candidates [42, 177].

There is a report on CO_2 capture from cement plants and steel mills [311]. MTR showed a simplified process flow diagram illustrating the impact of a selective recycle membrane on a two-stage membrane system. The recycle membrane raises the feed gas concentration from 14.8% to 18% simultaneously lowering the discharge CO_2 concentration from 6.7% to 3.7% CO_2 (80% CO_2 capture). Power consumption is reduced by 30 kW_e/ton of CO_2 captured (see Figure 3.47). It is believed that they adopted Polaris membrane for this process. Figure 3.48 shows a schematic diagram of a 1000 ton/day cement plant fitted with a membrane CO_2 capture system. The system captures 80% of the CO_2 emitted (800 tons/day) at an energy cost of 286 kW_e/ton CO_2 using 3400 $m^2/(ton\ h)^{-1}$ of membrane. The CO_2 condensation column operates at 30 bar and −25 °C (Figure 3.48). Several plants like this process diagram have been successfully installed globally to reduce the CO_2 emission from cement and steel industries.

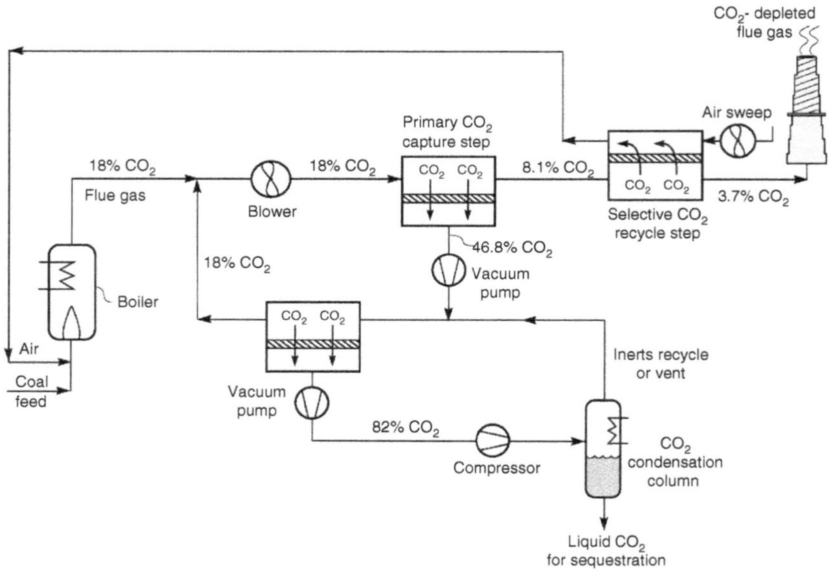

Figure 3.47 A simplified process flow diagram illustrating the impact of a selective recycle membrane on a two-stage membrane system. The recycle membrane raises the feed gas concentration from 14.8% to 18% simultaneously lowering the discharge CO_2 concentration from 6.7% to 3.7% CO_2 (80% CO_2 capture). Power consumption is reduced by 30 kW$_e$/ton of CO_2 captured. The CO_2 condensation column operates at 30 bar and −25 °C. Adapted from [311].

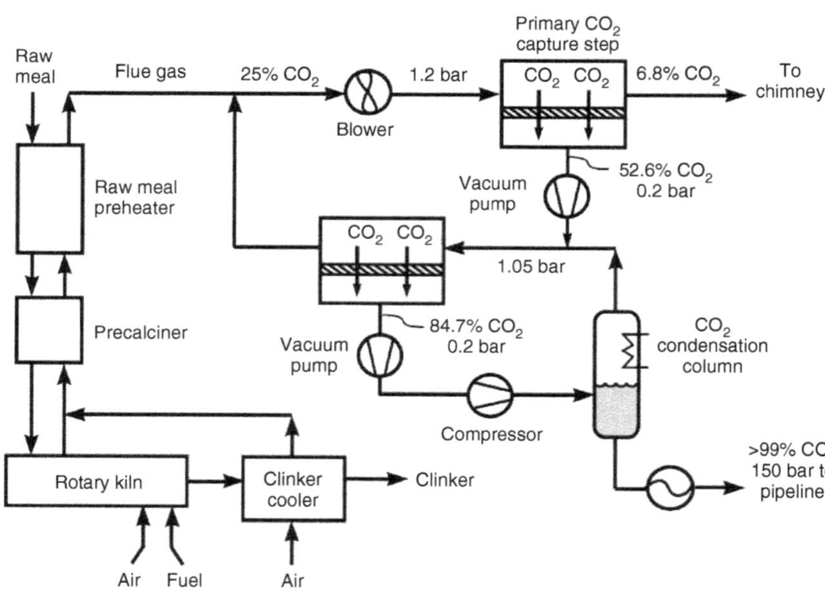

Figure 3.48 Schematic diagram of a 1000 ton/day cement plant fitted with a membrane CO_2 capture system. The system captures 80% of the CO_2 emitted (800 tons/day) at an energy cost of 286 kW$_e$/ton CO_2 using 3400 m^2/(ton h)$^{-1}$ of membrane. The CO_2 condensation column operates at 30 bar and −25 °C. Source: Adapted from [311].

3.5.4.2 CO_2 Removal from Natural Gas

The production of natural gas requires pretreatment before delivery to the pipeline. Typically, raw natural gas contains 75–90% methane. It also contains ethane, propane, butane, and 1–3% of other hydrocarbons. Impurities such as nitrogen, water, carbon dioxide, and hydrogen sulfide may be present and should be removed to upgrade the required natural gas purity. Membrane technology competes with the widely used amine absorption technique in small scale applications (<5 million scfd). Although amine absorption suffers from high capital costs and complex operations, membranes systems compete with the amine absorption techniques even if current membranes still exhibit low gas permeability and selectivity. Non-plasticized high CO_2-permable polymeric membranes should be developed and adopted for practical applications.

Natural gas sweetening processes remove CO_2 from high-pressure methane evolving from natural gas walls. Many natural gas streams are available in the pressure range of 330–1200 psig. In this case, membrane systems are frequently used in combination with conventional gas separation processes. The bulk of CO_2 can be separated from raw natural gas by using membrane system. The final purification to pipeline quality gas can be carried out by the amine absorption process. The capacity of an existing amine plant can be increased by integrating a membrane unit. Figure 3.49 shows a picture equipped with UOP Separex membrane systems to treat 680 MMscfd (217 $Nm^3\ s^{-1}$) to recover CO_2 from natural gas. Over 50 plants like this system are operating in the world. Figure 3.50 is a picture of Cynara membrane plant (Natco Group) to recover acid gas from petroleum refinery plant with treatment capacity of 700 MMscfd (223 $Nm^3\ s^{-1}$) installed about over 30 plants in USA, Canada, Southeast Asia and in the world. These large natural gas treatment plants adopt membranes made from cellulose acetate because of the economic reasons.

In enhanced oil recovery, the gas stream exits the field at pressures frequently as high as 2000 psig. CO_2 is used as injection medium in existing oil fields to increase the oil production. Because its viscosity is lowered by the dissolved CO_2, the oil is easily driven to the surface. Along with the oil, a casing head gas is produced. This gas

Figure 3.49 UOP Separex membrane systems to treat 680 MMscfd (217 $Nm^3\ s^{-1}$) to recover CO_2 from natural gas. Over 50 plants are operating in the world.

Figure 3.50 Cynara membrane plant (Natco Group) to recover acid gas from petroleum refinery plant with treatment capacity of 700 MMscfd (223 $Nm^3 s^{-1}$) installed about over 30 plants in USA, Canada Southeast Asia and in the world.

contains a variety of hydrocarbons including substantial amounts of methane. The casing head gas is mixed with CO_2 at the well. Concentration levels of 40–90 mol% carbon dioxide are achieved. The CO_2 must be removed to utilize the hydrocarbons. In this application, the natural gas as well as the carbon dioxide is the desired product. The natural gas is to be used as fuel and the CO_2 is to be used for re-injection. In the latter case, a purity of 95 mol% is often sufficient to maintain the solvent power of CO_2.

Further expansion of the natural gas purification market to include CO_2 removal, dehydration, and removal of higher hydrocarbons to control heating values and dewpoints will be possible using membrane systems.

3.5.4.3 CO_2 Recovery from Biogas

Biogas is a mixture of various gases produced by microorganisms in the process known as anaerobic digestion. Bio-methane produced by biogas upgrading is a promising source of green energy. Biogas is coming from bio-fermentation of biomass, sewage, landfill gas, wastewater etc. As with the natural gas sweetening, chemical absorption, physical adsorption, and cryogenic separation technologies are competing with membrane separation technology. Although water scrubbing was preferred because of the simplicity of the operation, during the last decade, membrane separation emerged due to its economic viability. Biogas production plants increased to over 18,000 in 2018 in Europe which was 73% increase compared to that in 2010 [312]. Biogas is mainly composed of methane and carbon dioxide and can be upgraded to bio-methane which can substitute traditional natural gas. Gas composition of biogas is tabulated in Table 3.12.

Landfill gas is produced at atmospheric pressures from the decomposition of organic materials under anaerobic conditions. Typically, this biogas contains 30–45 mol% CO_2, 35–65 mol% CH_4, 4% nitrogen, 1% oxygen, 1% water vapor, and traces of hydrogen sulfide and halogenated hydrocarbons (CFC). This biogas can be recovered for potential use when the landfill is covered. The upgrading of CH_4 for use in the local gas distribution system is possible by applying a membrane system.

Table 3.12 Biogas composition from different sources [312].

Compound	Biogas from sewage	Biogas from landfill	Biogas from wastewater	Natural gas
CH_4 (%)	60–70	35–65	55–58	90–95
CO_2 (%)	34–38	30–45	32–50	0.2–2
H_2O (%)	1–7	1–5	1–5	—
NH_3 (ppm)	50–100	0–5	0–100	—
H_2 (%)	Traces	0–5	Traces	—
O_2 (%)	Traces	0–1	Traces	—
N_2 (%)	0–2	0–15	Traces	0–0.5
H_2S (ppm)	0–4000	0–100	0–4000	0–10
Siloxanes (%)	0–0.2	0–0.2	0–0.5	—

Biogas composition varies depending of its origin source. In comparison with natural gas, where methane content is about 90%, biogas presents a wide range of 35–70% methane. Carbon dioxide represents about 30–50% of the biogas composition.

For this application, the toxic trace gases of H_2S and CFC are removed first by an adsorption process. The remaining gas stream is fed to a membrane system and is often compressed to higher pressure, up to 500 psig, to enhance membrane efficiency.

Figure 3.51 describes three modes of operation: (i) the driving force generation alternatives where feed is compressed and the process is operated in ambient mode, (ii) the feed gas and operation is in ambient pressure whereas the permeate is coming

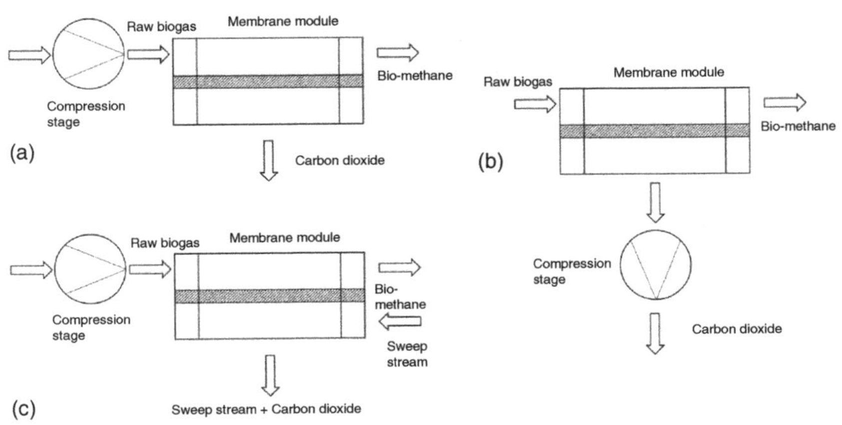

Figure 3.51 Driving force generation alternatives: (a) Feed compression, in which the feed gas must be pressurized all over the membrane surface, whereas the permeate can operate at ambient pressure; (b) vacuum in the permeate side, where the feed gas and operates at ambient pressure while vacuum is applied on the permeate side of the membrane; (c) sweep stream configuration, in which the presence of the sweep gas allows the creation of a driving force in the permeate side. Source: Adapted from Scholz et al. [312].

from the vacuum mode, and (iii) sweep stream configuration where the sweep gas is fed to create a driving force in the permeate side.

3.5.5 Vapor/Gas Separation

In this application, volatile organic compound (VOC) removal from air or nitrogen streams, such as ethylene or propylene recovery from polyolefin production, and gasoline vapor from air are representative membrane processes. Governmental regulations in the world are becoming stricter to limit the emissions of VOCs. To meet the worldwide release limitations of VOCs, many VOCs recovery technologies, such as absorption, condensation, and membrane technology have emerged. Among them, membrane technology has been regarded as one of the most energy-efficient process and thus different membrane materials have been developed. Both rubbery polymeric materials and glassy polymers have been developed for this purpose. Rubbery polymers, such as PDMS, showed permselectivity to VOCs and have been developed for removal of VOC from the waste streams. Glassy polymers are permeable to nitrogen rather than large VOCs. Therefore, nitrogen permeability is relatively larger than that of VOCs [313].

Main application of vapor separation membranes is the recovery of hydrocarbon monomers from ethylene and PE and PP production plants. MTR (Membrane Technology Research) installed the first propylene recovery membrane system (VaporSep®) in 1996. Since then, over 20 systems that recover propylene and nitrogen from resin degassing bins have been installed. Typically, during the production of PP, a considerable amount of the propylene feedstock is lost. The value of the lost feedstock is substantial, ranging from US $1 million to US $3 million per year for a typical PP plant. Usually, for vapor/gas separation, rubbery polymers such as cross-linked PDMS or reverse-selective glassy polymers such as PTMSP or PMP can be used. These polymers are mostly solubility-selective membranes. A schematic of process flow design of a typical membrane system integrated into a polyolefin production plant is shown in Figure 3.52. The vapor-enriched stream is recycled to the compressor inlet, while high purity N_2 is sent to the second stage membrane unit for further purification.

Usually, the emission of gasoline vapors at gas stations generates the formation of ground-level ozone due to high emissions of hydrocarbons. Membrane-based gas separations can recover and recycle gasoline vapors back to the underground storage tanks. For gasoline vapor recovery, a silicone-based membrane system was first considered. Silicone membranes for vapor recovery and recycling separate the hydrocarbon portion of the gasoline vapor, and the retained air is vented. However, silicone rubbers exhibit a weak resistance to gasoline vapors. If the system failed, further safeguards must be present to avoid uncontrolled release of gasoline vapors into the atmosphere. In addition, silicone-based membrane systems require many components such as pumps, condensers, valving, fans, and membrane cartridges to operate properly. About 95–99% of VOCs emission can be reduced.

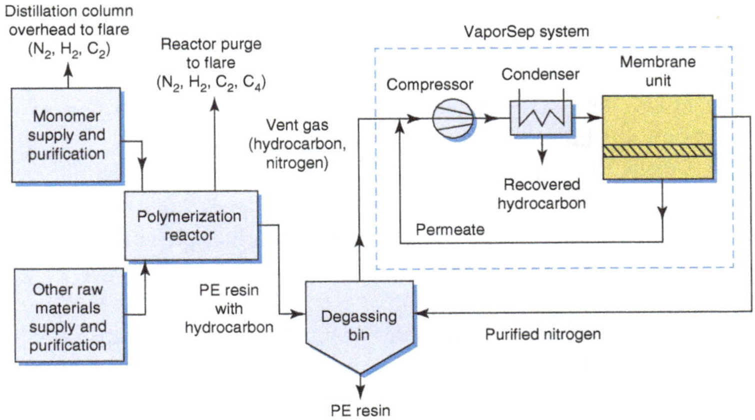

Figure 3.52 Propylene/nitrogen recovery unit. Silicone rubber propylene-selective membranes are used. https://www.mtrinc.com/our-business/petrochemical/polyethylene-production/.

3.6 Conclusions and Future Perspectives

Gas separation membranes have been advanced over the last several decades to separate hydrogen, methane, air, and CO_2. Polymer membranes have been mainly used for these applications. Although many high-performance polymers have been reported in the literature, it remains in the laboratory for further development. As the global climate changes have spurred the development of membranes for carbon capture and utilization for the carbon-neutral society, future development of gas separation requires the high permeability as well as high selectivity to gas molecules such as hydrogen and CO_2. Engineering of membrane modules and processes spurred the development of carbon capture technology which is demonstrated in several places in the world. Future development is expected to be realized in the large-scale and real size membrane modules and processes equipped in the power plants to battle with the climate change. Utilization of CO_2 from the carbon capture is another area to be considered.

References

1 Mitchell, J.K. (1995). On the penetrativeness of fluids. *Journal of Membrane Science* 100: 11–16.
2 Graham, T. (1867). On the absorption and dialytic separation of gases by colloid septa. *Journal of the Franklin Institute* 83: 39–41.
3 Fick, A. (1995). On liquid diffusion. *Journal of Membrane Science* 100: 33–38.

4 Graham, T. (1995). On the absorption and dialytic separation of gases by colloid septa Part I. Action of a septum of caoutchouc. *Journal of Membrane Science* 100: 27–31.

5 Knudsen, M. (1995). The laws of molecular flow and of inner friction flow of gases through tubes. *Journal of Membrane Science* 100: 23–25.

6 Daynes, H.A. (1920). The process of diffusion through a rubber membrane. *Proceedings of the Royal Society of London Series A* 97: 286–307.

7 Barrer, R. and Skirrow, G. (1948). Transport and equilibrium phenomena in gas–elastomer systems. I. Kinetic phenomena. *Journal of Polymer Science* 3: 549–563.

8 Loeb, S. and Sourirajan, S. (1963). Sea water demineralization by means of an osmotic membrane. In: *Saline Water Conversion—II*, 117–132. American Chemical Society.

9 Henis, J.M. and Tripodi, M.K. (1980).

10 Spillman, R. (1995). Economics of gas separation membrane processes. In: *Membrane Science and Technology*, 589–667. Elsevier.

11 Koros, W.J. and Fleming, G. (1993). Membrane-based gas separation. *Journal of Membrane Science* 83: 1–80.

12 Carslaw, H. and Jaeger, J. (1959). *Heat Conduction in Solids*. Oxford University Press.

13 Park, G.S. and Crank, J. (1968). *Diffusion in Polymers*. Academic Press.

14 Crank, J. (1979). *The Mathematics of Diffusion*. Oxford University Press.

15 Doolittle, A.K. (1951). Studies in Newtonian flow. I. The dependence of the viscosity of liquids on temperature. *Journal of Applied Physics* 22: 1031–1035.

16 Cohen, M.H. and Turnbull, D. (1959). Molecular transport in liquids and glasses. *The Journal of Chemical Physics* 31: 1164–1169.

17 Miyata, S., Sato, S., Nagai, K. et al. (2008). Relationship between gas transport properties and fractional free volume determined from dielectric constant in polyimide films containing the hexafluoroisopropylidene group. *Journal of Applied Polymer Science* 107: 3933–3944.

18 Brandrup, J., Immergut, E.H., and Grulke, E.A. (ed.) (1999). *Polymer Handbook*, VIe, 232. Wiley Interscience.

19 Hildebrand, J.H. and Scott, R.L. (1950). *The Solubility of Nonelectrolytes*. Reinhold Publishing Corporation.

20 Koros, W.J. and Chern, R.T. (1987). Separation of gaseous mixtures using polymer membranes. In: *Handbook of Separation Process Technology*, 862–953. New York, NY: Wiley-Interscience.

21 Kanehashi, S. and Nagai, K. (2005). Analysis of dual-mode model parameters for gas sorption in glassy polymers. *Journal of Membrane Science* 253: 117–138.

22 Petropoulos, J. (1970). Quantitative analysis of gaseous diffusion in glassy polymers. *Journal of Polymer Science, Part A-2: Polymer Physics* 8: 1797–1801.

23 Vieth, W. and Sladek, K. (1965). A model for diffusion in a glassy polymer. *Journal of Colloid Science* 20: 1014–1033.

24 Koros, W.J., Chan, A., and Paul, D. (1977). Sorption and transport of various gases in polycarbonate. *Journal of Membrane Science* 2: 165–190.

25 Koros, W. (1980). Model for sorption of mixed gases in glassy polymers. *Journal of Polymer Science Polymer Physics Edition* 18: 981–992.

26 Matteucci, S., Yampolskii, Y., Freeman, B.D., and Pinnau, I. (2006). Transport of gases and vapors in glassy and rubbery polymers. In: *Materials Science of Membranes for Gas and Vapor Separation*, 1–2. Wiley.

27 Kusakabe, K., Yoneshige, S., Murata, A., and Morooka, S. (1996). Morphology and gas permeance of ZSM-5-type zeolite membrane formed on a porous α-alumina support tube. *Journal of Membrane Science* 116: 39–46.

28 Huang, A., Bux, H., Steinbach, F., and Caro, J. (2010). Molecular-sieve membrane with hydrogen permselectivity: ZIF-22 in LTA topology prepared with 3-aminopropyltriethoxysilane as covalent linker. *Angewandte Chemie* 122: 5078–5081.

29 Chui, S.S.-Y., Lo, S.M.-F., Charmant, J.P. et al. (1999). A chemically functionalizable nanoporous material $[Cu_3 (TMA)_2 (H_2O)_3]n$. *Science* 283: 1148–1150.

30 Toda, K., Furue, R., and Hayami, S. (2015). Recent progress in applications of graphene oxide for gas sensing: a review. *Analytica Chimica Acta* 878: 43–53.

31 Liu, X.-H., Guan, C.-Z., Ding, S.-Y. et al. (2013). On-surface synthesis of single-layered two-dimensional covalent organic frameworks via solid–vapor interface reactions. *Journal of the American Chemical Society* 135: 10470–10474.

32 Fu, S., Sanders, E.S., Kulkarni, S.S., and Koros, W.J. (2015). Carbon molecular sieve membrane structure–property relationships for four novel 6FDA based polyimide precursors. *Journal of Membrane Science* 487: 60–73.

33 Hayashi, J.-I., Mizuta, H., Yamamoto, M. et al. (1997). Pore size control of carbonized BPDA-pp'ODA polyimide membrane by chemical vapor deposition of carbon. *Journal of Membrane Science* 124: 243–252.

34 Suda, H. and Haraya, K. (1997). Gas permeation through micropores of carbon molecular sieve membranes derived from Kapton polyimide. *The Journal of Physical Chemistry B* 101: 3988–3994.

35 Park, H. and Lee, Y.M. (2004). Relationship between chemical structure of aromatic polyimides and gas permeation properties of their carbon molecular sieve membranes. *Journal of Membrane Science* 229: 117–127.

36 Park, H.B. and Lee, Y.M. (2005). Fabrication and characterization of nanoporous carbon/silica membranes. *Advanced Materials* 17: 477–483.

37 Kim, S., Wang, H., and Lee, Y.M. (2019). 2D nanosheets and their composite membranes for water, gas, and ion separation. *Angewandte Chemie* 131: 17674–17689.

38 Shen, J., Liu, G., Huang, K. et al. (2015). Membranes with fast and selective gas-transport channels of laminar graphene oxide for efficient CO_2 capture. *Angewandte Chemie International Edition* 54: 578–582.

39 Wang, Y., Low, Z.X., Kim, S. et al. (2018). Functionalized boron nitride nanosheets: a thermally rearranged polymer nanocomposite membrane for hydrogen separation. *Angewandte Chemie* 130: 16288–16293.

40 Van Krevelen, D.W. (1990). Cohesive properties and solubility. In: *Properties of Polymers*, 189–225. Elsevier.

41 Breck, D.W. and Sieves, Z.M. (1974). Structure, chemistry and use. In: *Zeolite Molecular Sieves*, 593–724. New York: Wiley.

42 Lin, H. and Freeman, B.D. (2005). Materials selection guidelines for membranes that remove CO_2 from gas mixtures. *Journal of Molecular Structure* 739: 57–74.

43 Robeson, L.M. (1991). Correlation of separation factor versus permeability for polymeric membranes. *Journal of Membrane Science* 62: 165–185.

44 Robeson, L.M. (2008). The upper bound revisited. *Journal of Membrane Science* 320: 390–400.

45 Freeman, B.D. (1999). Basis of permeability/selectivity tradeoff relations in polymeric gas separation membranes. *Macromolecules* 32: 375–380.

46 Swaidan, R., Ghanem, B., and Pinnau, I. (2015). Fine-tuned intrinsically ultramicroporous polymers redefine the permeability/selectivity upper bounds of membrane-based air and hydrogen separations. *ACS Macro Letters* 4: 947–951.

47 Comesaña-Gándara, B., Chen, J., Bezzu, C.G. et al. (2019). Redefining the Robeson upper bounds for CO_2/CH_4 and CO_2/N_2 separations using a series of ultrapermeable benzotriptycene-based polymers of intrinsic microporosity. *Energy & Environmental Science* 12: 2733–2740.

48 Alentiev, A.Y. and Yampolskii, Y.P. (2000). Free volume model and tradeoff relations of gas permeability and selectivity in glassy polymers. *Journal of Membrane Science* 165: 201–216.

49 Loeb, S. and Sourirajan, S. (1962). High-flow semipermeable membranes for separation of water from saline solutions. *Advances in Chemistry Series* 38: 117–132.

50 Gantzel, P.K. and Merten, U. (1970). Gas separations with high-flux cellulose acetate membranes. *Industrial and Engineering Chemistry Process Design and Development* 9: 331–332.

51 Schell, W., Wensley, C., Chen, M. et al. (1989). Recent advances in cellulosic membranes for gas separation and pervaporation. *Gas Separation & Purification* 3: 162–169.

52 Puleo, A., Paul, D.R., and Kelley, S. (1989). The effect of degree of acetylation on gas sorption and transport behavior in cellulose acetate. *Journal of Membrane Science* 47: 301–332.

53 White, L.S. (2010). Evolution of natural gas treatment with membrane systems. In: *Membrane Gas Separation* (ed. Y. Yampolskii and B. Freeman), 313–332. Wiley.

54 Tang, M.-W., King, W.M., and Wensley, C.G. (1989). Air dried cellulose acetate membranes, US Patent 4855048A Google Patents.

55 Minhas, B.S., Matsuura, T., and Sourirajan, S. (1987). Formation of asymmetric cellulose acetate membranes for the separation of carbon dioxide-methane gas mixtures. *Industrial & Engineering Chemistry Research* 26: 2344–2348.

56 Stern, S. and Saxena, V. (1980). Concentration-dependent transport of gases and vapors in glassy polymers. *Journal of Membrane Science* 7: 47–59.

57 Duval, J., Kemperman, A., Folkers, B. et al. (1994). *Journal of Applied Polymer Science* 54: 409.

58 Bikson, B., Nelson, J., and Muruganandam, N. (1994). Composite cellulose acetate/poly (methyl methacrylate) blend gas separation membranes. *Journal of Membrane Science* 94: 313–328.

59 Henis, J.M. and Tripodi, M.K. (1980). A novel approach to gas separations using composite hollow fiber membranes. *Separation Science and Technology* 15: 1059–1068.

60 Henis, J.M. and Tripodi, M.K. (1981). Composite hollow fiber membranes for gas separation: the resistance model approach. *Journal of Membrane Science* 8: 233–246.

61 Henis, J.M.S. and Tripodi, M.K. (1983). The developing technology of gas separating membranes. *Science* 220: 11–17.

62 Sanaeepur, H., Amooghin, A.E., Bandehali, S. et al. (2019). Polyimides in membrane gas separation: monomer's molecular design and structural engineering. *Progress in Polymer Science* 91: 80–125.

63 Kim, T.H., Koros, W.J., Husk, G.R., and O'Brien, K.C. (1988). *Journal of Membrane Science* 37: 45–62.

64 Wang, Y.-C., Huang, S.-H., Hu, C.-C. et al. (2005). Sorption and transport properties of gases in aromatic polyimide membranes. *Journal of Membrane Science* 248: 15–25.

65 Budd, P.M., Elabas, E.S., Ghanem, B.S. et al. (2004). Solution-processed, organophilic membrane derived from a polymer of intrinsic microporosity. *Advanced Materials* 16: 456–459.

66 Vanherck, K., Koeckelberghs, G., and Vankelecom, I.F. (2013). Crosslinking polyimides for membrane applications: a review. *Progress in Polymer Science* 38: 874–896.

67 Jiang, L., Wang, Y., Chung, T. et al. (2009). Polyimides for pervaporation membranes and biofuels separation. *Progress in Polymer Science* 34: 1135–1160.

68 Zeigler, J.M. and Fearon, F.G. (1989). *Silicon-based Polymer Science: A Comprehensive Resource*. ACS Publications.

69 Bostick, E. (1969). *Ring Opening Polymerization*, 327. New York: Marcel Dekker.

70 Rochow, E.G. (1987). *Silicon and Silicones: About Stone-age Tools, Antique Pottery, Modern Ceramics, Computers, Space Materials, and How They All Got That Way*. Springer Verlag.

71 Schultz, J. and Peinemann, K.-V. (1996). Membranes for separation of higher hydrocarbons from methane. *Journal of Membrane Science* 110: 37–45.

72 Barrer, R. and Chio, H. (1965). Solution and diffusion of gases and vapors in silicone rubber membranes. *Journal of Polymer Science Part C: Polymer Symposia* 10: 111–138.

73 Stern, S., Shah, V., and Hardy, B. (1987). Structure-permeability relationships in silicone polymers. *Journal of Polymer Science Part B: Polymer Physics* 25: 1263–1298.

74 Baker, R.W., Yoshioka, N., Mohr, J.M., and Khan, A.J. (1987). Separation of organic vapors from air. *Journal of Membrane Science* 31: 259–271.

75 Lee, C.-L., Chapman, H.L., Cifuentes, M.E. et al. (1988). Effects of polymer structure on the gas permeability of silicone membranes. *Journal of Membrane Science* 38: 55–70.

76 Bhide, B. and Stern, S. (1991). Permeability of silicone polymers to hydrogen. *Journal of Applied Polymer Science* 42: 2397–2403.

77 Merkel, T., Bondar, V., Nagai, K. et al. (2000). Gas sorption, diffusion, and permeation in poly (dimethylsiloxane). *Journal of Polymer Science Part B: Polymer Physics* 38: 415–434.

78 Jordan, S. and Koros, W. (1990). Permeability of pure and mixed gases in silicone rubber at elevated pressures. *Journal of Polymer Science Part B: Polymer Physics* 28: 795–809.

79 Leemann, M., Eigenberger, G., and Strathmann, H. (1996). Vapour permeation for the recovery of organic solvents from waste air streams: separation capacities and process optimization. *Journal of Membrane Science* 113: 313–322.

80 Singh, A., Freeman, B., and Pinnau, I. (1998). Pure and mixed gas acetone/nitrogen permeation properties of polydimethylsiloxane [PDMS]. *Journal of Polymer Science Part B: Polymer Physics* 36: 289–301.

81 Hägg, M.-B. (2000). Membrane purification of Cl_2 gas: I. Permeabilities as a function of temperature for Cl_2, O_2, N_2, H_2 in two types of PDMS membranes. *Journal of Membrane Science* 170: 173–190.

82 Yeom, C., Lee, S., Song, H., and Lee, J. (2002). Vapor permeations of a series of VOCs/N_2 mixtures through PDMS membrane. *Journal of Membrane Science* 198: 129–143.

83 Pinnau, I., Casillas, C., Morisato, A., and Freeman, B. (1997). Long-term permeation properties of poly (1-trimethylsilyl-1-propyne) membranes in hydrocarbon—vapor environment. *Journal of Polymer Science Part B: Polymer Physics* 35: 1483–1490.

84 Pinnau, I. and He, Z. (2004). Pure- and mixed-gas permeation properties of polydimethylsiloxane for hydrocarbon/methane and hydrocarbon/hydrogen separation. *Journal of Membrane Science* 244: 227–233.

85 Masuda, T., Isobe, E., Higashimura, T., and Takada, K. (1983). Poly[1-(trimethylsilyl)-1-propyne]: a new high polymer synthesized with transition-metal catalysts and characterized by extremely high gas permeability. *Journal of the American Chemical Society* 105: 7473–7474.

86 Nagai, K., Masuda, T., Nakagawa, T. et al. (2001). Poly [1-(trimethylsilyl)-1-propyne] and related polymers: synthesis, properties and functions. *Progress in Polymer Science* 26: 721–798.

87 Srinivasan, R., Auvil, S., and Burban, P. (1994). Elucidating the mechanism (s) of gas transport in poly [1-(trimethylsilyl)-1-propyne](PTMSP) membranes. *Journal of Membrane Science* 86: 67–86.

88 Pinnau, I. and Toy, L.G. (1996). Gas and vapor transport properties of amorphous perfluorinated copolymer membranes based on 2,2-bistrifluoromethyl-4,5-difluoro-1,3-dioxole/tetrafluoroethylene. *Journal of Membrane Science* 109: 125–133.

89 Park, H.B. and Lee, Y.M. (2008). Polymeric membrane materials and potential use in gas separation. In: *Advanced Membrane Technology and Applications* (ed. N.N. Li), 633–669. Wiley.

90 Polyakov, A.M., Starannikova, L.E., and Yampolskii, Y.P. (2003). Amorphous Teflons AF as organophilic pervaporation materials transport of individual components. *Journal of Membrane Science* 216: 241–256.

91 Merkel, T.C., Bondar, V., Nagai, K. et al. (1999). Gas sorption, diffusion, and permeation in Poly(2,2-bis(trifluoromethyl)-4,5-difluoro-1,3-dioxole-co-tetrafluoroethylene). *Macromolecules* 32: 8427–8440.

92 Alentiev, A.Y., Shantarovich, V.P., Merkel, T.C. et al. (2002). Gas and vapor sorption, permeation, and diffusion in glassy amorphous Teflon AF1600. *Macromolecules* 35: 9513–9522.

93 Yampolskii, Y., Pinnau, I., and Freeman, B.D. (2006). *Materials Science of Membranes for Gas and Vapor Separation*. Hoboken, NJ, USA: John Wiley & Sons, Ltd.

94 Orme, C.J., Stone, M.L., Benson, M.T., and Peterson, E.S. (2003). Testing of polymer membranes for the selective permeability of hydrogen. *Separation Science and Technology* 38: 3225–3238.

95 Jansen, J.C., Tasselli, F., Tocci, E., and Drioli, E. (2006). High-flux composite perfluorinated gas separation membranes of Hyflon® AD on a hollow fibre ultrafiltration membrane support. *Desalination* 192: 207–213.

96 Jansen, J.C., Macchione, M., and Drioli, E. (2007). On the unusual solvent retention and the effect on the gas transport in perfluorinated Hyflon AD® membranes. *Journal of Membrane Science* 287: 132–137.

97 Macchione, M., Jansen, J.C., De Luca, G. et al. (2007). Experimental analysis and simulation of the gas transport in dense Hyflon® AD60X membranes: influence of residual solvent. *Polymer* 48: 2619–2635.

98 Pesiri, D.R., Jorgensen, B., and Dye, R.C. (2003). Thermal optimization of polybenzimidazole meniscus membranes for the separation of hydrogen, methane, and carbon dioxide. *Journal of Membrane Science* 218: 11–18.

99 Neuse, E.W. (1982). Aromatic polybenzimidazoles. Syntheses, properties, and applications. In: *Synthesis and Degradation Rheology and Extrusion*, 1–42. Springer.

100 Singh, R.P., Dahe, G.J., Dudeck, K.W., and Berchtold, K.A. (2020). Macrovoid-free high performance polybenzimidazole hollow fiber membranes for elevated temperature H_2/CO_2 separations. *International Journal of Hydrogen Energy* 45: 27331–27345.

101 Berchtold, K.A., Singh, R.P., Young, J.S., and Dudeck, K.W. (2012). Polybenzimidazole composite membranes for high temperature synthesis gas separations. *Journal of Membrane Science* 415: 265–270.

102 Merkel, T., Freeman, B., Spontak, R. et al. (2002). Ultrapermeable, reverse-selective nanocomposite membranes. *Science* 296: 519–522.

103 Barrer, R. (1968). Diffusion and permeation in heterogeneous media. In: *Diffusion in Polymers*, 165–217. Academic Press.

104 Maxwell, J.C. (1873). *A Treatise on Electricity and Magnetism*. Oxford: Clarendon Press.

105 Gomes, D., Nunes, S.P., and Peinemann, K.-V. (2005). Membranes for gas separation based on poly(1-trimethylsilyl-1-propyne)–silica nanocomposites. *Journal of Membrane Science* 246: 13–25.

106 Merkel, T., Freeman, B., Spontak, R. et al. (2003). Sorption, transport, and structural evidence for enhanced free volume in poly (4-methyl-2-pentyne)/fumed silica nanocomposite membranes. *Chemistry of Materials* 15: 109–123.

107 Merkel, T.C., He, Z., Pinnau, I. et al. (2003). Effect of nanoparticles on gas sorption and transport in poly (1-trimethylsilyl-1-propyne). *Macromolecules* 36: 6844–6855.

108 Park, H.B., Kim, J.K., Nam, S.Y., and Lee, Y.M. (2003). Imide-siloxane block copolymer/silica hybrid membranes: preparation, characterization and gas separation properties. *Journal of Membrane Science* 220: 59–73.

109 Kim, S., Wang, H., and Lee, Y.M. (2019). 2D nanosheets and their composite membranes for water, gas, and ion separation. *Angewandte Chemie (International Ed. in English)* 58: 17512–17527.

110 Kim, S. and Lee, Y.M. (2023). Two-dimensional nanosheets and membranes for their emerging technologies. *Current Opinion in Chemical Engineering* 39.

111 Furukawa, H., Cordova, K.E., O'Keeffe, M., and Yaghi, O.M. (2013). The chemistry and applications of metal-organic frameworks. *Science* 341.

112 Zhao, M., Huang, Y., Peng, Y. et al. (2018). Two-dimensional metal–organic framework nanosheets: synthesis and applications. *Chemical Society Reviews* 47: 6267–6295.

113 Lei, W., Mochalin, V.N., Liu, D. et al. (2015). Boron nitride colloidal solutions, ultralight aerogels and freestanding membranes through one-step exfoliation and functionalization. *Nature Communications* 6: 8849.

114 Wang, Y., Wang, X., and Antonietti, M. (2012). Polymeric graphitic carbon nitride as a heterogeneous organocatalyst: from photochemistry to multipurpose catalysis to sustainable chemistry. *Angewandte Chemie International Edition* 51: 68–89.

115 Gao, M.-R., Chan, M.K., and Sun, Y. (2015). Edge-terminated molybdenum disulfide with a 9.4-Å interlayer spacing for electrochemical hydrogen production, Nature. *Communications* 6: 7493.

116 Mao, X., Xu, Y., Xue, Q. et al. (2013). Ferromagnetism in exfoliated tungsten disulfide nanosheets. *Nanoscale Research Letters* 8: 430.

117 Ding, L., Wei, Y., Li, L. et al. (2018). MXene molecular sieving membranes for highly efficient gas separation. *Nature Communications* 9: 155.

118 Xu, M., Liang, T., Shi, M., and Chen, H. (2013). Graphene-like two-dimensional materials. *Chemical Reviews* 113: 3766–3798.

119 Naguib, M., Mochalin, V.N., Barsoum, M.W., and Gogotsi, Y. (2014). 25th anniversary article: MXenes: a new family of two-dimensional materials. *Advanced Materials* 26: 992–1005.

120 Bhimanapati, G.R., Lin, Z., Meunier, V. et al. (2015). Recent advances in two-dimensional materials beyond graphene. *ACS Nano* 9: 11509–11539.

121 Huang, H., Ying, Y., and Peng, X. (2014). Graphene oxide nanosheet: an emerging star material for novel separation membranes. *Journal of Materials Chemistry A* 2: 13772–13782.

122 Liu, G., Jin, W., and Xu, N. (2015). Graphene-based membranes. *Chemical Society Reviews* 44: 5016–5030.

123 Liu, G., Jin, W., and Xu, N. (2016). Two-dimensional-material membranes: a new family of high-performance separation membranes. *Angewandte Chemie International Edition* 55: 13384–13397.

124 Cheng, X.Q., Wang, Z.X., Jiang, X. et al. (2018). Towards sustainable ultrafast molecular-separation membranes: from conventional polymers to emerging materials. *Progress in Materials Science* 92: 258–283.

125 Jiang, D.-E., Cooper, V.R., and Dai, S. (2009). Porous graphene as the ultimate membrane for gas separation. *Nano Letters* 9: 4019–4024.

126 Kim, H.W., Yoon, H.W., Yoon, S.-M. et al. (2013). Selective gas transport through few-layered graphene and graphene oxide membranes. *Science* 342: 91–95.

127 Wang, S., Wu, Y., Zhang, N. et al. (2016). A highly permeable graphene oxide membrane with fast and selective transport nanochannels for efficient carbon capture. *Energy & Environmental Science* 9: 3107–3112.

128 Yan, N., Capezzuto, F., Lavorgna, M. et al. (2016). Borate cross-linked graphene oxide–chitosan as robust and high gas barrier films. *Nanoscale* 8: 10783–10791.

129 Yang, J., Gong, D., Li, G. et al. (2018). Self-assembly of thiourea-crosslinked graphene oxide framework membranes toward separation of small molecules. *Advanced Materials* 30: 1705775.

130 Kim, S., Hou, J., Wang, Y. et al. (2018). Highly permeable thermally rearranged polymer composite membranes with a graphene oxide scaffold for gas separation. *Journal of Materials Chemistry A* 6: 7668–7674.

131 Rodenas, T., Luz, I., Prieto, G. et al. (2015). Metal–organic framework nanosheets in polymer composite materials for gas separation. *Nature Materials* 14: 48.

132 Pustovarenko, A., Goesten, M.G., Sachdeva, S. et al. (2018). Nanosheets of non-layered aluminum metal–organic frameworks through a surfactant-assisted method. *Advanced Materials* 30: 1707234.

133 Peng, Y., Li, Y., Ban, Y. et al. (2014). Metal-organic framework nanosheets as building blocks for molecular sieving membranes. *Science* 346: 1356–1359.

134 Wang, X., Chi, C., Zhang, K. et al. (2017). Reversed thermo-switchable molecular sieving membranes composed of two-dimensional metal-organic nanosheets for gas separation. *Nature Communications* 8: 14460.

135 Hu, Y., Wei, J., Liang, Y. et al. (2016). Zeolitic imidazolate framework/graphene oxide hybrid nanosheets as seeds for the growth of ultrathin molecular sieving membranes. *Angewandte Chemie International Edition* 128: 2088–2092.

136 Hou, J., Wei, Y., Zhou, S. et al. (2018). Highly efficient H_2/CO_2 separation via an ultrathin metal-organic framework membrane. *Chemical Engineering Science* 182: 180–188.

137 Shan, M., Liu, X., Wang, X. et al. (2018). Facile manufacture of porous organic framework membranes for precombustion CO_2 capture, Science. *Advances* 4: eaau1698.

138 Wang, D., Wang, Z., Wang, L. et al. (2015). Ultrathin membranes of single-layered MoS2 nanosheets for high-permeance hydrogen separation. *Nanoscale* 7: 17649–17652.

139 Shen, J., Liu, G., Ji, Y. et al. (2018). 2D MXene nanofilms with tunable gas transport channels. *Advanced Functional Materials* 28: 1801511.

140 Geng, Z., Song, Q., Zhang, X. et al. (2018). Mixed matrix membranes composed of WS2 nanosheets and fluorinated poly(2,6-dimethyl-1,4-phenylene oxide) via Suzuki reaction for improved CO_2 separation. *Journal of Membrane Science* 565: 226–232.

141 Wang, Y., Low, Z.X., Kim, S. et al. (2018). Functionalized boron nitride nanosheets: a thermally rearranged polymer nanocomposite membrane for hydrogen separation. *Angewandte Chemie International Edition* 130: 1–7.

142 Rezac, M.E., John, T., and Pfromm, P.H. (1997). Effect of copolymer composition on the solubility and diffusivity of water and methanol in a series of polyether amides. *Journal of Applied Polymer Science* 65: 1983–1993.

143 Bondar, V., Freeman, B., and Pinnau, I. (2000). Gas transport properties of poly (ether-b-amide) segmented block copolymers. *Journal of Polymer Science Part B: Polymer Physics* 38: 2051–2062.

144 Kim, J.H., Ha, S.Y., and Lee, Y.M. (2001). Gas permeation of poly (amide-6-b-ethylene oxide) copolymer. *Journal of Membrane Science* 190: 179–193.

145 Kim, J.H. and Lee, Y.M. (2001). Gas permeation properties of poly (amide-6-b-ethylene oxide)–silica hybrid membranes. *Journal of Membrane Science* 193: 209–225.

146 McKeown, N.B., Gahnem, B., Msayib, K.J. et al. (2006). Towards polymer-based hydrogen storage materials: engineering ultramicroporous cavities within polymers of intrinsic microporosity. *Angewandte Chemie International Edition* 45: 1804–1807.

147 Budd, P.M., Ghanem, B.S., Makhseed, S. et al. (2004). Polymers of intrinsic microporosity (PIMs): robust, solution-processable, organic nanoporous materials. *Chemical Communications* 10: 230–231.

148 McKeown, N.B. (2020). Polymers of intrinsic microporosity (PIMs). *Polymer* 202: 122736.

149 McKeown, N.B. (2022). The structure-property relationships of polymers of intrinsic microporosity (PIMs). *Current Opinion in Chemical Engineering* 36: 100785.

150 Staiger, C.L., Pas, S.J., Hill, A.J., and Cornelius, C.J. (2008). Gas separation, free volume distribution, and physical aging of a highly microporous spirobisindane polymer. *Chemistry of Materials* 20: 2606–2608.

151 Budd, P.M., McKeown, N.B., and Fritsch, D. (2005). Free volume and intrinsic microporosity in polymers. *Journal of Materials Chemistry* 15: 1977–1986.

152 Shrimant, B., Shaligram, S.V., Kharul, U.K., and Wadgaonkar, P.P. (2018). Synthesis, characterization, and gas permeation properties of adamantane-containing polymers of intrinsic microporosity. *Journal of Polymer Science Part A: Polymer Chemistry* 56: 16–24.

153 Ghanem, B.S., McKeown, N.B., Budd, P.M., and Fritsch, D. (2008). Polymers of intrinsic microporosity derived from bis(phenazyl) monomers. *Macromolecules* 41: 1640–1646.

154 Du, N., Robertson, G.P., Song, J. et al. (2008). Polymers of intrinsic microporosity containing trifluoromethyl and phenylsulfone groups as materials for membrane gas separation. *Macromolecules* 41: 9656–9662.

155 Du, N., Robertson, G.P., Pinnau, I., and Guiver, M.D. (2009). Polymers of intrinsic microporosity derived from novel disulfone-based monomers. *Macromolecules* 42: 6023–6030.

156 Du, N., Robertson, G.P., Pinnau, I. et al. (2009). Copolymers of intrinsic microporosity based on 2,2′,3,3′- Tetrahydroxy-1,1′-dinaphthyl. *Macromolecular Rapid Communications* 30: 584–588.

157 McKeown, N.B. and Budd, P.M. (2010). Exploitation of intrinsic microporosity in polymer-based materials. *Macromolecules* 43: 5163–5176.

158 Ghanem, B.S., Swaidan, R., Ma, X. et al. (2014). Energy-efficient hydrogen separation by AB-type ladder-polymer molecular sieves. *Advanced Materials* 26: 6696–6700.

159 Rose, I., Bezzu, C.G., Carta, M. et al. (2017). Polymer ultrapermeability from the inefficient packing of 2D chains. *Nature Materials* 16: 932–937.

160 Fuoco, A., Comesaña-Gándara, B., Longo, M. et al. (2018). Temperature dependence of gas permeation and diffusion in triptycene-based ultrapermeable polymers of intrinsic microporosity. *ACS Applied Materials & Interfaces* 10: 36475–36482.

161 Longo, M., De Santo, M.P., Esposito, E. et al. (2019). Correlating gas permeability and Young's modulus during the physical aging of polymers of intrinsic microporosity using atomic force microscopy. *Industrial & Engineering Chemistry Research* 59: 5381–5391.

162 Corrado, T. and Guo, R. (2020). Macromolecular design strategies toward tailoring free volume in glassy polymers for high performance gas separation membranes. *Molecular Systems Design & Engineering* 5: 22–48.

163 Bezzu, C.G., Carta, M., Tonkins, A. et al. (2012). A spirobifluorene-based polymer of intrinsic microporosity with improved performance for gas separation. *Advanced Materials* 24: 5930–5933.

164 Bezzu, C.G., Carta, M., Ferrari, M.-C. et al. (2018). The synthesis, chain-packing simulation and long-term gas permeability of highly selective spirobifluorene-based polymers of intrinsic microporosity. *Journal of Materials Chemistry A* 6: 10507–10514.

165 Emmler, T., Heinrich, K., Fritsch, D. et al. (2010). Free volume investigation of polymers of intrinsic microporosity (PIMs): PIM-1 and PIM1 copolymers incorporating ethanoanthracene units. *Macromolecules* 43: 6075–6084.

166 Zhang, J., Jin, J., Cooney, R. et al. (2015). Synthesis of perfectly alternating copolymers for polymers of intrinsic microporosity. *Polymer Chemistry* 6: 5003–5008.

167 Starannikova, L., Belov, N., Shantarovich, V. et al. (2018). Effective increase in permeability and free volume of PIM copolymers containing ethanoanthracene unit and comparison between the alternating and random copolymers. *Journal of Membrane Science* 548: 593–597.

168 Carta, M., Malpass-Evans, R., Croad, M. et al. (2013). An efficient polymer molecular sieve for membrane gas separations. *Science* 339: 303–307.

169 Guiver, M.D. and Lee, Y.M. (2013). Polymer rigidity improves microporous membranes. *Science* 339: 284–285.

170 Carta, M., Croad, M., Malpass-Evans, R. et al. (2014). Triptycene induced enhancement of membrane gas selectivity for microporous Tröger's base polymers. *Advanced Materials* 26: 3526–3531.

171 Rose, I., Carta, M., Malpass-Evans, R. et al. (2015). Highly permeable benzotriptycene-based polymer of intrinsic microporosity. *ACS Macro Letters* 4: 912–915.

172 Zhang, J., Kang, H., Martin, J. et al. (2016). The enhancement of chain rigidity and gas transport performance of polymers of intrinsic microporosity via intramolecular locking of the spiro-carbon. *Chemical Communications* 52: 6553–6556.

173 Ghanem, B.S., McKeown, N.B., Budd, P.M. et al. (2008). High-performance membranes from polyimides with intrinsic microporosity. *Advanced Materials* 20: 2766–2771.

174 Ghanem, B.S., McKeown, N.B., Budd, P.M. et al. (2009). Synthesis, characterization, and gas permeation properties of a novel group of polymers with intrinsic microporosity: PIM-polyimides. *Macromolecules* 42: 7881–7888.

175 Song, J., Du, N., Dai, Y. et al. (2008). Linear high molecular weight ladder polymers by optimized polycondensation of tetrahydroxytetramethylspirobisindane and 1,4-dicyanotetrafluorobenzene. *Macromolecules* 41: 7411–7417.

176 Park, H.B., Jung, C.H., Lee, Y.M. et al. (2007). Polymers with cavities tuned for fast selective transport of small molecules and ions. *Science* 318: 254–258.

177 Park, H.B., Han, S.H., Jung, C.H. et al. (2010). Thermally rearranged (TR) polymer membranes for CO_2 separation. *Journal of Membrane Science* 359: 11–24.

178 Han, S.H., Lee, J.E., Lee, K.-J. et al. (2010). Highly gas permeable and microporous polybenzimidazole membrane by thermal rearrangement. *Journal of Membrane Science* 357: 143–151.

179 Calle, M. and Lee, Y.M. (2011). Thermally rearranged (TR) poly (ether–benzoxazole) membranes for gas separation. *Macromolecules* 44: 1156–1165.

180 Soo, C.Y., Jo, H.J., Lee, Y.M. et al. (2013). Effect of the chemical structure of various diamines on the gas separation of thermally rearranged poly (benzoxazole-co-imide)(TR-PBO-co-I) membranes. *Journal of Membrane Science* 444: 365–377.

181 Do, Y.S., Seong, J.G., Kim, S. et al. (2013). Thermally rearranged (TR) poly (benzoxazole-co-amide) membranes for hydrogen separation derived from 3, 3′-dihydroxy-4, 4′-diamino-biphenyl (HAB), 4, 4′-oxydianiline (ODA) and isophthaloyl chloride (IPCl). *Journal of Membrane Science* 446: 294–302.

182 Han, S.H., Kwon, H.J., Kim, K.Y. et al. (2012). Tuning microcavities in thermally rearranged polymer membranes for CO_2 capture. *Physical Chemistry Chemical Physics* 14: 4365–4373.

183 Zhuang, Y., Seong, J.G., Do, Y.S. et al. (2014). Effect of isomerism on molecular packing and gas transport properties of poly (benzoxazole-co-imide)s. *Macromolecules* 47: 7947–7957.

184 Jo, H.J., Soo, C.Y., Dong, G. et al. (2015). Thermally rearranged poly (benzoxazole-co-imide) membranes with superior mechanical strength for gas separation obtained by tuning chain rigidity. *Macromolecules* 48: 2194–2202.

185 Kim, S., Seong, J.G., Do, Y.S., and Lee, Y.M. (2015). Gas sorption and transport in thermally rearranged polybenzoxazole membranes derived from polyhydroxylamides. *Journal of Membrane Science* 474: 122–131.

186 Jung, C.H., Lee, J.E., Han, S.H. et al. (2010). Highly permeable and selective poly(benzoxazole-co-imide) membranes for gas separation. *Journal of Membrane Science* 350: 301–309.

187 Han, S.H., Misdan, N., Kim, S. et al. (2010). Thermally rearranged (TR) polybenzoxazole: effects of diverse imidization routes on physical properties and gas transport behaviors. *Macromolecules* 43: 7657–7667.

188 Brunetti, A., Tocci, E., Cersosimo, M. et al. (2019). Mutual influence of mixed-gas permeation in thermally rearranged poly (benzoxazole-co-imide) polymer membranes. *Journal of Membrane Science* 580: 202–213.

189 Comesaña-Gándara, B., Hernández, A., Jose, G. et al. (2015). Thermally rearranged polybenzoxazoles and poly (benzoxazole-co-imide) s from ortho-hydroxyamine monomers for high performance gas separation membranes. *Journal of Membrane Science* 493: 329–339.

190 Zhuang, Y., Seong, J.G., Lee, W.H. et al. (2015). Mechanically tough, thermally rearranged (TR) random/block poly (benzoxazole-co-imide) gas separation membranes. *Macromolecules* 48: 5286–5299.

191 Smith, S.J., Hou, R., Lau, C.H. et al. (2019). Highly permeable thermally rearranged mixed matrix membranes (TR-MMM). *Journal of Membrane Science* 585: 260–270.

192 Kim, S., Shamsaei, E., Lin, X. et al. (2018). The enhanced hydrogen separation performance of mixed matrix membranes by incorporation of two-dimensional ZIF-L into polyimide containing hydroxyl group. *Journal of Membrane Science* 549: 260–266.

193 Hu, X., Lee, W.H., Zhao, J. et al. (2020). Thermally rearranged polymer membranes containing highly rigid biphenyl ortho-hydroxyl diamine for hydrogen separation. *Journal of Membrane Science* 604: 118053.

194 Kim, S., Han, S.H., and Lee, Y.M. (2012). Thermally rearranged (TR) polybenzoxazole hollow fiber membranes for CO_2 capture. *Journal of Membrane Science* 403: 169–178.

195 Lee, J.H., Lee, J., Jo, H.J. et al. (2017). Wet CO_2/N_2 permeation through a crosslinked thermally rearranged poly (benzoxazole-co-imide)(XTR-PBOI) hollow fiber membrane module for CO_2 capture. *Journal of Membrane Science* 539: 412–420.

196 Lee, J., Kim, J.S., Kim, J.F. et al. (2019). Densification-induced hollow fiber membranes using crosslinked thermally rearranged (XTR) polymer for CO_2 capture. *Journal of Membrane Science* 573: 393–402.

197 Moon, S.J., Kim, J.H., Seong, J.G. et al. (2020). Thin film composite on fluorinated thermally rearranged polymer nanofibrous membrane achieves power density of $87\,W\,m^{-2}$ in pressure retarded osmosis, improving economics of osmotic heat engine. *Journal of Membrane Science* 607: 118120.

198 Aguilar-Lugo, C., Álvarez, C., Lee, Y.M. et al. (2018). Thermally rearranged polybenzoxazoles containing bulky adamantyl groups from ortho-substituted precursor copolyimides. *Macromolecules* 51: 1605–1619.

199 Yeong, Y.F., Wang, H., Pramoda, K.P., and Chung, T.-S. (2012). Thermal induced structural rearrangement of cardo-copolybenzoxazole membranes for enhanced gas transport properties. *Journal of Membrane Science* 397: 51–65.

200 Li, S., Jo, H.J., Han, S.H. et al. (2013). Mechanically robust thermally rearranged (TR) polymer membranes with spirobisindane for gas separation. *Journal of Membrane Science* 434: 137–147.

201 Ma, X., Salinas, O., Litwiller, E., and Pinnau, I. (2014). Pristine and thermally-rearranged gas separation membranes from novel o-hydroxyl-functionalized spirobifluorene-based polyimides. *Polymer Chemistry* 5: 6914–6922.

202 Shamsipur, H., Dawood, B.A., Budd, P.M. et al. (2014). Thermally rearrangeable PIM-polyimides for gas separation membranes. *Macromolecules* 47: 5595–5606.

203 Meckler, S.M., Bachman, J.E., Robertson, B.P. et al. (2018). Thermally rearranged polymer membranes containing tröger's base units have exceptional performance for air separations. *Angewandte Chemie International Edition* 57: 4912–4916.

204 Calle, M., Jo, H.J., Doherty, C.M. et al. (2015). Cross-linked thermally rearranged poly (benzoxazole-co-imide) membranes prepared from ortho-hydroxycopolyimides containing pendant carboxyl groups and gas separation properties. *Macromolecules* 48: 2603–2613.

205 Calle, M., Doherty, C.M., Hill, A.J., and Lee, Y.M. (2013). Cross-linked thermally rearranged poly (benzoxazole-co-imide) membranes for gas separation. *Macromolecules* 46: 8179–8189.

206 Do, Y.S., Lee, W.H., Seong, J.G. et al. (2016). Thermally rearranged (TR) bismaleimide-based network polymers for gas separation membranes. *Chemical Communications* 52: 13556–13559.

207 Seong, J.G., Lee, W.H., Lee, J. et al. (2021). Microporous polymers with cascaded cavities for controlled transport of small gas molecules Science. *Advances* 7: eabi9062.

208 Li, Y., Wang, X., Ding, M., and Xu, J. (1996). Effects of molecular structure on the permeability and permselectivity of aromatic polyimides. *Journal of Applied Polymer Science* 61: 741–748.

209 Schmidhauser, J.C. and Longley, K.L. (1990). The effect of bisphenol monomer structure on the gas permeability of aromatic polycarbonates. *Journal of Applied Polymer Science* 39: 2083–2096.

210 Tanaka, K., Kita, H., Okano, M., and Okamoto, K.-I. (1992). Permeability and permselectivity of gases in fluorinated and non-fluorinated polyimides. *Polymer* 33: 585–592.

211 Mi, Y., Stern, S., and Trohalaki, S. (1993). Dependence of the gas permeability of some polyimide isomers on their intrasegmental mobility. *Journal of Membrane Science* 77: 41–48.

212 Langsam, M. and Burgoyne, W.F. (1993). Effects of diamine monomer structure on the gas permeability of polyimides. I. Bridged diamines. *Journal of Polymer Science Part A: Polymer Chemistry* 31: 909–921.

213 Ghanem, B.S., Swaidan, R., Litwiller, E., and Pinnau, I. (2014). Ultramicroporous triptycene-based polyimide membranes for high-performance gas separation. *Advanced Materials* 26: 3688–3692.

214 Swaidan, R., Al-Saeedi, M., Ghanem, B. et al. (2014). Rational design of intrinsically ultramicroporous polyimides containing bridgehead-substituted triptycene for highly selective and permeable gas separation membranes. *Macromolecules* 47: 5104–5114.

215 Wiegand, J.R., Smith, Z.P., Liu, Q. et al. (2014). Synthesis and characterization of triptycene-based polyimides with tunable high fractional free volume for gas separation membranes. *Journal of Materials Chemistry A* 2: 13309–13320.

216 Alghunaimi, F., Ghanem, B., Alaslai, N. et al. (2015). Gas permeation and physical aging properties of iptycene diamine-based microporous polyimides. *Journal of Membrane Science* 490: 321–327.

217 Luo, S., Liu, Q., Zhang, B. et al. (2015). Pentiptycene-based polyimides with hierarchically controlled molecular cavity architecture for efficient membrane gas separation. *Journal of Membrane Science* 480: 20–30.

218 Ghanem, B., Alghunaimi, F., Ma, X. et al. (2016). Synthesis and characterization of novel triptycene dianhydrides and polyimides of intrinsic microporosity based on 3, 3′-dimethylnaphthidine. *Polymer* 101: 225–232.

219 Luo, S., Stevens, K.A., Park, J.S. et al. (2016). Highly CO_2-selective gas separation membranes based on segmented copolymers of poly (ethylene oxide) reinforced with pentiptycene-containing polyimide hard segments. *ACS Applied Materials & Interfaces* 8: 2306–2317.

220 Luo, S., Wiegand, J.R., Kazanowska, B. et al. (2016). Finely tuning the free volume architecture in iptycene-containing polyimides for highly selective and fast hydrogen transport. *Macromolecules* 49: 3395–3405.

221 Weidman, J.R. and Guo, R. (2017). The use of iptycenes in rational macromolecular design for gas separation membrane applications. *Industrial & Engineering Chemistry Research* 56: 4220–4236.

222 Weidman, J.R., Luo, S., Breier, J.M. et al. (2017). Triptycene-based copolyimides with tailored backbone rigidity for enhanced gas transport. *Polymer* 126: 314–323.

223 Weidman, J.R., Luo, S., Zhang, Q., and Guo, R. (2017). Structure manipulation in triptycene-based polyimides through main chain geometry variation and its effect on gas transport properties. *Industrial & Engineering Chemistry Research* 56: 1868–1879.

224 Ghanem, B.S., Alghunaimi, F., Wang, Y. et al. (2018). Synthesis of highly gas-permeable polyimides of intrinsic microporosity derived from 1, 3, 6, 8-tetramethyl-2, 7-diaminotriptycene. *ACS Omega* 3: 11874–11882.

225 Maya, E., Garcia-Yoldi, I., Lozano, A. et al. (2011). Synthesis, characterization, and gas separation properties of novel copolyimides containing adamantyl ester pendant groups. *Macromolecules* 44: 2780–2790.

226 Plaza-Lozano, D., Comesaña-Gándara, B., de la Viuda, M. et al. (2015). New aromatic polyamides and polyimides having an adamantane bulky group. *Materials Today Communications* 5: 23–31.

227 Pixton, M. and Paul, D.R. (1995). Gas transport properties of adamantane-based polysulfones. *Polymer* 36: 3165–3172.

228 Bera, D., Bandyopadhyay, P., Ghosh, S., and Banerjee, S. (2014). Gas transport properties of aromatic polyamides containing adamantyl moiety. *Journal of Membrane Science* 453: 175–191.

229 Kazama, S., Teramoto, T., and Haraya, K. (2002). Carbon dioxide and nitrogen transport properties of bis (phenyl) fluorene-based cardo polymer membranes. *Journal of Membrane Science* 207: 91–104.

230 Han, J.Y., Lee, J.Y., Kim, H.J. et al. (2014). Synthesis and characterization of fluorene-based polybenzimidazole copolymer for gas separation. *Journal of Applied Polymer Science* 131: https://doi.org/10.1002/app.40521.

231 Park, J. and Paul, D.R. (1997). Correlation and prediction of gas permeability in glassy polymer membrane materials via a modified free volume based group contribution method. *Journal of Membrane Science* 125: 23–39.

232 Recio, R., Lozano, A.E., Prádanos, P. et al. (2008). Effect of fractional free volume and Tg on gas separation through membranes made with different glassy polymers. *Journal of Applied Polymer Science* 107: 1039–1046.

233 Ma, X., Ghanem, B., Salines, O. et al. (2015). Synthesis and effect of physical aging on gas transport properties of a microporous polyimide derived from a novel spirobifluorene-based dianhydride. *ACS Macro Letters* 4: 231–235.

234 Yi, S., Ma, X., Pinnau, I., and Koros, W.J. (2015). A high-performance hydroxyl-functionalized polymer of intrinsic microporosity for an environmentally attractive membrane-based approach to decontamination of sour natural gas. *Journal of Materials Chemistry A* 3: 22794–22806.

235 Rogan, Y., Malpass-Evans, R., Carta, M. et al. (2014). A highly permeable polyimide with enhanced selectivity for membrane gas separations. *Journal of Materials Chemistry A* 2: 4874–4877.

236 Zhuang, Y., Seong, J.G., Do, Y.S. et al. (2014). Intrinsically microporous soluble polyimides incorporating Tröger's base for membrane gas separation. *Macromolecules* 47: 3254–3262.

237 Zhuang, Y., Seong, J.G., Do, Y.S. et al. (2016). Soluble, microporous, Tröger's base copolyimides with tunable membrane performance for gas separation. *Chemical Communications* 52: 3817–3820.

238 Seong, J.G., Zhuang, Y., Kim, S. et al. (2015). Effect of methanol treatment on gas sorption and transport behavior of intrinsically microporous polyimide membranes incorporating Tröger′ s base. *Journal of Membrane Science* 480: 104–114.

239 Zhuang, Y., Seong, J.G., Do, Y.S. et al. (2016). High-strength, soluble polyimide membranes incorporating Tröger's Base for gas separation. *Journal of Membrane Science* 504: 55–65.

240 Staudt-Bickel, C. and Koros, W.J. (1999). Improvement of CO_2/CH_4 separation characteristics of polyimides by chemical crosslinking. *Journal of Membrane Science* 155: 145–154.

241 Qiu, W., Zhang, K., Li, F.S. et al. (2014). Gas separation performance of carbon molecular sieve membranes based on 6FDA-mPDA/DABA (3:2) polyimide. *ChemSusChem* 7: 1186–1194.

242 Dose, M.E., Chwatko, M., Hubacek, I. et al. (2019). Thermally cross-linked diaminophenylindane (DAPI) containing polyimides for membrane based gas separations. *Polymer* 161: 16–26.

243 Askari, M., Xiao, Y., Li, P., and Chung, T.-S. (2012). Natural gas purification and olefin/paraffin separation using cross-linkable 6FDA-Durene/DABA co-polyimides grafted with α, β, and γ-cyclodextrin. *Journal of Membrane Science* 390: 141–151.

244 MacLean, D., Bollinger, W., King, D., and Narayan, R. (1986). Gas separation design with membranes. In: *Recent Developments in Separation Science* (ed. N.N. Li and J.M. Calo). Boca Raton, FL: CRC Press.

245 Spillman, R.W. (1989). Economics of gas separation membranes. *Chemical Engineering Progress* 85: 41–62.

246 Ren, Y., Liang, X., Dou, H. et al. (2020). Membrane-based olefin/paraffin separations. *Advanced Science (Weinheim, Germany)* 7: 2001398.

247 Das, M. and Koros, W.J. (2010). Performance of 6FDA–6FpDA polyimide for propylene/propane separations. *Journal of Membrane Science* 365: 399–408.

248 Faiz, R. and Li, K. (2012). Polymeric membranes for light olefin/paraffin separation. *Desalination* 287: 82–97.

249 Rungta, M., Zhang, C., Koros, W.J., and Xu, L. (2013). Membrane-based ethylene/ethane separation: the upper bound and beyond. *AICHE Journal* 59: 3475–3489.

250 Smith, Z.P., Hernández, G., Gleason, K.L. et al. (2015). Effect of polymer structure on gas transport properties of selected aromatic polyimides, polyamides and TR polymers. *Journal of Membrane Science* 493: 766–781.

251 Swaidan, R.J., Ghanem, B., Swaidan, R. et al. (2015). Pure- and mixed-gas propylene/propane permeation properties of spiro- and triptycene-based microporous polyimides. *Journal of Membrane Science* 492: 116–122.

252 Swaidan, R.J., Ma, X., Litwiller, E., and Pinnau, I. (2015). Enhanced propylene/propane separation by thermal annealing of an intrinsically microporous hydroxyl-functionalized polyimide membrane. *Journal of Membrane Science* 495: 235–241.

253 Do, X.H., Nguyen, Q.T., Kim, S. et al. (2021). Effect of thermal processing on brominated 6FDA-DAM for effective propylene/propane separation. *Separation and Purification Technology* 262: 118331.

254 Lee, M.J., Kwon, H.T., and Jeong, H.K. (2018). High-flux zeolitic imidazolate framework membranes for propylene/propane separation by postsynthetic linker exchange. *Angewandte Chemie International Edition* 57: 156–161.

255 Lee, M.J., Hamid, M.R.A., Lee, J. et al. (2018). Ultrathin zeolitic-imidazolate framework ZIF-8 membranes on polymeric hollow fibers for propylene/propane separation. *Journal of Membrane Science* 559: 28–34.

256 Kwon, H.T., Jeong, H.K., Lee, A.S. et al. (2015). Heteroepitaxially grown zeolitic imidazolate framework membranes with unprecedented propylene/propane separation performances. *Journal of the American Chemical Society* 137: 12304–12311.

257 Zhang, C., Dai, Y., Johnson, J.R. et al. (2012). High performance ZIF-8/6FDA-DAM mixed matrix membrane for propylene/propane separations. *Journal of Membrane Science* 389: 34–42.

258 Askari, M. and Chung, T.-S. (2013). Natural gas purification and olefin/paraffin separation using thermal cross-linkable co-polyimide/ZIF-8 mixed matrix membranes. *Journal of Membrane Science* 444: 173–183.

259 Japip, S., Wang, H., Xiao, Y., and Shung-Chung, T. (2014). Highly permeable zeolitic imidazolate framework (ZIF)-71 nano-particles enhanced polyimide membranes for gas separation. *Journal of Membrane Science* 467: 162–174.

260 Chi, W.S., Kim, S.J., Lee, S.J. et al. (2015). Enhanced performance of mixed-matrix membranes through a graft copolymer-directed interface and interaction tuning approach. *ChemSusChem* 8: 650–658.

261 Zhang, Q., Li, H., Chen, S. et al. (2020). Mixed-matrix membranes with soluble porous organic molecular cage for highly efficient C3H6/C3H8 separation. *Journal of Membrane Science* 611: 118288.

262 Bachman, J.E., Smith, Z.P., Li, T. et al. (2016). Enhanced ethylene separation and plasticization resistance in polymer membranes incorporating metal-organic framework nanocrystals. *Nature Materials* 15: 845–849.

263 Ma, X., Swaidan, R.J., Wang, Y. et al. (2018). Highly compatible hydroxyl-functionalized microporous polyimide-ZIF-8 mixed matrix membranes for energy efficient propylene/propane separation. *ACS Applied Nano Materials* 1: 3541–3547.

264 Liu, Y., Chen, Z., Liu, G. et al. (2019). Conformation-controlled molecular sieving effects for membrane-based propylene/propane separation. *Advanced Materials* 31: e1807513.

265 Park, S., Abdul Hamid, M.R., and Jeong, H.K. (2019). Highly propylene-selective mixed-matrix membranes by in situ metal-organic framework formation using a polymer-modification strategy. *ACS Applied Materials & Interfaces* 11: 25949–25957.

266 Shen, Q., Cong, S., He, R. et al. (2019). SIFSIX-3-Zn/PIM-1 mixed matrix membranes with enhanced permeability for propylene/propane separation. *Journal of Membrane Science* 588: 117201.

267 Liu, Y., Kita, H., Tanaka, K. et al. (2020). Mechanochemically synthesized ZIF-8 nanoparticles blended into 6FDA-TrMPD membranes for C3H6/C3H8 separation. *Journal of Applied Polymer Science* 138: 50251.

268 Moghadam, F., Lee, T.H., Park, I., and Park, H.B. (2020). Thermally annealed polyimide-based mixed matrix membrane containing ZIF-67 decorated porous graphene oxide nanosheets with enhanced propylene/propane selectivity. *Journal of Membrane Science* 603: 118019.

269 An, H., Cho, K.Y., Back, S. et al. (2021). The significance of the interfacial interaction in mixed matrix membranes for enhanced propylene/propane separation performance and plasticization resistance. *Separation and Purification Technology* 261: 118279.

270 Chen, G., Chen, X., Pan, Y. et al. (2021). M-gallate MOF/6FDA-polyimide mixed-matrix membranes for C2H4/C2H6 separation. *Journal of Membrane Science* 620: 118852.

271 Salinas, O., Ma, X., Litwiller, E., and Pinnau, I. (2016). Ethylene/ethane permeation, diffusion and gas sorption properties of carbon molecular sieve membranes derived from the prototype ladder polymer of intrinsic microporosity (PIM-1). *Journal of Membrane Science* 504: 133–140.

272 Qian, Q., Asinger, P.A., Lee, M.J. et al. (2020). MOF-based membranes for gas separations. *Chemical Reviews* 120: 8161–8266.

273 Salinas, O., Ma, X., Litwiller, E., and Pinnau, I. (2016). High-performance carbon molecular sieve membranes for ethylene/ethane separation derived from an intrinsically microporous polyimide. *Journal of Membrane Science* 500: 115–123.

274 Liao, K.-S., Japip, S., Lai, J.-Y., and Chung, T.-S. (2017). Boron-embedded hydrolyzed PIM-1 carbon membranes for synergistic ethylene/ethane purification. *Journal of Membrane Science* 534: 92–99.

275 Salinas, O., Ma, X., Wang, Y. et al. (2017). Carbon molecular sieve membrane from a microporous spirobisindane-based polyimide precursor with enhanced ethylene/ethane mixed-gas selectivity. *RSC Advances* 7: 3265–3272.

276 Qiu, W., Xu, L., Liu, Z. et al. (2021). Surprising olefin/paraffin separation performance recovery of highly aged carbon molecular sieve hollow fiber membranes by a super-hyperaging treatment. *Journal of Membrane Science* 620: 118271.

277 Wang, Q., Huang, F., Cornelius, C.J., and Fan, Y. (2021). Carbon molecular sieve membranes derived from crosslinkable polyimides for CO_2/CH_4 and C_2H_4/C_2H_6 separations. *Journal of Membrane Science* 621.

278 Campos, A.C.C., dos Reis, R.A., Ortiz, A. et al. (2018). A perspective of solutions for membrane instabilities in olefin/paraffin separations: a review. *Industrial & Engineering Chemistry Research* 57: 10071–10085.

279 Chae, I.S., Kang, S.W., Park, J.Y. et al. (2011). Surface energy-level tuning of silver nanoparticles for facilitated olefin transport. *Angewandte Chemie (International Ed. in English)* 50: 2982–2985.

280 Fallanza, M., Ortiz, A., Gorri, D., and Ortiz, I. (2013). Polymer–ionic liquid composite membranes for propane/propylene separation by facilitated transport. *Journal of Membrane Science* 444: 164–172.

281 Liao, K.-S., Lai, J.-Y., and Chung, T.-S. (2016). Metal ion modified PIM-1 and its application for propylene/propane separation. *Journal of Membrane Science* 515: 36–44.

282 Dou, H., Jiang, B., Xiao, X. et al. (2018). Novel protic ionic liquid composite membranes with fast and selective gas transport nanochannels for ethylene/ethane separation. *ACS Applied Materials & Interfaces* 10: 13963–13974.

283 Dou, H., Jiang, B., Xiao, X. et al. (2018). Ultra-stable and cost-efficient protic ionic liquid based facilitated transport membranes for highly selective olefin/paraffin separation. *Journal of Membrane Science* 557: 76–86.

284 Dou, H., Jiang, B., Xu, M. et al. (2019). Boron nitride membranes with a distinct nanoconfinement effect for efficient ethylene/ethane separation. *Angewandte Chemie (International Ed. in English)* 58: 13969–13975.

285 Dou, H., Xu, M., Jiang, B. et al. (2019). Bioinspired graphene oxide membranes with dual transport mechanisms for precise molecular separation. *Advanced Functional Materials* 29: 1905229.

286 Powell, C.E. and Qiao, G.G. (2006). Polymeric CO_2/N_2 gas separation membranes for the capture of carbon dioxide from power plant flue gases. *Journal of Membrane Science* 279: 1–49.

287 Maya, E.M., Muñoz, D.M., de la Campa, J.G. et al. (2006). Thermal effect on polyethyleneoxide-containing copolyimide membranes for CO_2/N_2 separation. *Desalination* 199: 188–190.

288 Sijbesma, H., Nymeijer, K., van Marwijk, R. et al. (2008). Flue gas dehydration using polymer membranes. *Journal of Membrane Science* 313: 263–276.

289 Kratochvil, A.M. and Koros, W.J. (2008). Decarboxylation-induced cross-linking of a polyimide for enhanced CO_2 plasticization resistance. *Macromolecules* 41: 7920–7927.

290 J.R. Klaehn, C.J. Orme, E.S. Peterson, et al., *CO_2 Separation Using Thermally Optimized Membranes: A Comprehensive Project Report (2000–2007)*, 2008.

291 Low, B.T., Chung, T.S., Chen, H. et al. (2009). Tuning the free volume cavities of polyimide membranes via the construction of pseudo-interpenetrating networks for enhanced gas separation performance. *Macromolecules* 42: 7042–7054.

292 Prasetya, N., Himma, N.F., Sutrisna, P.D. et al. (2020). A review on emerging organic-containing microporous material membranes for carbon capture and separation. *Chemical Engineering Journal* 391: 123575.

293 Elwell, L. and Grant, W.S. (2006). Technology options for capturing CO_2. *Power* 150: 60–65.

294 He, X., Lie, J.A., Sheridan, E., and Hägg, M.-B. (2009). CO_2 capture by hollow fibre carbon membranes: experiments and process simulations. *Energy Procedia* 1: 261–268.

295 Merkel, T.C., Lin, H., Wei, X., and Baker, R. (2010). Power plant post-combustion carbon dioxide capture: an opportunity for membranes. *Journal of Membrane Science* 359: 126–139.

296 Baker, R.W., Freeman, B., Kniep, J. et al. (2017). CO_2 capture from natural gas power plants using selective exhaust gas recycle membrane designs. *International Journal of Greenhouse Gas Control* 66: 35–47.

297 Pohlmann, J., Bram, M., Wilkner, K., and Brinkmann, T. (2016). Pilot scale separation of CO_2 from power plant flue gases by membrane technology. *International Journal of Greenhouse Gas Control* 53: 56–64.

298 Merkel, T.C., Wei, X., He, Z. et al. (2013). Selective exhaust gas recycle with membranes for CO_2 capture from natural gas combined cycle power plants. *Industrial & Engineering Chemistry Research* 52: 1150–1159.

299 Merkel, T., Kniep, J., Wei, X. et al. (2015). *Pilot Testing of a Membrane System for Postcombustion CO_2 Capture*. Incorporated, Newark, CA (United States): Membrane Technology and Research.

300 Han, Y. and Ho, W.W. (2020). Recent advances in polymeric facilitated transport membranes for carbon dioxide separation and hydrogen purification. *Journal of Polymer Science* 58: 2435–2449.

301 Salim, W., Vakharia, V., Chen, Y. et al. (2018). Fabrication and field testing of spiral-wound membrane modules for CO_2 capture from flue gas. *Journal of Membrane Science* 556: 126–137.

302 Vakharia, V., Salim, W., Wu, D. et al. (2018). Scale-up of amine-containing thin-film composite membranes for CO_2 capture from flue gas. *Journal of Membrane Science* 555: 379–387.

303 Han, Y., Salim, W., Chen, K.K. et al. (2019). Field trial of spiral-wound facilitated transport membrane module for CO_2 capture from flue gas. *Journal of Membrane Science* 575: 242–251.

304 Chen, K.K., Salim, W., Han, Y. et al. (2020). Fabrication and scale-up of multi-leaf spiral-wound membrane modules for CO_2 capture from flue gas. *Journal of Membrane Science* 595: 117504.

305 Hussain, A. and Hägg, M.-B. (2010). A feasibility study of CO_2 capture from flue gas by a facilitated transport membrane. *Journal of Membrane Science* 359: 140–148.

306 Ramasubramanian, K., Verweij, H., and Ho, W.W. (2012). Membrane processes for carbon capture from coal-fired power plant flue gas: a modeling and cost study. *Journal of Membrane Science* 421: 299–310.

307 Roussanaly, S., Anantharaman, R., Lindqvist, K. et al. (2016). Membrane properties required for post-combustion CO_2 capture at coal-fired power plants. *Journal of Membrane Science* 511: 250–264.

308 Roussanaly, S., Anantharaman, R., Lindqvist, K., and Hagen, B. (2018). A new approach to the identification of high-potential materials for cost-efficient membrane-based post-combustion CO_2 capture Sustainable. *Energy & Fuels* 2: 1225–1243.

309 Hägg, M.-B., Lindbråthen, A., He, X. et al. (2017). Pilot demonstration-reporting on CO_2 capture from a cement plant using hollow fiber process. *Energy Procedia* 114: 6150–6165.

310 He, X. (2018). A review of material development in the field of carbon capture and the application of membrane-based processes in power plants and energy-intensive industries. *Energy, Sustainability and Society* 8: 34.

311 Baker, R.W., Freeman, B., Kniep, J. et al. (2018). CO_2 capture from cement plants and steel mills using membranes. *Industrial & Engineering Chemistry Research* 57: 15963–15970.

312 Baena-Moreno, F.M., le Saché, E., Pastor-Pérez, L., and Reina, T.R. (2020). Membrane-based technologies for biogas upgrading: a review. *Environmental Chemistry Letters* 18: 1649–1658.

313 Dai, S., Liao, R., Zhou, H., and Jin, W. (2020). Synthesis of triptycene-based linear polyamide membrane for molecular sieving of N_2 from the VOC mixture. *Separation and Purification Technology* 252: 117355.

4

Membranes for Fuel Cell

4.1 Introduction

Polymers have been used in hydrogen fuel cells, both as a membrane and a binder or ionomer for proton exchange membrane (PEM) fuel cells, direct methanol fuel cells (DMFCs), and anion exchange membrane fuel cells (AEMFCs). Perfluorosulfonic acid (PFSA) was developed for fuel cells and used in fuel cell vehicles. With the enormous industrial development in environmentally friendly electrical vehicles and low-carbon emissions in mobile applications, hydrogen fuel cells using polymer membranes and binders are continuously growing and expanding. The progress in low-cost catalysts and binders as well as membranes thereof is required to spread the wide use of fuel cells for mobile and stationary applications. In this chapter, fluoropolymers used in hydrogen fuel cell technology are reviewed in detail, from the basic principles of electrochemistry to the selection and application of materials for fuel cells.

The first fuel cell was demonstrated by William Grove in 1839, using an experiment along the lines of that shown in Figure 4.1a,b. In Figure 4.1a, water is being electrolyzed into hydrogen and oxygen by passing an electric current through it. In Figure 4.1b, the power supply has been replaced with an ammeter, and a small current flows. The electrolysis is being reversed – the hydrogen and oxygen recombined, and an electric current is being produced. The experiment shown in Figure 4.1 makes a reasonable demonstration of the basic principle of the fuel cell, but the current produced is very small. The two main reasons for the small current are (i) the low contact area between the gas, the electrode, and the electrolyte – basically just a small ring where the electrode emerges from the electrolyte and (ii) the large distance between the electrodes – the electrolyte resists the flow of electric current.

To overcome these problems, the electrodes are usually made flat, with a thin layer of electrolyte, as illustrated in Figure 4.2. The structure of the electrode is porous so that both the electrolyte from one side and the gas from the other can penetrate it. This is to give the maximum possible contact between electrodes, electrolyte, and gas.

At this point, the reaction between hydrogen and oxygen to produce an electric current from each electrode needs to be understood. The detailed information

Membranes for Energy Applications, First Edition. Young Moo Lee.

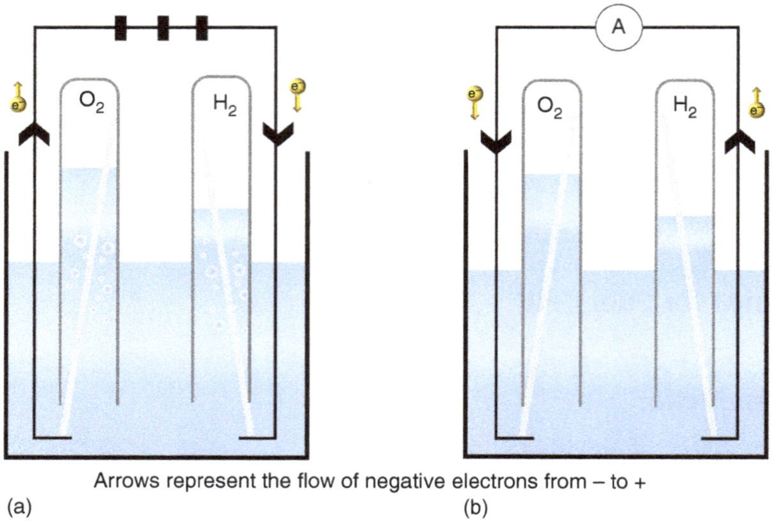

Figure 4.1 First fuel cell experiments by William Grove in 1839.

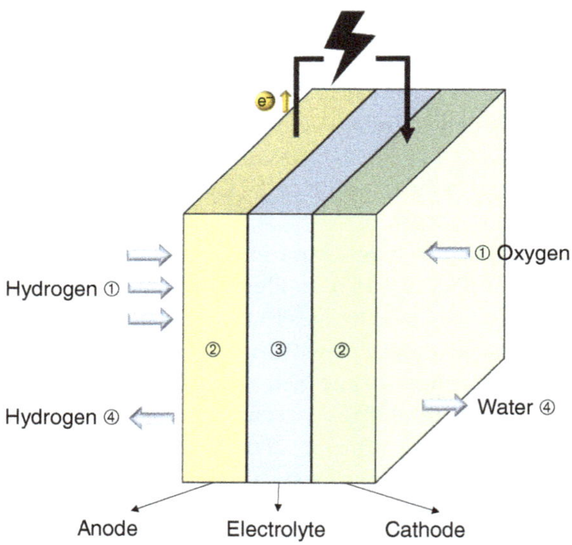

Figure 4.2 Basic scheme of anode–electrolyte–cathode construction of a fuel cell. A fuel cell is like a factory that takes fuel as input and produces electricity as output. Similar to a factory, fuel cells produce electricity as long as fuel is fed into.

depends on the type of fuel cells. Grove used the most common and simple fuel cell based on acid electrolytes.

At the anode of an acid electrolyte fuel cell, the hydrogen gas ionizes, releasing electrons and creating H^+ ions (or protons), as described in Eq. (4.1).

$$2H_2 \rightarrow 4H^+ + 4e^- \tag{4.1}$$

This reaction produces energy. At the cathode, oxygen reacts with electrons taken from the anode and H^+ ions from the electrolyte to form water, as illustrated in Eq. (4.2).

$$O_2 + 4e^- + 4H^+ \rightarrow 2H_2O \tag{4.2}$$

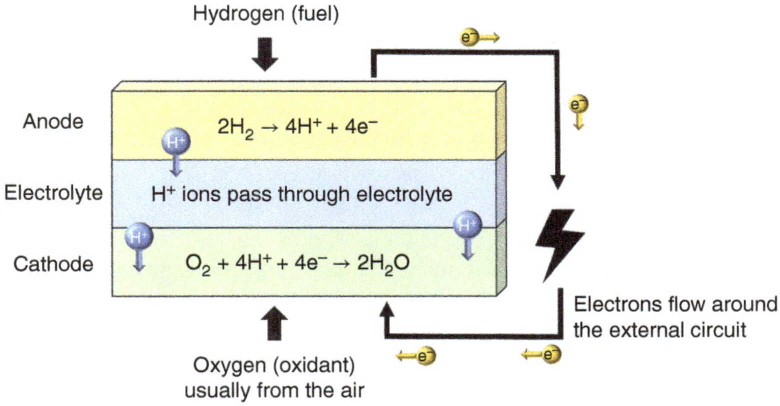

Figure 4.3 Schematic of acid electrolyte fuel cell. Proton ions pass through electrolyte from anode to cathode side.

For both these reactions to continue, electrons taken from the anode must pass through the electric circuit to the cathode. H^+ ions must pass through the electrolyte as well. Polymers that contain mobile H^+ ions are called cation exchange membranes (CEMs) or PEMs.

Therefore, the overall reaction in a fuel cell is as follows:

$$2H_2 + O_2 \rightarrow 2H_2O \tag{4.3}$$

In Eq. (4.3), we can see that two hydrogen molecules are required for each oxygen molecule to maintain a balance in the system. In Figure 4.3, we can see the electrode reactions and charge flow for an acid electrolyte fuel cell. Note that electrolyte only allows H^+ ions to pass through it and not electrons. If the electrons pass through the electrolyte, all electrons are lost, which leads to nonfunctional external motor or circuits.

In an alkaline electrolyte membrane fuel cell, the overall reaction is the same as in Eq. (4.3), but the reactions at each electrode are different (Figure 4.4). In an alkaline

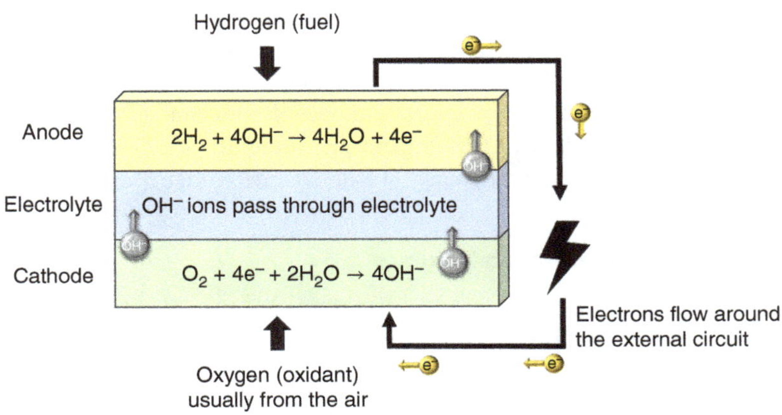

Figure 4.4 Schematic of alkaline electrolyte fuel cell. Hydroxyl ions pass through electrolyte from cathode to anode side due to electro-osmotic drag. Water back diffusion occurs from anode to cathode.

condition, hydroxyl (OH^-) ions are present and mobile. At the anode, these react with hydrogen, releasing energy and electrons and producing water.

$$2H_2 + 4OH^- \rightarrow 4H_2O + 4e^- \tag{4.4}$$

At the cathode, oxygen reacts with electrons taken from the electrode and water in the electrolyte, forming new OH^- ions, as shown in Eq. (4.5).

$$O_2 + 4e^- + 2H_2O \rightarrow 4OH^- \tag{4.5}$$

For these reactions to proceed continuously, OH^- ions must pass through the electrolyte, and there should be an electrical circuit for the electrons to move from the anode to the cathode. As with acid electrolyte membrane fuel cells, twice as much hydrogen is needed as oxygen to produce water. Note that four water molecules are produced in the anode, whereas only two water molecules are consumed at the cathode, resulting in imbalance of water in alkaline fuel cells (AFCs). Therefore, water management in this case is very important.

In fuel cell, the rate of reaction at both electrodes determines the amount of energy produced. If the probability of a molecule having enough energy is low, then the reaction only proceeds slowly. This is the case in a fuel cell, except for operation at high temperatures. The three main ways to circumvent this slow reaction rate are as follows:

- The use of catalyst
- Increasing the reaction temperature
- Increasing the electrode area

The first two ways can be applied to any chemical reaction. However, the third is specific to fuel cells and is very important. Since hydrogen fuel (or oxygen as oxidant gas), electrode, and electrolyte are in contact with each other, the surface area of the electrode determines the rate of the reaction. In fuel cell, *three-phase boundary* between these three components (porosity phase or binder, electrolyte, and catalyst) is very important to increase the contact area. This issue is very important in fuel cell design. Another important thing in the catalyst layer (CL) is the porosity, as porous structure of electrode increases the surface area of the electrode. Modern fuel cell electrodes have a microstructure that gives thousands of times their straightforward "length X width." Therefore, the microporous structure of the CL and the manufacturing of a fuel cell electrode are very important topics for practical fuel cells. Other than this surface area issue, high-temperature durability of a catalyst in the electrodes should be considered, particularly in a harsh environment.

4.1.1 Fuel Cells as Electrochemical Engines

Fuel cell technologies are attractive, as a fuel cell is a "factory" that takes fuel as input and produces electricity as output. Similar to a factory, a fuel cell produces electricity as long as fuel is fed. The difference between a fuel cell and a battery is that a fuel cell continuously produces electricity, whereas a battery produces electricity from stored reactants. A battery is like a "warehouse" or "storehouse." A sealed battery depletes

Figure 4.5 Schematic of a battery. A battery is like "warehouse," depleting reactants and products to produce electricity.

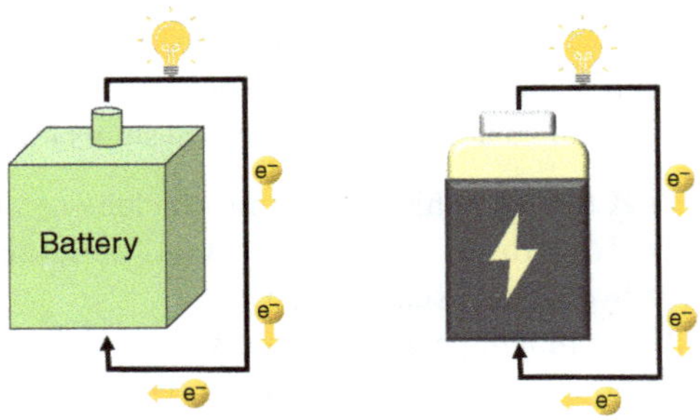

stored reactants and products, whereas a fuel cell has a constant flow of reactants and products. Therefore, a fuel cell can operate as a true steady-state device. Instead, a battery does not achieve a strict steady-state operation.

Unlike a fuel cell, a primary battery is non-rechargeable. A secondary battery is rechargeable, but the process or recharging involves controlled reversal of electrochemical reactions and takes significantly longer than refilling the flow of oxidizer and fuel in a fuel cell (Figure 4.5).

A fuel cell acts as an engine as it is a device that converts chemical energy into useful work, whereas a combustion engine converts the chemical energy of the fuel and oxidizer into mechanical work. Therefore, a fuel cell and a battery can be considered electrochemical engines. Both combustion and electrochemical engines use a fuel and an oxidizer as reactants. Both systems derive the desired output of useful work from the chemical bond energy released via the oxidation of the fuel. The overall chemical reaction and the potential energy released by the reaction are identical for the same fuel and oxidizer. The difference between combustion and electrochemical engines is the process of conversion of the enthalpy of reaction to useful work (Figure 4.6). In a combustion engine, the fuel and oxidizer react via combustion to generate heat, which is then converted into useful work via some mechanical process. An internal combustion engine in a car is a good example. In an electrochemical engine, the same enthalpy of reaction is directly converted into electrical current via an electrochemical oxidation process. The direct conversion of energy from chemical to electrical energy has a significant impact on the maximum

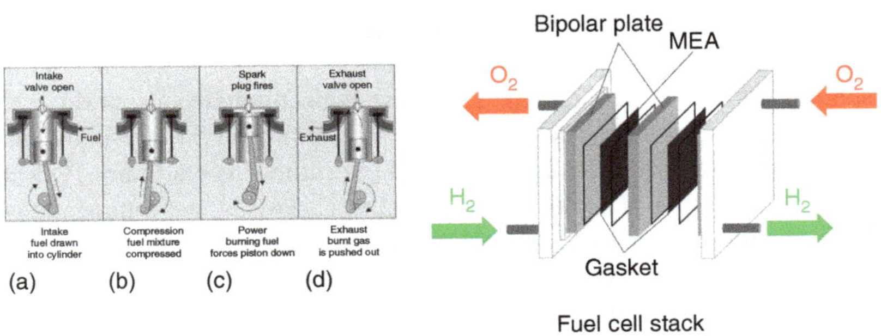

Figure 4.6 Schematic of combustion engine and fuel cell stacks.

theoretical efficiency of electrochemical devices. The waste heat from combustion is much more than that from electrochemical engines, as the former is too hot to touch, whereas the latter can be touched by hand.

Several advantages of fuel cell systems include:

(1) A relatively high operating efficiency scalable to all power plants.
(2) If hydrogen is used as a fuel, pollution emissions are strictly a product of hydrogen, usually water.
(3) No moving parts are necessary, with the exception of pumps, compressors, and blowers to drive fuel and oxidizer.
(4) There are multiple fuel choices, from natural resources, such as natural gas, petroleum, or coal, to renewable biofuels, such as ethanol and hydrogen from biomass.
(5) No limit of recharging capacity compared to batteries.

4.1.2 Classification of Fuel Cells

Fuel cells can be classified into the following five types depending on the types of electrolytes they use:

(1) Alkaline fuel cell (AFC): AFC uses KOH solution with high concentration (85 wt%) for high-temperature operation (250 °C) and a less concentrated solution (35–50 wt%) for lower-temperature operation (<120 °C). The electrolyte is retained in a matrix, such as asbestos, and a wide range of electrocatalysts can be used, such as Ni, Ag, and metal oxides. AFC is intolerant to CO_2 present in either fuel or oxidant. AFC has been used in the Apollo and Space Shuttle space programs since the 1960s. Anion exchange membrane fuel cell (AEMFC) uses anion exchange membrane (AEM) as a solid electrolyte that has been actively studied recently. Usually, AEMFCs are reported to operate below 80 °C.
(2) Polymer electrolyte membrane (PEM) or proton exchange membrane fuel cell (PEMFC): These use a thin (less than 30 μm) proton-conductive polymer membrane, such as a PFSA polymer as the electrolyte. The electrocatalyst is typically platinum supported on carbon with loadings of about 0.2 $mg\,cm^{-2}$, or if the hydrogen contains minor amounts of CO, Pt–Ru alloys supported on carbon are used. The operating temperature is usually 60–89 °C. PEMFCs are widely used in hydrogen fuel-cell operated automobiles, stationary power generation, and portable power applications as well.
(3) Phosphoric acid fuel cells (PAFC); These use concentrated phosphoric acid (~100%) as the electrolyte. The matrix used to retain the acid is usually SiC and the electrocatalyst in both the anode and the cathode is platinum. Typical operating temperature is between 150 and 220 °C. PAFCs are commercially available in container packages of 200 kW for stationary electricity generation. Hundreds of units have been installed all over the world.
(4) Molten carbonate fuel cells (MCFC): These use the electrolyte composed of a combination of alkali (Li, Na, and K) carbonates, which is retained in a ceramic matrix of $LiAlO_2$. The typical operating temperature is between 600 and 700 °C,

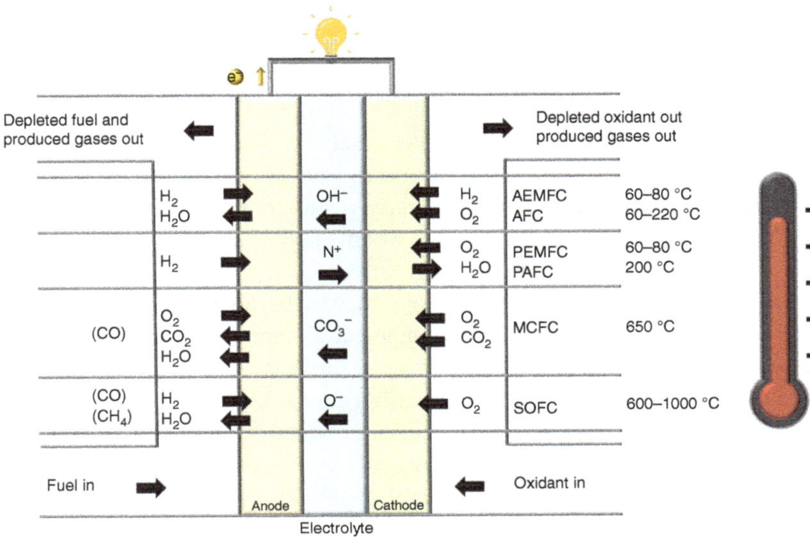

Figure 4.7 Schematic of fuel cell types depending on the electrochemical reactions and fuels and typical operating temperatures.

where the carbonates form a highly conductive molten salt, with carbonate ions that provide the ionic conduction. High-cost noble metal catalysts are not required at that high operating temperature. MCFCs have been installed in the stationary power generation plants.

(5) Solid oxide fuel cells (SOFC): These use a solid, nonporous metal oxide, usually Y_2O_3-stabilized ZrO_2 (YSZ) as the electrolyte. SOFCs are typically operated between 800 and 1000 °C, where ionic conduction by oxygen ions takes place. Similar to MCFCs, SOFCs are in the commercial/demonstration stage for stationary power generation.

Other than these five typical fuel cells, a microbial fuel cell uses bio-organisms to generate electricity using membranes. Among these five types of fuel cells, we will mainly focus on the first two types for this book, as we are dealing with membranes (typically made of polymers) for fuel cells: AEMFC and PEMFC. Figure 4.7 summarizes the basic principles and electrochemical reactions in several fuel cell types.

4.1.3 History of Fuel Cell Development

Sir William Grove made the first discovery of the fuel cell principle in 1839 [1], although Swiss scientist Christian F. Shoenbein independently discovered the same effect at about the same time. As the development of fuel cells from 1983 until the mid-1990s is summarized [2], fuel cells have been a subject of scientific attempts throughout the century from 1839 to 1939 to make a practical device. Francis T. Bacon developed the first practical 6 kW fuel cell in the late 1950s. However, the first

PEM fuel cells were developed by General Electric (GE) in the United States for the US Gemini space program in the early 1960s [3]. The fuel cell was expected to provide electricity and water for an animal orbiting in space in a satellite called the BIOS satellite program. At the beginning, GE applied membranes made from poly(styrene sulfonic acid) (PSSA) as the electrolyte in these fuel cells. However, these fuel cells broke down prematurely because the membranes were chemically unstable in the oxidative environment of the fuel cell. The early version of the PEMFC, as used in the NASA Gemini spacecraft, had a lifetime of only about 500 hours, but that was sufficient for those limited early space missions.

This was followed by the Apollo Space Program, which utilized alkaline fuel cells to generate electricity for life support, guidance, and communications. Pratt and Whitney built these fuel cells based on licenses taken from Bacon's patents. In the mid-1960s, General Motors experimented with a fuel cell-powered van based on fuel cell from Union Carbide. In 1989, fuel cell-powered submarine was successfully demonstrated, followed by fuel cell-powered buses and passenger cars in 1993 by Ballard Power Systems.

The development of the space program continued with the incorporation of the new polymer, Nafion. In 1967, DuPont contacted GE in an effort to search for applications of new polymer. Nafion received much attention due to its use as an electrolyte in the PEMFC. The catalyst loading required about 28 mg of platinum per cm^2 at that time, compared to $0.2\,mg\,cm^{-2}$ or less. The development of Nafion membranes and polymer dispersions (or ionomers) advanced fuel cell design, including a 100-fold reduction in catalyst loading and improvements in the gas diffusion layer (GDL) performance and bipolar plates.

From the 1970s to the 1990s, the development of fuel cells was a renaissance of interest in PEMFCs. Among them, Ballard Power Systems of Vancouver, Canada, and Los Alamos National Laboratory in the United States can get credit due to their active involvement in the development. Recent developments in fuel cells have brought the current densities over $1\,A\,cm^{-2}$ or more.

The development of Nafion began in the early 1960s when DuPont Co. was expanding fluorine technology that had previously resulted in the development of Teflon fluorocarbon resins and Viton fluoroelastomers [4, 5]. They investigated new monomers for copolymerization with tetrafluoroethylene (TFE). They also developed a general approach to synthesize perfluorinated vinyl ether monomers, the starting point for Nafion and other novel commercial fluoropolymers. Studies revealed that one of the acid fluorides, based on the reaction product of TFE and sulfur trioxide, led to an unusual TFE copolymer containing branches with pendant sulfonic acid groups [6]. Considering the properties that make Teflon valuable for many applications, in particular, its unequaled dielectric properties, exceptionally low coefficient of friction, anti-stick properties, and chemical barrier properties – one would expect that all of these properties would be adversely affected by the introduction of sulfonic acid groups. On the other hand, this newly prepared copolymer was believed to be basically different from all other TFE copolymers made before, having the potential to interact with the environment instead of merely resisting severe chemical or thermal environments.

Nafion perfluorinated membranes have a general chemical formula [7], as described in detail in Section 4.3.1, where the value of m can be as low as 1 and the value of n ranges between 6 and 13. The sulfonyl fluoride group is easily hydrolyzed to form the strongly acidic sulfonic acid ion exchange site. The equivalent weight (EW) of the ionized polymer can be defined as the mass, in grams, of the dry polymer in the acid form that would neutralize one equivalent of base. This polymer possesses outstanding thermal, chemical, and mechanical stability. It has been employed as a membrane separator in electrochemical applications.

The initial Nafion extrusion-cast membranes were N-115 and N-117 (5 and 7 mil [or 125 and 178 μm] thick films, 1 mil = 25.4 μm), which satisfied the high catalyst loadings and low current density operating modes of the pioneering fuel cell efforts. Researchers began to reduce catalyst loadings and increase the current density while the thick membranes struggled with water management issues and performance decline problems. The high current density operation tended to dehydrate the membrane, resulting in reduced proton conductivity. Therefore, N-112, a 2 mil (or ~50 μm) thick membrane, showed the performance improvement achieved with thinner membranes.

Fuel-cell passenger cars are expected to grow and can be seen easily on the road nowadays. The analysis of the recent fuel cell vehicle market and trends is described in the later part of this chapter.

4.2 Basic Electrochemical Principles

4.2.1 Electrochemical Reactions

Electrochemical reactions occur when the fuels (or reactants) are delivered to the electrodes. The current generated by the fuel cell is directly associated with how fast the electrochemical reactions undergo. Fast electrochemical reactions result in a high current output from the fuel cell. Slow reactions result in low current output. Obviously, high current output is desired. Therefore, catalysts are usually used to increase the electrochemical reaction rate and efficiency of the electrochemical reactions. Fuel cell performance mainly depends on the choice of the right catalyst and the careful design of reaction zones. Electrochemical reaction kinetics critically affect the limitation of fuel cell performance.

Ionic conduction (or electronic conduction) occurs because electrochemical reactions either produce or consume ions and electrons. Ions produced at one electrode must be consumed at the other electrode. The same is true for electrons. These ions and electrons must be transported from locations where they are generated to the locations where they are consumed to keep the charge balance. For electron transport, an electric conductive wire provides a path between the two electrodes, as illustrated in Figures 4.2–4.4. For ions, however, an electrolyte must be used for ions to provide a path between the two electrodes for the ions to flow, as ions are larger than electrons. Compared to electron transport, ion transport process is less efficient. Therefore, ionic transport represents a significant resistance loss,

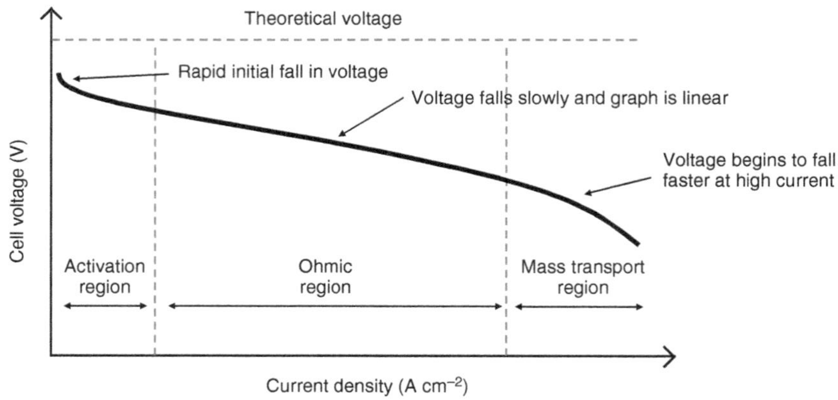

Figure 4.8 Graph showing the current-voltage (i–V) curve. Dashed line represents ideal or theoretical voltage of a fuel cell. Ideal voltage is also called "No loss" voltage, which is usually 1.2 V for low-temperature air-pressure fuel cell and 1.0 V for an air-pressure fuel cell operated at high temperature.

reducing fuel cell performance. To battle this issue, the electrolytes in fuel cells are fabricated as thin as possible to minimize the distance over which ionic conduction must occur.

Fuel cell performance can be depicted in a graph of current–voltage characteristics. This graph is called i–V curve and shows the voltage output of the fuel cell for a given current output, as shown in Figure 4.8. Because the current has been normalized by the area of the fuel cell, the current density is expressed in amperes per square centimeter. The actual voltage output of a real fuel cell is less than the ideal theoretically (or thermodynamically) predicted voltage.

The current supplied by a fuel cell is directly proportional to the amount of fuel consumed. Therefore, as fuel cell voltage decreases, the electric power produced per unit of fuel also decreases. Therefore, fuel cell voltage can be seen as a measure of fuel cell efficiency. We can, therefore, see the cell voltage in the y-axis as fuel cell efficiency. It is difficult to maintain a high fuel cell voltage under a current load. The voltage output of a real fuel cell is less than ideal or thermodynamically predicted voltage output due to irreversible losses. Three major losses in a fuel cell can occur in three different steps, depicted in Figure 4.8. They are (i) activation losses due to electrochemical reactions, (ii) Ohmic losses due to ionic and electronic conduction, and (iii) concentration losses due to mass transport.

The electrical power (P) delivered by a fuel cell is given by the product of current and voltage, and the energy output is the following well-known Eq. (4.6):

$$P = i\,V \tag{4.6}$$

$$E = i\,V\,t$$

The real voltage output for a real fuel cell can thus be written by starting with thermodynamically predicted voltage output of the fuel cell and then subtracting

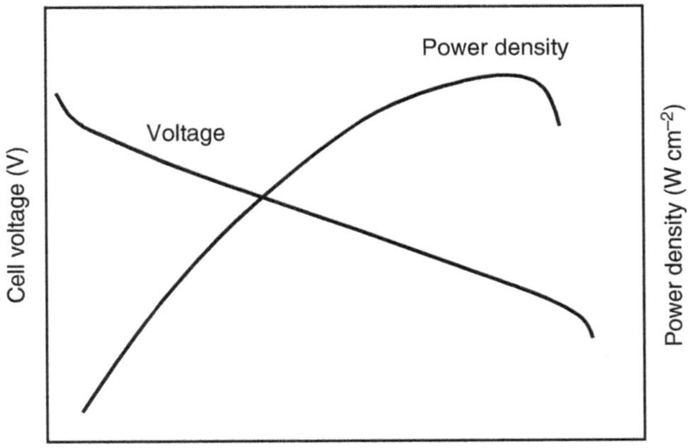

Figure 4.9 Typical *i*–*V*–*P* curve. A maximum power density curve is call peak power density.

the voltage drops due to the voltage losses from three major regions:

$$V = E_{\text{thermo}} - \Delta V_{\text{act}} - \Delta V_{\text{ohmic}} - \Delta V_{\text{conc}} \tag{4.7}$$

where V is the real voltage output of fuel cell, E_{thermo} is the thermodynamically predicted fuel cell voltage output (or the irreversible open circuit voltage [OCV]), ΔV_{act} is the activation losses due to reaction kinetics, ΔV_{ohmic} is the Ohmic losses from ionic and electronic conduction, and ΔV_{conc} is the concentration losses due to mass transport. As seen in Figure 4.8, activation losses are seen at the initial section of the curve; the Ohmic losses are evident in the middle part of the curve; the concentration losses are apparent in the last part of the curve.

A fuel cell power density curve can be drawn from the information in a fuel cell *i*–*V* curve. The power density curve is produced by multiplying the measured voltage at each point on the *i*–*V* curve by the corresponding current density. Usually, the cell voltage is given on the left-hand *y*-axis while power density is given on the right-hand *y*-axis, as seen in Figure 4.9.

4.2.2 Basic Principles of Fuel Cells

In the case of a fuel cell, "Gibbs free energy" is important. Gibbs free energy is defined as "the energy available to do external work neglecting any work done by changes in pressure and/or volume." In a fuel cell, the external work involves moving electrons around an external circuit – any work done by a change in volume between the input and output is not implemented by the fuel cell. Enthalpy is the Gibbs free energy plus the energy connected with the entropy. All these forms of chemical energy are similar to ordinary mechanical potential energy. In a fuel cell, the change in Gibbs free energy of formation, ΔG_f, is the energy released. This change is the difference between the Gibbs free energy of the products and the

Table 4.1 $\Delta\bar{g}_f$ for the reaction $H_2 + \frac{1}{2}O_2 \rightarrow H_2O$ at various temperatures.

Form of water product	Temperature (°C)	$\Delta\bar{g}_f$ (kJ mol^{-1})
Liquid	25	−237.2
Liquid	80	−228.2
Gas	80	−226.1
Gas	100	−225.2
Gas	200	−220.4
Gas	400	−210.3
Gas	600	−199.6
Gas	800	−188.6
Gas	1000	−177.4

Source: Taken from Larmine and Dicks [8].

Gibbs free energy of the inputs or reactants.

$$\Delta G_f = G_f \text{ of products} - G_f \text{ of reactants}$$

It is convenient to consider these quantities in their "per mole" form, indicated by over the lower-case letter, for example, $(\bar{g}_f)_{H_2O}$ is the molar-specific Gibbs free energy of formation for water (see Table 4.1).

Consider the basic reaction for the hydrogen fuel cell:

$$2H_2 + O_2 \rightarrow 2H_2O$$

which is equivalent to the following Eq. (4.8):

$$H_2 + \frac{1}{2}O_2 \rightarrow H_2O \tag{4.8}$$

The product is one mole of H_2O, and the reactants are one mole of H_2 and half a mole of O_2. Thus,

$$\Delta\bar{g}_f = \bar{g}_f \text{ of products} - \bar{g}_f \text{ of reactants}$$

So we have,

$$\Delta\bar{g}_f = (\bar{g}_f)_{H_2O} - (\bar{g}_f)_{H_2} - \frac{1}{2}(\bar{g}_f)_{O_2} \tag{4.9}$$

Since Gibbs free energy of formation is not constant, it changes with temperature and state (liquid or gas).

If there are no losses in the fuel cell, or if the process is "reversible," then all this Gibbs free energy is converted into electrical energy. In practice, however, some energy is released as heat.

In the basic operation of a hydrogen fuel cell, two electrons pass around the external circuit for each water molecule produced and each hydrogen molecule used.

So, for 1 mol of hydrogen used, $2N$ electrons pass round the external circuit – where N is Avogadro's number. If $-e$ is the charge on one electron, then the charge that flows is

$$-2Ne = -2F \text{ coulombs}$$

where F is the Faraday constant (96,485 J) or the charge on 1 mol of electrons.

If E is the voltage (or electrical energy) of the fuel cell, then the electrical work done moving this charge round the external circuit is

$$\text{Electrical work done} = \text{charge} \times \text{voltage} = -2FE \quad \text{Joules}$$

If the system is reversible or has no losses, then this electrical work done equals the Gibbs free energy released, $\Delta\overline{g}_f$. So

$$\Delta\overline{g}_f = -2F \cdot E$$

Thus,

$$E = \frac{-\Delta\overline{g}_f}{2F} \tag{4.10}$$

The fundamental equation gives the electromotive force (EMF) or reversible OCV of the hydrogen fuel cell.

For example, a hydrogen fuel cell operating at 80 °C has $\Delta\overline{g}_f = -226.1$ kJ, so

$$E = \frac{226{,}100}{2 \times 96{,}485} = 1.17\,\text{V}$$

Note that this value assumes no "irreversibilities" and assumes pure hydrogen and oxygen at standard pressure (0.1 MPa). In practice, the voltage would be lower than this because of the voltage drops. Some of these irreversibilities apply even when no current is drawn, so even the OCV of a fuel cell will usually be lower than the value given in Eq. (4.5).

Since a fuel cell uses reactants and oxidants burnt to release their energy, it would make sense to compare the electrical energy produced with the heat that would be produced by burning the fuel. This is called calorific value or the change in "enthalpy of formation," $\Delta\overline{h}_f$. As with the Gibbs free energy, it is the convention that $\Delta\overline{h}_f$ is negative when energy is released. So, the efficiency of the fuel cell is usually defined as:

$$\frac{\text{Electrical energy produced per mole of fuel}}{\Delta\overline{h}_f} \tag{4.11}$$

We have two different values of $\Delta\overline{h}_f$ depending on the state of the water. For the burning of hydrogen

$$H_2 + \frac{1}{2}O_2 \rightarrow H_2O\,\text{(steam)}$$

$$\Delta\overline{h}_f = -241.83\,\text{kJ mol}^{-1}$$

Table 4.2 $\Delta\bar{g}_f$, maximum EMF (or reversible open circuit voltage), and efficiency limit (HHV basis) for hydrogen fuel cells.

Form of water product	Temperature (°C)	$\Delta\bar{g}_f$ (kJ mol^{-1})	Max EMF (V)	Efficiency limit (%)
Liquid	25	−237.2	1.23	83
Liquid	80	−228.2	1.18	80
Gas	100	−225.2	1.17	79
Gas	200	−220.4	1.14	77
Gas	400	−210.3	1.09	74
Gas	600	−199.6	1.04	70
Gas	800	−188.6	0.98	66
Gas	1000	−177.4	0.92	62

Source: Taken from Larmine and Dicks [8].

whereas if the product water is condensed back to liquid, the reaction is

$$H_2 + \frac{1}{2}O_2 \rightarrow H_2O\ (\text{liquid})$$

$$\Delta\bar{h}_f = -285.84\ \text{kJ mol}^{-1}$$

The difference between these two values for $\Delta\bar{h}_f$ (44.01 kJ mol^{-1}) is the molar enthalpy of vaporization or molar "latent heat." The higher value is called the higher heating value (HHV), and the lower figure is called the lower heating value (LHV). According to Eq. (4.9), there is a limit to the efficiency. The maximum electrical energy available is equal to the change in Gibbs free energy, so

$$\text{Maximum efficiency possible} = \frac{\Delta\bar{g}_f}{\Delta\bar{h}_f} \times 100\% \tag{4.12}$$

This maximum efficiency limit is sometimes known as the "thermodynamic efficiency." Table 4.2 provides the values of the efficiency limit relative to the HHV for a hydrogen fuel cell. The maximum voltage from Eq. (4.9) is also given.

If all the energy ($\Delta\bar{h}_f$) from the hydrogen fuel were transformed into electrical energy, then the EMF would be $E = 1.48$ V if using the HHV, or $E = 1.25$ V if using the LHV, according to Eq. (4.8) and replacing Gibbs free energy with $\Delta\bar{h}_f$. These are the values that would be obtained from a 100% efficient system with reference to the HHV or LHV. Since the actual cell voltage is defined as V_c, then the actual cell efficiency is,

$$\text{Cell efficiency} = \frac{V_c}{1.48} 100\%\ (\text{with reference to HHV})$$

However, in practice, it is found that all the fuel that is fed into a fuel cell can be used. Some fuel usually pass through unreacted. A fuel utilization coefficient can be defined as

$$\mu_f = \frac{\text{Mass of fuel reacted in cell}}{\text{Mass of fuel input to cell}}$$

This is the same as the ratio of fuel cell current and the current obtained if all the fuel were reacted. The fuel cell efficiency based on HHV is, therefore, is given by

$$\text{Efficiency} = \mu_f \frac{V_c}{1.48} 100\% \tag{4.13}$$

If a figure relative to the LHV is required, replace 1.25 with 1.48. A good estimate for μ_f is 0.95, which allows the efficiency of a fuel cell to be accurately estimated from the very simple measurement of its voltage.

The Nernst equation is the basis of a fuel cell estimation. Now the Gibbs free energy changes in a chemical reaction vary with temperature. Also noticed is that the changes in Gibbs free energy also vary with reactant pressure and concentration.

Since the activity of reactants and products is related to pressure in an ideal gas,

$$\text{Activity } a = \frac{P}{P^o}$$

where P is the pressure or partial pressure of the gas, and P^o is the standard pressure at 0.1 MPa. Since fuel cells as gas reactors utilize gas mixtures, the activity is proportional to partial pressure. Fuel cell produces water in liquid form and steam or vapor as well, which makes it a difficult situation. For steam, we can define the activity as:

$$a_{H_2O} = \frac{P}{P^o_{H_2O}}$$

where $P^o_{H_2O}$ is the vapor pressure of the steam at the temperature concerned, which can be found in the steam tables. If liquid water is a product, it is reasonable to estimate that $a_{H_2O} = 1$.

For the water produced in the fuel cell, as in the overall reaction in Eq. (4.8), the Gibbs free energy Eq. (4.9) becomes:

$$\Delta \overline{g}_f = \Delta \overline{g}_f{}^o - RT \ln \left(\frac{a_{H_2} \cdot a_{O_2}^{\frac{1}{2}}}{a_{H_2O}} \right) \tag{4.14}$$

$\Delta \overline{g}_f{}^o$ is the value given in Tables 4.1 and 4.2. If the activity of the reactants increases, $\Delta \overline{g}_f$ becomes more negative, that is, more energy is released. On the other hand, if the activity of the product increases, $\Delta \overline{g}_f$ increases, thus becoming less negative, and less energy is released. To see how this equation affects voltage, we can substitute Eq. (4.14) into Eq. (4.10),

$$E = \frac{-\Delta \overline{g}_f{}^o}{2F} + RT \ln \left(\frac{a_{H_2} \cdot a_{O_2}^{\frac{1}{2}}}{a_{H_2O}} \right)$$

$$= E^o + RT \ln \left(\frac{a_{H_2} \cdot a_{O_2}^{\frac{1}{2}}}{a_{H_2O}} \right) \tag{4.15}$$

where E^o is the EMF or reversible OCV at standard pressure and is the number given in column 4 of Table 4.2. The equation shows precisely how raising the activity of the reactants increases the voltage. The Eq. (4.15) is called *Nernst equations*, which give an EMF in terms of product and/or reactant activity. The EMF or reversible OCV calculated from such equations is called the "*Nernst voltage*" and is the reversible cell voltage that would exist at a given temperature and pressure.

At high temperatures (e.g. in a SOFC at 1000 °C), we can assume that the steam behaves like an ideal gas, all the pressures are in bars, and activity of each component becomes,

$$a_{H_2} = \frac{P}{P^o_{H_2}}, \quad a_{O_2} = \frac{P}{P^o_{O_2}}, \quad a_{H_2O} = \frac{P}{P^o_{H_2O}}$$

At high temperatures, $P^o = 1$, and activity will become partial pressure of gases. Then the Eq. (4.15) becomes,

$$E = E^o + \frac{RT}{2F} \ln\left(\frac{P_{H_2} \cdot P_{O_2}^{\frac{1}{2}}}{P_{H_2O}}\right) \tag{4.16}$$

The pressure in Eq. (4.16) is partial pressure, as the gases will be part of a mixture. For example, hydrogen gas can be part of a mixture of H_2 and CO_2 from a fuel reformer and product steam. The oxygen will be part of the air. Fuel cell design will be simpler if the pressure on both the anode and the cathode is approximately the same. If this system pressure is P, then:

$$P_{H_2} = \alpha P, P_{O_2} = \beta P \quad \text{and} \quad P_{H_2O} = \delta P$$

where α, β, and δ are constants depending on the molar masses and concentrations of H_2, O_2, and H_2O. Then Eq. (4.16) becomes

$$\begin{aligned} E &= E^o + \frac{RT}{2F} \ln\left(\frac{\alpha \cdot \beta^{\frac{1}{2}}}{\delta} \cdot P^{\frac{1}{2}}\right) \\ &= E^o + \frac{RT}{2F} \ln\left(\frac{\alpha \cdot \beta^{\frac{1}{2}}}{\delta}\right) + \frac{RT}{4F} \ln(P) \end{aligned} \tag{4.17}$$

Equations (4.16) and (4.17) are forms of the Nernst equation. They provide a theoretical basis and a quantitative indication for many variables in fuel cell design and operation. As air passes through the fuel cell, oxygen is used, and so the partial pressure of oxygen will reduce. Similarly, the fuel partial pressure will often decline as the proportion of fuel reduces and reaction products increase. In Eq. (4.17), usually α and β decrease, whereas δ increases. Therefore, the second term in Eq. (4.17) becomes smaller, and so the EMF or OCV will fall. This varies within the cell. It will be worst near the fuel outlet as the fuel is used. According to the third term in Eq. (4.17), the EMF of a fuel cell is increased by the system pressure P. Voltage changes in high-temperature fuel cells agree well with the estimated values from the third term in Eq. (4.17), while the agreement is not as good in low-temperature fuel cell cases. In other words, at lower temperatures, the benefits of raising system pressure are much greater than what the Nernst equation predicts. This is because increasing the pressure also reduces the losses at the electrodes, particularly at the cathode.

4.2.3 Voltage Losses

As mentioned in Figure 4.4 on the shape of the voltage/current density graph, there are major irreversibilities: activation losses, ohmic losses, and mass transport or concentration losses. These three aspects are considered in the following discussion.

4.2.3.1 Activation Losses

Activation losses are caused by the slowness of the reactions taking place on the surface of the electrodes. A part of the voltage generated is lost in driving the chemical reaction that transfers the electrons to or from the electrode. This voltage drop is highly nonlinear.

In 1905, Tafel observed that the overvoltage at the surface of an electrode followed a similar pattern in a variety of electrochemical reactions. This pattern shows that if overvoltage is plotted against log of current density, the graph approximates to a straight line. Such plots of overvoltage against log of current density are known as "Tafel plots." Therefore, the voltage drop due to activation loss can be expressed as *Tafel equation*, as described in Eq. (4.18).

$$\Delta V_{act} = a\log\left(\frac{i}{i_o}\right) \text{ or } \Delta V_{act} = A\ln\left(\frac{i}{i_o}\right) \tag{4.18}$$

The constant i_o is higher if the reaction is faster. The current density i_o can be considered the current density at which the overvoltage begins to move from zero and is sometimes called the *exchange current density*. Note that the Tafel equation only holds true when $i > i_o$.

Here, constant A can be expressed as the following relation:

$$A = \frac{RT}{2\alpha F} \tag{4.19}$$

Constant A is higher for a slow electrochemical reaction. The constant α is called the *charge transfer coefficient* and is the proportion of the electrical energy applied that is harnessed in changing the rate of an electrochemical reaction.

If we combine Eqs. (4.18) and (4.19) and rearrange the equation,

$$i = i_o \exp\left(\frac{2\alpha F\Delta V_{act}}{RT}\right)$$

This equation is called *Buttler–Volmer* equation, and it is often used as an equivalent alternative to the Tafel equation. If the fuel cell has no losses at all, except for activation overvoltage on one electrode, its voltage can be given by the equation:

$$V = E - A\ln\left(\frac{i}{i_o}\right) \tag{4.20}$$

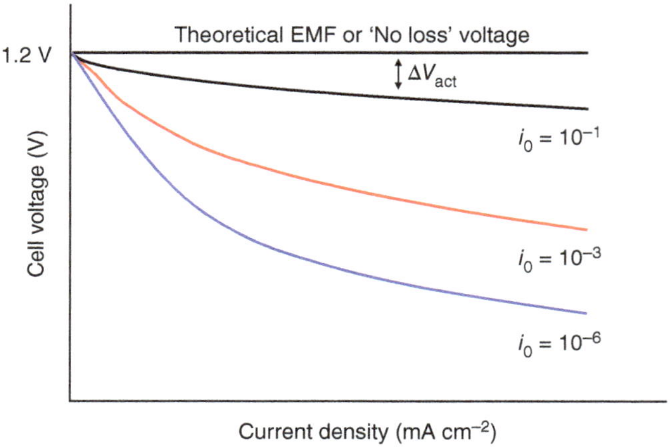

Figure 4.10 Effect of activation overvoltage on fuel cell performance. Reaction kinetics typically inflicts exponential loss on a fuel cell's *i*–*V* curve as determined by the Butler–Volmer equation. The magnitude of the loss is influenced by the size of i_0.

The importance of i_0 is to reduce the cell voltage by a fixed amount, as predicted by the Tafel equation. The smaller the i_0, the greater this voltage drop. The extent of i_0 depends on the type of catalyst, indicating a strong catalytic effect. In general, i_0 at the anode is negligible compared to that of the cathode, at least in the case of hydrogen fuel cells. For a low-temperature hydrogen fuel cell operating at ambient pressure in air, a typical value for i_0 would be about 0.1 mA cm^{-2} at the cathode and about 200 mA cm^{-2} at the anode. A typical graph showing the effect of i_0 can be seen in Figure 4.10.

Note that activation overvoltage can be reduced, and increasing the exchange current density i_0 improves the fuel cell performance, particularly at the cathode. There are several ways to reduce the activation overvoltage:

(1) *Increasing the cell temperature*: This will accelerate the reaction kinetics and increase the i_0 value.
(2) *Using more effective catalysts*: Effective catalysts should have higher i_0 values, improving the cell performance.
(3) *Increasing the roughness of the electrodes*: This increases the active surface area of each nominal area of 1 cm^2 and increases i_0.
(4) *Increasing the reactant concentration*: For example, by using pure oxygen rather than air, catalyst sites are more effectively occupied by reactants.
(5) *Increasing the pressure*: This increases the catalyst sites available for reaction, particularly at the cathode. This also increases the reversible OCV.

In summary, in low- and medium-temperature fuel cells, activation overvoltage is the most important irreversibility and cause of voltage drop and occurs mainly at the cathode. Activation overvoltage at both electrodes is important in cells using fuels other than hydrogen, such as methanol. At higher temperatures and pressures, the activation overvoltage becomes less important.

4.2.3.2 Fuel Crossover and Internal Currents

Fuel crossover is defined as the amount of fuel (e.g. hydrogen) that diffuses through the anode through the electrolyte to the cathode. Fuel crossover and internal currents are essentially equivalent. The crossover of one hydrogen molecule from the anode to the cathode wastes two electrons, which is the same amount of crossover over that of two electrons from the anode to the cathode. In the case of hydrogen fuel cells, the major loss in the fuel cell is the transfer of electrons at the cathode interface, and thus, the effect of crossover of fuel and electrons on the cell voltage is the same.

Because of this internal current density, the cell current density is not zero even if the cell is open circuit. In Eq. (4.19), a PEMFC operated at RT and air, reasonable values for this equation are $E = 1.2\,\text{V}$, $A = 0.06\,\text{V}$, and $i_o = 0.04\,\text{mA cm}^{-2}$. At lower voltages, the internal current density i can be obtained from this equation. For example, at $i = 1.0\,\text{mA cm}^{-2}$, the OCV is 0.97 V, which is about 0.2 V (or 20%) less than the theoretical OCV of 1.2 V.

If i_n is the value of internal current density, then Eq. (4.20) for cell voltage can be redefined to Eq. (4.21),

$$V = E - A \ln\left(\frac{i + i_n}{i_o}\right) \tag{4.21}$$

Typical values for a low-temperature fuel cell are again $E = 1.2\,\text{V}$, $A = 0.06\,\text{V}$, $i_o = 0.04\,\text{mA cm}^{-2}$, and $i_n = 3\,\text{mA cm}^{-2}$, and we can see the current–voltage curve, as shown in Figure 4.6, which is similar to Figure 4.5. Here, for a high-temperature cell, an initial cell voltage drop is not so much higher as i_o is higher because an exchange current density i_o is much higher, whereas for low-temperature fuel cells, an initial fall in cell voltage is usually at least 20% less than the theoretical OCV. This drop in cell voltage is mainly due to the initial fall in voltage in hydrogen fuel cells, where the cathode activation overvoltage dominates. The importance of this internal current density is less in the case of high-temperature fuel cells because the exchange current density, i_o, is so much higher, and so the initial fall in voltage is not nearly so marked. For low-temperature fuel cells, however, it has a significant effect on the open-circuit voltage.

4.2.3.3 Ohmic Losses

As the amount of voltage drop is proportional to the current,

$$V = IR$$

In most fuel cells, the resistance is mainly caused by the electrolyte. We can express this equation in terms of current density i in mA cm^{-2}. If we can define the resistance corresponding to 1 cm^2 of the cell, then *area-specific resistance* (or ASR) r ($\text{k}\Omega\,\text{cm}^2$) can be used to convert the above mentioned equation.

$$\Delta V_{\text{ohm}} = ir \tag{4.22}$$

This ohmic voltage loss is important in all types of cells, especially important in SOFC. Making the membranes as thin as possible can reduce the internal resistance of the cell. The use of electrodes with the highest possible conductivity is another way of reducing the resistance.

4.2.3.4 Mass Transport and Concentration Losses

If the oxygen is supplied in the cathode in the form of air, the concentration of oxygen will cause a reduction in the partial pressure of the oxygen, and thus it is less in the electrode reducing the fuel cell performance. The reduction in gas pressure will result in a reduction in voltage. In the following Eq. (4.23), we can see that the change in voltage caused by a change in gas pressure is given by

$$\Delta V = \frac{RT}{2F} \ln \left(\frac{P_2}{P_1} \right) \tag{4.23}$$

If the pressure increases from P_1 to P_2, then the effect of pressure on the i–V curve can be seen in Figure 4.11.

If P_1 is the pressure when the current density is zero, and we assume that the pressure falls linearly down to zero at the current density i_1, then the pressure P_2 at any current density i is given by the following equation:

$$P_2 = P_1 \left(1 - \frac{i}{i_1} \right)$$

If we substitute this with Eq. (4.23),

$$\Delta V = \frac{RT}{2F} \ln \left(1 - \frac{i}{i_1} \right) \tag{4.24}$$

This is the voltage change due to the mass transport losses. Equations (4.23) and (4.24) are the voltage gain, and if we would like to express this equation in terms of voltage drop, then we need to put minus sign here.

$$\Delta V_{\text{trans}} = -\frac{RT}{2F} \ln \left(1 - \frac{i}{i_1} \right)$$

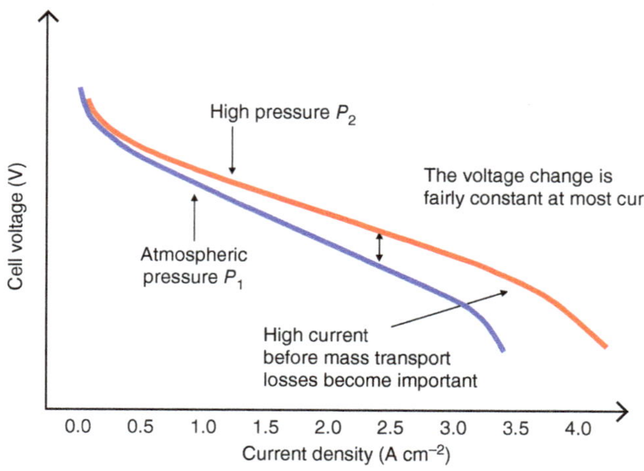

Figure 4.11 The effect of increasing pressure on the voltage–current graph of a typical fuel cell.

Since $RT/2F$ is a constant and is different for different reactants. For example, for oxygen, it will be $RT/4F$. The general equation for concentration or mass transport losses is given by the following equation:

$$\Delta V_{\text{trans}} = -B \ln\left(1 - \frac{i}{i_1}\right) \quad (4.25)$$

where B is a constant that depends on the fuel cell and its operating state.

$$\Delta V_{\text{trans}} = m \exp(ni) \quad (4.26)$$

where m will typically be about 3×10^{-5} V, and n about 8×10^{-3} cm^2 mA^{-1}. If the constants are chosen correctly, the results between Eqs. (4.25) and (4.26) are similar. The mass transport or concentration overvoltage is important in cases where the hydrogen is supplied from reformer.

Combining the four irreversibilities mentioned in Eqs. (4.18), (4.22), and (4.26), we can arrive at the following equation for the operating voltage of a fuel cell at a current density i.

$$V = E - \Delta V_{\text{act}} - \Delta V_{\text{ohm}} - \Delta V_{\text{trans}}$$

or

$$V = E - A \ln\left(\frac{i + i_n}{i_o}\right) - ir - m \exp(ni) \quad (4.27)$$

where E is the reversible OCV, i_n is the internal and fuel crossover equivalent current density, which is usually very small, A is the slope of the Tafel line, i_o is either the exchange current density at the cathode if the cathodic overvoltage is much greater than the anodic or it is a function of both exchange current densities, m and n are the constants in the mass-transfer overvoltage Eq. (4.26), and r is the area-specific resistance or ASR. Again, it is very small and has little impact on operating losses of fuel cells at working currents, making it difficult to measure. If we assume that current is always greater than the exchange current, i_o is much more countable. Then the equation for the activation overvoltage is rearranged to

$$\Delta V_{\text{act}} = A \ln\left(\frac{i}{i_o}\right) = A \ln(i) - A \ln(i_o) \quad (4.28)$$

Since practical OCV E_{oc} is a constant as described below:

$$E_{oc} = E + A \ln(i_o) \quad (4.29)$$

where E is the theoretical and reversible OCV given by Eq. (4.10). $E_{oc} < E$ because i_o will generate negative logarithms. If we replace Eqs. (4.28) and (4.29) into (4.27) and neglect i_n, we can obtain,

$$V = E_{oc} - ir - A \ln(i) + m \exp(ni) \quad (4.30)$$

This equation is simple but has been found to give an excellent fit with the results obtained from real fuel cells.

4.2.4 Water Management

Water is produced not as steam but as a liquid in PEMFC and AEMFC. A critical requirement of these cells is to maintain a high water content in the electrolyte to ensure high ionic conductivity. The ionic conductivity depends on the water content of the electrolyte membrane, and a fully saturated membrane offers a low resistance to current flow, increasing overall efficiency. Water content in the cell is determined by the balance of water or its transport during the reactive mode of operation. As depicted in Figure 4.12 for AEMFC, water transport is influenced by the water drag through the cell, back diffusion from the anode, and the diffusion of any water in the fuel stream through the anode. Water transport is a function of cell current and the characteristics of the membrane and the electrodes. Water drag refers to the amount of water pulled by osmotic action along the hydroxide. Between 1 and 5 molecules are dragged with each hydroxide ion. The water drag increases at high current density, and this makes the water balance a potential concern. During actual operation, however, back diffusion of water from the anode to the cathode through the thin membrane results in a net water transport of nearly zero.

Water management has a significant impact on cell performance, specifically at high current densities, where mass transport issues associated with water formation and distribution can limit cell output. Without adequate water management, an imbalance will occur between water production and evaporation within the cell, as depicted in the case of AEMFC in Figure 4.12 and PEMFC in Figure 4.13. Adverse effects include dilution of reactant gases by water vapor, flooding of the electrodes, and dehydration of the polymer membrane. The adherence of the membrane to the electrode will also be adversely affected if dehydration occurs. The intimate contact between the electrodes and the polyelectrolyte membrane is important because there is no free liquid electrolyte to form a conducting bridge. If more water is exhausted than produced, it is important to humidify the incoming anode gas. If there is too much humidification, however, the electrode floods, which causes problems in gas

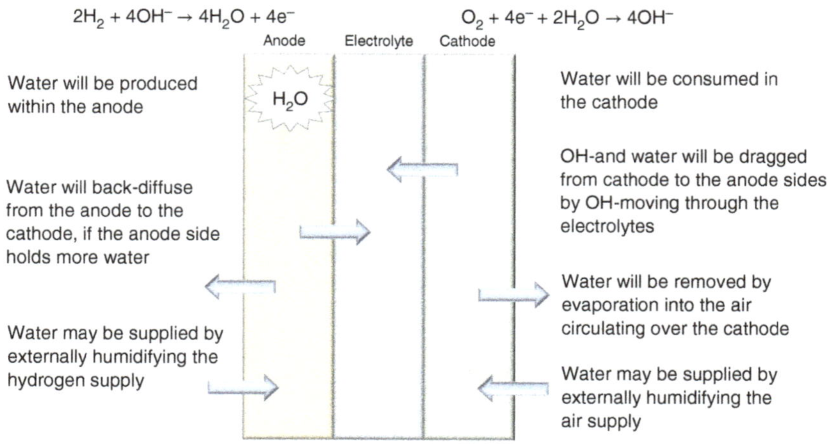

Figure 4.12 Different water movements to, within, and from the electrolyte of an AEMFC.

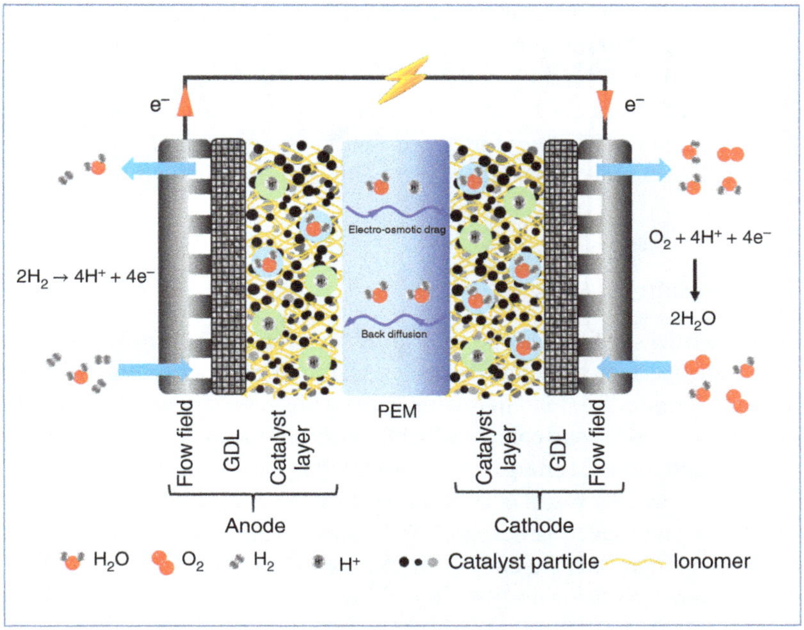

Figure 4.13 Schematic of a proton exchange membrane fuel cell.

diffusion to the electrode. A smaller current, larger reactant flow, lower humidity, higher temperature, or lower pressure will result in a water deficit and dry-out problem. A higher current, smaller reactant flow, higher humidity, lower temperature, or higher pressure will lead to a water surplus. More reliable forms of water management are also being developed based on continuous flow field design and appropriate operating adjustments.

4.3 Membranes in Proton Exchange Membrane Fuel Cell

The PEM is a semipermeable membrane that selectively transports protons via functional groups of polymers and transport media (e.g. water, acid), while preventing permeation of fuel and oxidants. Figure 4.13 depicts a proton exchange membrane fuel cell. Ideal characteristics of PEMs include high ion conductivity, low electrical conductivity, low fuel and oxidant permeability, good chemical and mechanical properties, and competitive cost.

The most widely used class of CEM material is PFSA polymers such as Nafion®, Hyflon®, Selemion®, and Aquivion®. These are composed of polytetrafluoroethylene (PTFE) hydrophobic main chains and super acidic sulfonic acid groups containing perfluoroether hydrophilic side chains. This unique structural class of PEMs forms hydrophilic ionic clusters connected by short and narrow channels within the hydrophobic matrix of PTFE. PFSAs have high proton conductivity below 90 °C despite of their low ion exchange capacity (IEC) due to well-connected ionic pathways. In addition, the fluorocarbon-based polymer structure enhances chemical and mechanical properties. However, the perfluoro aliphatic polymer chains of PFSAs have relatively low glass transition temperatures (T_gs), and the ionic clusters dehydrate above 80 °C. For this reason, proton conductivity and chemical and mechanical stabilities of PFSAs are drastically reduced above 100 °C, leading to a decrease in fuel cell performance. Furthermore, high fuel permeability and high cost are impediments to the commercialization of PFSAs for medium and high-temperature fuel cells.

4.3.1 Perfluorosulfonic Acids

As mentioned earlier, Nafion was developed by DuPont and is a PFSA polymer synthesized by a free radical initiated copolymerization of a perfluorinated vinyl ether sulfonyl fluoride co-monomer with TFE. This reaction provides a poly (tetrafluoroethylene) backbone connected with perfluoroether pendant side chains terminated with sulfonic acid groups (Figure 4.14a). Polymers with similar chemical structures are produced by Asahi Glass Company, Asahi Kasei, and Fuma-Tech as different commercial product names, such as Flemion, Aciplex, and Fumion. Since Nafion contains a longer side chain, it is called a long side chain (LSC) type PFSA, compared with shorter side chain (SSC) type introduced from different producers as in Figure 4.14b,c. The SSC-type perfluorinated ionomer was first introduced by Dow Chemical with no fluoroether group in the pendant side chain and comprising only two CF_2 groups during the 1980s. The industrial development of the corresponding SSC ionomer was manufactured by Solvay Specialty Polymers by applying its fluoroether process to produce SSC monomer in Figure 4.14b. During the same period, 3M Corporation also developed a 3M ionomer in Figure 4.14c with a fluoroether-free pendant chain with four $-CF_2-$ groups by the electrochemical fluorination of a hydrocarbon starting material.

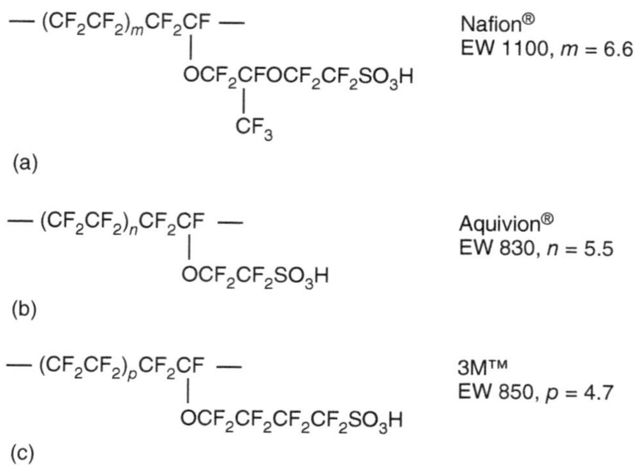

Figure 4.14 Nafion, Aquivion, and 3 M perfluorosulfonic acid membranes and ionomers from DuPont, Solvay Specialty Polymers, and 3 M, respectively. EW = equivalent weight.

The EW of an ionomer is the weight of the polymer required to provide 1 mol of exchangeable protons, i.e. it is the inverse of IEC. These properties are related to proton conductivity and the swelling tendency. EW and IEC of PFSA depend on the ratio of TFE and side-chain functionalized TFE. LSC-type membranes typically consist of an ionomer of EW of 1100–900 g mol^{-1} or IEC of 0.91–1.11 mmol g^{-1}.

The SSC-type Aquivion membrane shows semi-crystalline characteristics even at low EW. Glass transition temperature of SSC-type membranes (Aquivion and 3M ionomer) exhibits higher T_g (T_g Aquivion = 140 °C and 3M ionomer = 125 °C), compared with LSC Nafion (T_g = 100 °C). This is due to the absence of the pendant CF_3 group and higher crystallinity with SSC. Studies showed that significantly lower fluoride release from the degradation of PFSA was noticed from the SSC-type membranes. This means that relative stability to radical attack was superior for SSC-type PFSA compared with LSC-type ones.

4.3.2 Characteristics of Nafion

PFSA-based PEMFCs have received much attention because they provide high power density at relatively low operating temperatures (typically 80 °C). These fuel cells have been used in portable and stationary power sources and for electric vehicle applications. Banerjee and Curtin summarized the details of Nafion membrane development in Du Pont in fuel cell applications [4].

In 1964, DuPont realized that the major commercial application of Nafion was as a membrane in chlor-alkali processes. In this application, concentrated brine is electrolyzed in the anode to produce chlorine, and water is decomposed in the cathode compartment to produce caustic soda and hydrogen. The role of the membrane is to transport Na^+ ions and to restrict the transport of OH^- ions. According to the classical Donnan equilibrium theory for homogeneous ion exchange membranes, the membrane is effective in the cell only if the concentration of the fixed ions in the membrane considerably exceeds the OH^- ion concentration in the cathode compartment. This condition is generally not met in Nafion membranes; instead, their unusual properties arise primarily from the heterogeneous structure found in wet resins.

Because the proton conductivity of Nafion is high only in the wet state and fades when it begins to dry, water management to protect it from drying is a major concern during fuel cell operation [9]. Therefore, the performance of the PEMFC is strongly dependent on the humidification of the membrane since membrane resistance changes drastically with water content. The water balance of the fuel cell is also extremely complicated. During the cell reaction in the PEMFC, the anode side (hydrogen fuel side) of the membrane is readily dried up by the electro-osmotic drag of the proton, while the water produced in the cathode reaction diffuses back to the anode side. Analysis of the performance of a PEMFC requires reliable conductivity for the membrane electrolyte at different humidities in combination with water uptake, water diffusion, gas diffusion properties, etc.

The cathode CL of PEMFC contains Nafion binder, carbon particles, and platinum particles. Raising the Nafion ionomer content in the CL increases proton transport but reduces void volume, thus lowering gaseous oxygen transport. Increasing Pt loading enhances the rate of the electrochemical reaction but increases the cost [10]. Due to limited availability and high cost of Pt, most investigations concentrate on lowering the Pt loading while keeping an acceptable performance. Optimal Nafion loading contents range from 27 to 40 wt%, depending on the type of the electrodes: different Pt loading, Pt/carbon support, and different ink preparations. Although Nafion is the state-of-the-art PEM for fuel cells, several PFSAs with different side chain lengths have been developed by several producers as listed in Figure 4.14.

In many ion-containing polymers, the ion exchange sites in Nafion resin aggregate into clusters (Figure 4.15). Such ionic clustering has significant effects on the transport properties [7]. Haubold et al. [11] studied Nafion membranes in methanol and water using small-angle X-ray scattering and concluded that pure methanol replaces the water and dehydrates the ionic shell. The side-chain swelling was observed at the cost of available volume for the core region. Mauritz and Moore studied the microscopic morphological investigation of Nafion ionomers using X-rays and neutron scattering and correlated it with the nature of ions and water in Nafion [12].

The transport of water and ions through Nafion has been an important topic for study [13–17]. Chiou and Paul studied the gas transport through dry Nafion membranes [18]. Nafion membrane showed high separation factors for He/CH_4, He/H_2, and N_2/CH_4 relative to other polymer membranes. Broka et al. studied the water uptake from the vapor phase by commercial Nafion 117 membranes, and recast Nafion film was also determined. It increased with increasing temperature and was slightly lower for the recast film. Water uptake by a heat-treated Nafion membrane was lower than that by the untreated sample and the recast film. Oxygen permeability increased with temperature and relative humidity for both forms of Nafion. Hydrogen permeability of Nafion 117 was about two times higher than the oxygen permeability. Oxygen transport in Nafion was highest at 80–85 °C when the

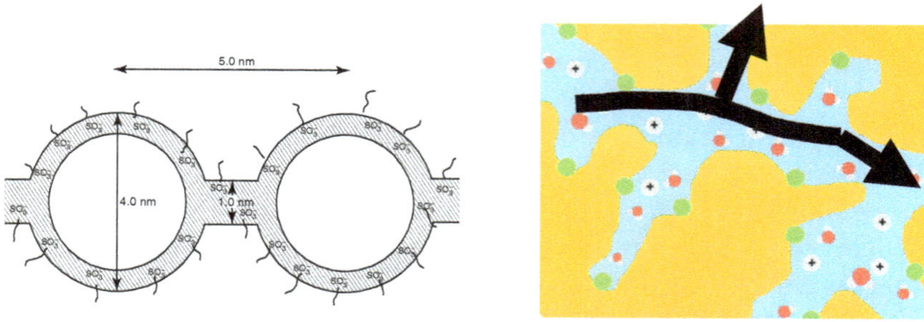

Figure 4.15 Cluster-network model for Nafion membranes. The polymeric ions and absorbed electrolyte phase-separate from the fluorocarbon backbone into approximately spherical clusters connected by short narrow channels. The shaded areas around the interface and inside a channel are the double-layer regions from which the hydroxyl ions are excluded electrostatically. Source: Adapted from Hsu and Gierke [7]. Scheme on the right describes the ion transport through hydrophilic water channels.

membrane water content was lowest, and, at the same time, permeability increased with relative humidity (and membrane water content) at a given temperature [19].

Morris and Sun [20] reported on the water sorption and transport properties of Nafion 117 by measuring its density and dimensional changes as a function of water content. Swelling of Nafion starts at λ (mole ratio of water to proton ion) = 1.9. Some anisotropy of dimensional changes was observed. Water sorption isotherms obey Henry's Law with a non-zero intercept, indicating some water retention. The diffusion coefficient of water in Nafion and its electrical conductivity of Nafion are strong functions of the water content.

Motupally et al. [16] studied the diffusion of water in the Nafion 115 membrane as a function of the water activity gradient by changing the flow rate and pressure of nitrogen gas on one side of the membrane, whereas the other side of the membrane was equilibrated with liquid water. The model was developed using the Fickian diffusion coefficient extracted from self-diffusion measurements. Majsztrik et al. [21] reported the water sorption, desorption, and transport in Nafion membranes at 30 and 90 °C. Obviously, water permeation increases with temperature. Liquid water permeation increased with the difference in water activity across the membrane. Water permeation from humidified gas into dry nitrogen went through a maximum with difference in water activity difference across the membrane. These results suggested that the membrane was less swollen in the presence of water vapor and that a thin skin formed on the dry side of the membrane that reduced permeability to water. Permeation was only weakly dependent on membrane thickness; results indicated that interfacial mass transport at the membrane/gas interface was the limiting resistance.

The diffusivity of water in Nafion deduced from water sorption into a dry Nafion film was almost two orders of magnitude slower than the diffusivity determined from permeation experiments. Water desorption from a water-saturated film was an order of magnitude faster than water sorption. Water desorption appeared to be limited by the rate of interfacial transport across the membrane/gas interface. They [22] later reported that diffusivity of water, interfacial mass transport, and volume available for transport are three critical factors that influence the transport of water in Nafion membranes. The self-diffusion coefficients increased with water activity, showing two regimes. Diffusion coefficients increased rapidly with increased water for the first four molecules per sulfonic acid and then increased more slowly for water activities corresponding to more than the four water molecules per sulfonic acid. The water sorption and diffusion results suggested four water molecules form a first solvation shell around the sulfonic acid groups. Beyond the first four water molecules, the diffusivity increases more rapidly with water sorption. Interfacial mass transfer coefficients evaluated for water transport across the Nafion/vapor interface indicated that at high membrane water activity, diffusion is fast, and water transport is limited by interfacial transport at the membrane/vapor interface.

Although there has been success in the area of fuel cells, stability enhancement and durability of the PEMs are critical for the lifetime and commercial viability of the PEMFCs. It has been found that the degradation of the fuel cell performance is primarily due to the decay of the membrane electrode assembly (MEA). The degradation of the PEM associated with the loss of fluoride ions and the decrease in conductivity is also a problem facing the long-term stability of the fuel cell [23].

4.3.3 Degradation of Nafion

The degradation behaviors of Nafion NR111 membrane were studied under various mechanical, chemical, and polarization conditions [24]. It was found that the fatigue strength or the safety limit of the cyclic stress for Nafion NR111 membrane is ~1.5 MPa, which is 1/10 of the tensile strength of the membrane. The cyclic stress and dimensional changes (or strain) induced by water uptake can be substantial and are the main causes of mechanical degradation and membrane failure. For example, in the case of RH cycling of water-soaked state to 25% RH state, the cyclic stress of the Nafion membrane was as high as 2.23 MPa, and the dimensional change was ~11%. The decomposition of the Nafion polymer in the H_2O_2 solution, in the presence of trace Fe, Cr, and Ni ions, started from the ends of the main chain, resulting in the loss of repeat units and the formation of voids and pinholes in the membrane. The high degradation rate of the membrane at the OCV most likely results from the attack of H_2O_2 formed between O_2 and H_2 crossover through the membrane.

Degradation of Nafion membrane has been studied by Yu et al. using quantum mechanics (QM) in a PEMFC [25]. Experiments suggest that Nafion degradation occurs due to the generation of trace radical species (such as OH radical and H radical) only in the presence of H_2, O_2, and Pt. It was found that OH radicals might be present during the oxygen reduction reaction (ORR), particularly when the fuel cell is turned on and off. Based on QM, two chemical mechanisms for OH radical attack on the Nafion polymer were proposed: (i) OH radical attack on the S—C bond to form H_2SO_4 plus a carbon radical, followed by decomposition of the carbon radical to form an epoxide. (ii) OH radical attack on H_2 crossover gas to form hydrogen radical, which subsequently attacks a C—F bond to form HF plus carbon radicals.

Dehydration in the Nafion membrane has been studied in detail [9]. Data showed that for stacks with humidified cathodes, the resistance is small and relatively insensitive to the presence of anodic humidification. On the contrary, the membrane resistance is high for stacks with non-humidified cathodes and strongly dependent on current and anodic humidification.

The accelerated stability testing (AST) was conducted using MEA from Nafion/PTFE composite membrane in studies combining RH and load cycling for over 1000 hours [26]. It was observed that edge protection should be adopted to avoid immediate mechanical failure for excessive or nonuniform mechanical stress, mainly caused by RH cycling in the early period of ASTs. TEM and SEM cross-sectional images presented that severe chemical degradation occurred on the membrane without hot-pressing with ionomer layers thinning, ionomer disappearing in the cathode side, and delamination or cracks along the interfaces of ionomer, catalyst, and PTFE layers. In addition, the agglomeration of Pt and possible carbon corrosion could accelerate the chemical decomposition for the generation of H_2O_2 in the MEAs.

4.3.4 Composite PEM

There have been numerous studies related to composites with Nafion membranes [26–35]. Satterfield et al. [27] added TiO_2 into Nafion and measured the mechanical properties and PEMFC performance. Nafion/TiO_2 composites have slightly higher elastic moduli and exhibit less strain hardening than Nafion. Composites also show a reduction in the long-time creep of ~40% compared to Nafion. Water uptake was faster in Nafion membranes recast from solution compared to extruded Nafion. The addition of 3–20% TiO_2 particles has a minimal effect on the rate of water uptake. The results demonstrate how the dynamics of water uptake and loss from membranes are dependent on physical constraints, and these constraints can impact fuel cell performance. Amjadi et al. [33] added TiO_2 nanoparticles to Nafion and showed increased water uptake and fuel cell performance, particularly at 110 °C.

Kannan et al. [28] incorporated sulfonic acid functionalized single-walled carbon nanotubes (S-SWCNT) in Nafion for PEMFC. As S-SWCNT was added, the composite membranes showed an increase in ionic conductivity, mechanical stability, and enhanced fuel cell performance.

Mesoporous silica was also filled into Nafion, increasing proton conductivity by more than 50%, probably due to the large surface area of the silica particles inducing water adsorption [29]. The addition of sulfated zirconia (S-ZrO_2) in Nafion increased the thermal stability of the composite membranes that performed better, particularly at high temperatures [30]. PEMFC performance of a single cell was the best when the S-ZrO_2 content was 15 wt%, and achieved 1.35 W cm^{-2} at 80 °C and 0.99 W cm^{-2} at 120 °C based on H_2/O_2 and at a pressure of 2 atm at 80 °C, the single-cell performance of S-ZrO_2/Nafion was far more than that of Nafion at the same condition (e.g. 1.28 W cm^{-2} at 80 °C and 0.75 W cm^{-2} at 120 °C). Nafion can be blended with polybenzimidazole (PBI) for high-temperature PEMFC operation when doped with phosphoric acid [31]. H_2/O_2 single-cell performance of the composite membrane, tested at 150 °C without humification, showed that the presence of Nafion enhanced the durability of the composites. Porous PTFE can be used as a support to impregnate the Nafion membrane. The composite membrane enhanced the mechanical durability as well as the durability of fuel cell performance [32].

4.3.5 Radiation-Grafted Fluoropolymer PEM

Although fluoropolymers are known to be chemically inert materials with good temperature resistance, low dielectric constants, and low surface energy, they are often the materials of choice for harsh chemical environments [36]. These properties arise from the C—F bond, which is the strongest of all bonds between carbon and other elements, and due to its large size, F atoms can protect the carbon backbone of polymers, such as PTFE, from chemical attack. However, although the C—F bond is much stronger than the C—H bond, the G values for radical formation on high-energy radiolysis of fluoropolymers are roughly comparable to those of their protonated counterparts. Therefore, efficient radiation grafting of fluoropolymers is practically possible, and this process can be applied to modify either the surface or bulk of a fluoropolymer. In fact, radiation-grafted fluoropolymers are currently being used as polyelectrolyte membranes for fuel cells, both in PEM and AEM, as binders and, as separation membranes. Table 4.3 summarizes the fluoropolymers used in fuel cell technology and base polymers for radiation grafting reported for the last two decades.

Cui et al. [36] summarized the recent progress in the development of novel fluoropolymers for membrane applications, including fuel cells and LIBs and their fabrication and modification methods.

Souzy and Ameduri [37] summarized various routes to synthesize functional fluoropolymers used in fuel cell membranes. They can be categorized into two types: (i) The direct radical copolymerization of fluoroalkenes with fluorinated functional monomers, for example, fluorinated vinyl ethers, α,β,β-trifluorostyrenes, or trifluorovinyl oxy aromatic monomers bearing sulfonic or phosphonic acids. The resulting membranes are well-known as Nafion, Flemion, Hyflon, Dow, Aciplex, or BAM3G. (ii) Synthesis of FP-g-poly(M) where FP and M stand for fluoropolymers and monomer, respectively, obtained by activation (e.g. irradiation using an electron beam, γ-rays, or ozone) of FP polymers followed by grafting of M monomers. In PEMFC, the most widely used M monomer is styrene, followed by sulfonation on FP-g-PS that leads to FP-g-PS sulfonic acid graft copolymers. In AEMFC, amination can be conducted on FP-g-PS instead.

Modification of fluoropolymers [38] can be achieved chemically by etching or using high-energy radiation.

Radiation-grafted ion exchange fluoropolymers [39–46] have been prepared using both simultaneous (mutual) and post-irradiation methods. Table 4.4 lists the parameters and conditions for either mutual (or simultaneous) and pre-irradiation methods. Typical radiation grafting of fluoropolymers such as poly(ethylene-co-tetraflyoroethylene) (ETFE) and poly(vinylidene fluoride) (PVDF) is depicted in Figure 4.16 to prepare PEMs.

A majority of fuel cell membrane research investigated the radiation-grafting of styrene into nonfluorinated (e.g. LDPE [low-density polyethylene]), partially fluorinated (e.g. PVDF), and fully fluorinated membranes (e.g. poly(tetrafluoroethene-co-hexafluoropropylene) [FEP]), with subsequent sulfonation to produce PEMs

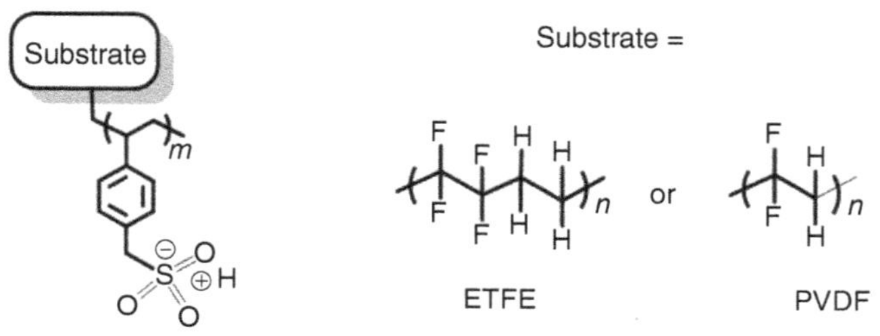

Figure 4.16 Radiation-grafted polymer membranes.

Table 4.3 Typical fluoropolymers and hydrocarbon polymer for radiation grafting toward fuel cell application and their physical properties.

Polymer	Abbreviation	Density ($g\,cm^{-3}$)	Thickness (μm)	Melting point (°C)	Crystallinity (%)	Structure
Poly(tetrafluoroethylene-co-sulfonyl vinyl ether)	Nafion	2.1	50	350	18	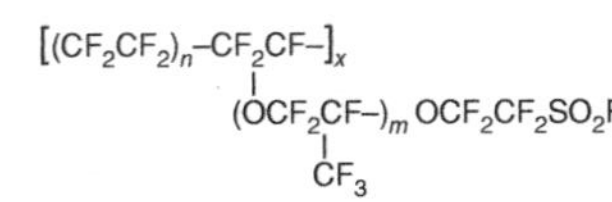
Poly(vinylidenefluoride)	PVDF	1.76	25	177	46	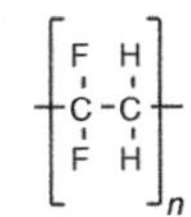
Poly(ethylene-tetrafluoroethylene)	ETFE	1.73	25	257	34	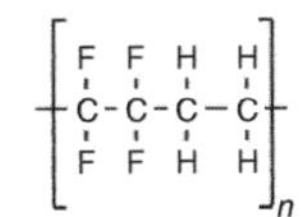
Low density polyethylene	LDPE	0.94	—	115	35	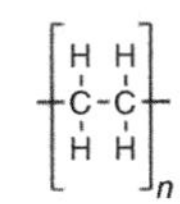

[38, 47]. The properties and compositions of the final modified membranes can be easily controlled with this method [48]. Radiation-grafted PVDF-based cation- and AEMs have been successfully applied to fuel cell technologies. Radiation-grafting methodology produces ion exchange membranes economically and has two major advantages: (i) preformed commercial polymer films are modified, alleviating the need for film formation steps; (ii) the availability of a wealth of adjustable experimental parameters (e.g. radiation dose/type, temperature, and film thickness), allowing a large degree of tailorability and customization.

Brack et al. [49] reported on the radiation grafting of ETFE with styrene by the pre-radiation method and compared that with FEP or PVDF. Film orientation and fluorination have an important impact on their grafting behavior. Horsfall and Lovell [50] reported on the radiation-grafted ETFE and PVDF and compared them with Nafion membrane in H_2/O_2 fuel cell (Figure 4.17). ETFE-grafted membrane MEA showed the peak power density (PPD) of up to $1\,W\,cm^{-2}$ that is comparable to Nafion115. Longer-term test showed that these fluoropolymer-grafted membranes are physically and chemically stable up to 135 hours. However, problems with MEA fabrication must be remarked in order to reduce the delamination problem when MEAs are fabricated using procedures developed for Nafion membranes.

Chen et al. [51] reported on the ETFE-based fuel cell membranes using UV-induced photo-grafting of styrene and electron beam-induced cross-linking technique, followed by sulfonation of the multiple-crosslinked membranes as PEMs. Photo-grafted polystyrene chains can completely penetrate into the base ETFE film; the resultant sulfonated electrolyte membranes show in-plane proton conductibility of up to about $120\,mS\,cm^{-1}$. On the one hand, the proton conductivity of the liquid-phase photo-grafted electrolyte membranes is higher than that of the vapor-phase one and is anisotropic in the surface and thickness directions. On the other hand, radiation-induced cross-linking greatly improves the chemical stability

Table 4.4 Conditions of radiation grafting for ion exchange membrane from commercially available films such as polyethylene and ETFE.

Parameter	Mutual (or simultaneous method)	Pre-irradiation method
Type of radiation	γ-ray, EB	γ-ray, EB
Absorbed dose	Low (10 kGy or less)	High (100 kGy and more)
Dose rate	Low ($kGy\,h^{-1}$)	High ($kGy\,s^{-1}$)
Irradiation time	Long (h)	Short (min) or (s)
Atmosphere	Inert gas	Air/inert gas/vacuum
Side reactions: homopolymerization	High	Low
Temperature	Ambient	Irradiation: Ambient/low temperature, graft Polymerization: High (peroxide decomposition)

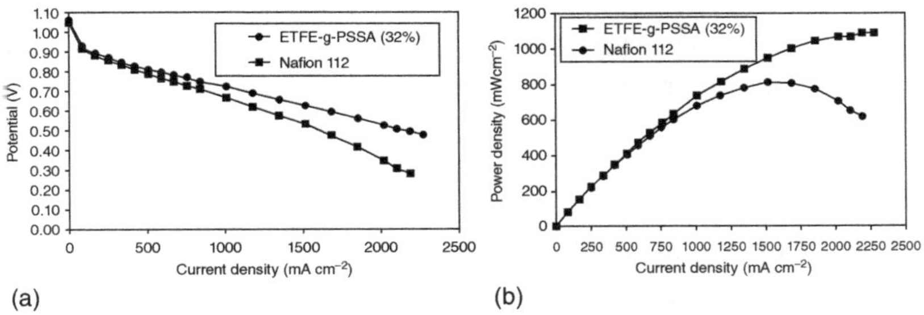

Figure 4.17 (a) Polarization curve and (b) power density on H_2/O_2 for ETFE-*g*-PSSA membrane and Nafion 112. Source: Taken from Horsfall and Lovell [50].

of the resultant membranes and increases the surface concentration of sulfonic acid groups to its higher level, which is very important for the performance of the relevant MEA.

They [52] also investigated the effect of three cross-linkers on the properties of ETFE-based radiation-grafted PEMs. The cross-linkers studied included divinyl benzene (DVB), 1,2-bis(*p,p*-vinyl-phenyl)ethane (BVPE), and triallyl cyanurate (TAC).

Although cross-linked, ETFE membranes showed lower ion conductivity, swelling, and water uptakes; they showed significantly higher chemical stability in 3% H_2O_2 at 50 °C. Among the three cross-linkers, DVB shows the most pronounced efficiency in cross-linking of the radiation-grafted membranes, while the TAC has no significant influence, and the BVPE is a mid and effective cross-linker, showing the moderate influence between DVB and TAC cross-linkers.

Chen et al. [53] investigated six different commercially available fluoropolymer films (PTFE, FEP, PFA, ETFE, PVDF, and PVF) for radiation grafting and sulfonation for PEMs. They concluded that PTFE and PVF are not suitable for the preparation of fuel cell membranes, as PTFE is quite unstable for radiation and PVF undergoes only surface grafting. The FEP-based membrane shows the highest proton conductivity over the entire IEC range and the lowest water uptake over the entire conductivity range, while the ETFE-based membrane shows the opposite results. Perfluorinated film-based membranes are considerably more chemically stable than partially fluorinated film-based membranes. However, partially fluorinated membranes show significantly better mechanical properties than perfluorinated membranes. They concluded that the perfluorinated film-based membranes are promising materials for hydrogen fuel cells, and the partially fluorinated film-based membranes are appropriate for liquid methanol-fed fuel cells.

Gubler et al. [54] studied the effect of cross-linker in ETFE-based radiation-grafted proton-conducting membranes. Different extents of the cross-linker DVB as a co-monomer of styrene were grafted onto 25 μm ETFE film. As the amount of DVB increases, water uptake, hydration number λ, proton conductivity, and fuel crossover decrease. Their subsequent study [55] reported that ETFE-based cross-linked membranes were far better than FEP-based proton-conducting membranes. An *in situ* long-term stability test showed an H_2/O_2 single-cell could last for 2180 hours using ETFE-based grafted membranes with a grafting level of 25%.

Sohn et al. [56] grafted binary mixture of styrene sulfonic acid (SSA) and poly (vinyltrimethoxysilane) (PVTMS) to ETFE, resulting in silane-cross-linked PEMs through sol–gel and sulfonation reactions. ETFE-g-PSSA showed high H^+ ion conductivity at 90% RH but exhibited low conductivity at 50% RH compared with cross-linked ETFE-g-PSSA/PVTMS membranes.

4.3.6 Hydrocarbon-Based Cation Exchange Membranes

To overcome these drawbacks of PFSAs, many researchers have developed hydrocarbon-based polymers based on sulfonated derivatives of poly(arylene ether)s (SPAEs), poly(arylene sulfide)s (SPASs), polyimides (SPIs), and non-sulfonated PBIs, and polyphenylenes (PPs). These sulfonated polymers can be synthesized via various synthetic routes from diverse monomers to obtain a variety of structural architectures. Most sulfonated polymers are thermally and mechanically stable up to 200 °C due to their more rigid polymer chain structures and have excellent fuel barrier properties, which lend themselves to CEM materials for medium temperature (<120 °C) PEMFC application. For a given IEC, sulfonated polymers generally have lower proton conductivities and higher dimensional swelling compared with PFSAs because of their lower acidity, less hydrophobicity of polymer backbone, and weaker phase separation between hydrophilic and hydrophobic moieties, resulting in less effective micro-phase-separated morphology for water channel formation. Chemical stability of sulfonated polymers appears to be lower than PFSAs, particularly those with electron-rich structures containing ether linkages. Thus, there are considerable efforts to solve these problems for medium-temperature fuel cell operation, especially for understanding proton conduction mechanisms in CEMs and the effects of morphology on CEM properties.

Ions are conducted through PEMs via a combination of vehicle, Grotthuss (or hopping), and surface mechanisms. The vehicle mechanism is the transport of acidic or basic media-solvated ions. While the ion diffuses with the carrier, there is a counterflow of nonionic carriers that continue the ion transport. The vehicle mechanism dominates ion conduction when the temperature increases because the diffusion rate of the ion carrier increases with temperature. Carrier content is also important for fast ion transport. Ions are also transported by forming and breaking hydrogen bonds with ion-accepting/donating media, such as water molecules and concentrated potassium hydroxide solution (i.e. Grotthuss mechanism). In addition, ions are conducted along with the cation or anion exchange groups from one moiety to another by the Grotthuss mechanism. Ion transport is influenced by the degree of media content in the hydrophilic ionic domains of the polymer and their connectivity. The vehicle mechanism is more dominant for ion conduction than the Grotthuss mechanism at low concentrations of ion carriers due to the bound counter ion (anion in CEM and cation in AEM) groups. The transport of ions via the Grotthuss mechanism is much faster than the diffusion of hydronium ions (vehicle mechanism) in the fully hydrated state (Figure 4.18) [57]. In general, the dependence of ion conductivity is influenced not only by the ion carrier content but also by the interconnectivity of

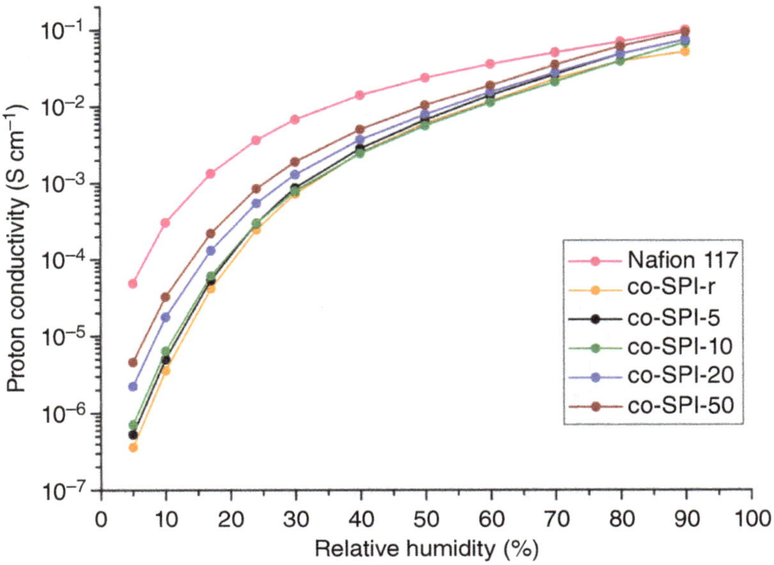

Figure 4.18 Proton conductivity as a function of relative humidity at 50 °C of the synthesized co-SPI membranes and Nafion 117 determined from dielectric spectroscopy in a two-electrode in-plane geometry [57].

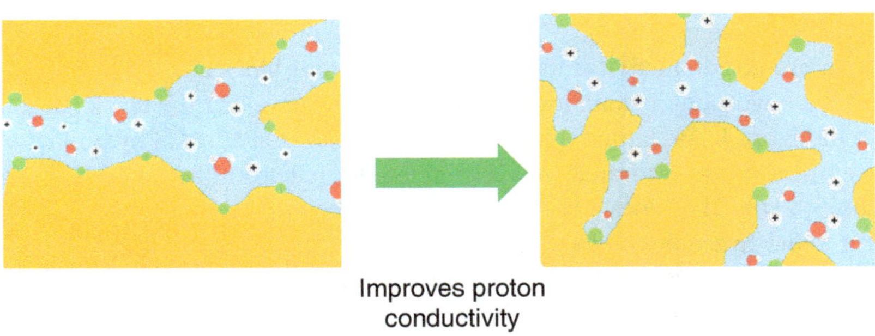

Figure 4.19 Schematic of the importance of ion-conducting channels to improve the proton conductivity.

the hydrophilic ionic nanochannels and the acidity or basicity of the ion-conducting functional groups. The membrane morphology determines the state of water, such as size, shape, and interface within hydrophilic ionic domains. For these reasons, well-connected hydrophilic ionic nanochannels facilitate the transportation of ions through the polymer membranes (Figure 4.19). In addition, such morphology is effective in controlling the IEC values and achieving high ion conductivities as well as chemical and mechanical stabilities because these properties are dependent on the nature of water within the polymer membrane.

In general, sulfonated random copolymers have homogeneous morphology because hydrophobic and hydrophilic moieties are randomly distributed.

In addition, hydrocarbon-based structures are less hydrophobic, and the sulfonic acid groups are less acidic than PFSAs, which also poses a greater barrier for hydrophilic–hydrophobic phase separation. However, the morphology of acid-bearing random copolymers is dependent on the chemical nature of the polymer structure. It is also affected by external processing conditions (e.g. solution casting, acidification, and hot pressing). Although random copolymers contain ionic clusters, their proton conductivity is much lower than that of PFSAs because ionic clusters are randomly dispersed in hydrophobic domains and are less interconnected.

Block copolymers are composed of hydrophilic and hydrophobic homo sequences. Therefore, block copolymers tend to have well-defined micro-phase separation of hydrophilic and hydrophobic regions. These polymer systems adopt various morphologies controlled by factors such as the degree of dissimilarity, chain length, and volume ratio of each sequence. Block copolymers often show anisotropic dimensional water swelling due to distinct morphological orientations. In addition, their proton conductivity may outperform PFSAs and random copolymers, especially at medium temperatures or low relative humidity. Therefore, amphiphilic block copolymers are regarded as attractive candidates for advanced PEM materials.

The block length of the hydrophilic and hydrophobic repeating unit is one of critical factors affecting morphology formation behavior. Figure 4.20 shows chemical structures of two representative block copolymers. McGrath and coworkers synthesized SPAE-based block copolymers (14a) with different hydrophilic and hydrophobic block lengths [58]. While all the block copolymers had similar IEC values, their proton conductivity increased with increasing block length because the connectivity of hydrophilic domains was improved by increasing the block length, as shown in Figure 4.21a. Jannasch's group demonstrated the effect of hydrophobic block length on phase-separated microstructure (see chemical structure for 14b and Figure 4.21b). The block copolymers had similar hydrophilic domain sizes due to the same hydrophilic block length. However, the size of the hydrophobic domains gradually increased relative to the hydrophilic chain length. Consequently, the short hydrophobic block increased the proton conductivity by increasing the IEC value and decreasing the volume fraction of the hydrophobic domain. Block length also influenced defects in the phase-separated domains as well as their size and connectivity. Kins et al. investigated the relationship between phase separation behavior and block length of imide-based block copolymers (see chemical structure for 14c and Figure 4.21c).

Acid-bearing graft copolymers, structurally mimicking PFSAs, are synthesized by attaching hydrophilic aromatic or aliphatic side chains to a hydrophobic main chain or *vice versa*. Similar to block copolymers, graft copolymers are composed of different homo sequences of ion-conducting and nonconducting units, which facilitate the formation of phase-separated morphology. The morphology of graft copolymers can be controlled by varying the length and periodicity of the side chains. The side-chain length and rigidity affect the size of hydrophilic ionic clusters or nanochannels, whereas the frequency of side chains over the polymer main-chain length influences the number of ionic domains per unit volume, allowing graft copolymers to be tailored for the desired electrochemical performances.

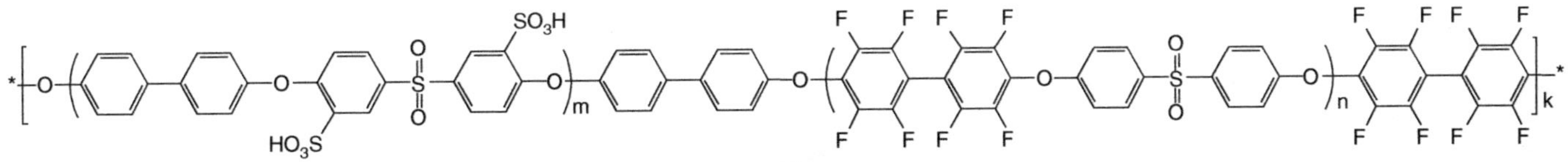

14a

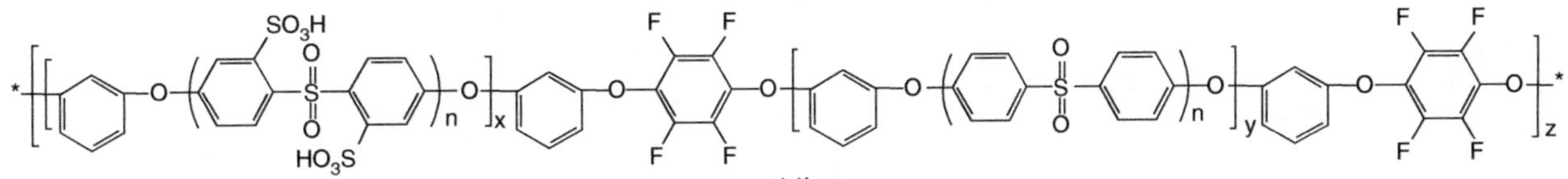

14b

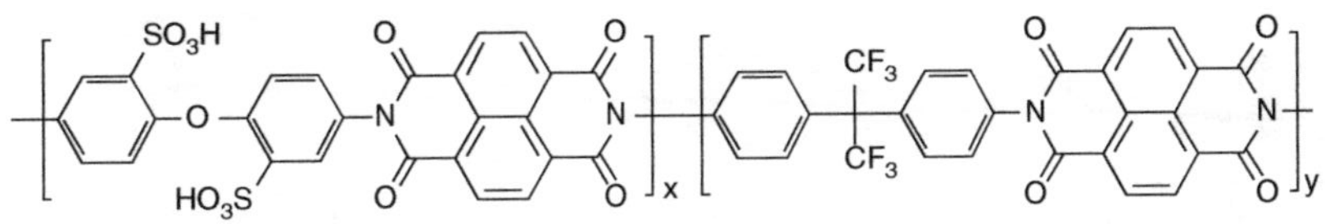

14c

Figure 4.20 Chemical structure of sulfonated poly(arylene ether sulfone)-based and sulfonated polyimide-based block copolymers [58].

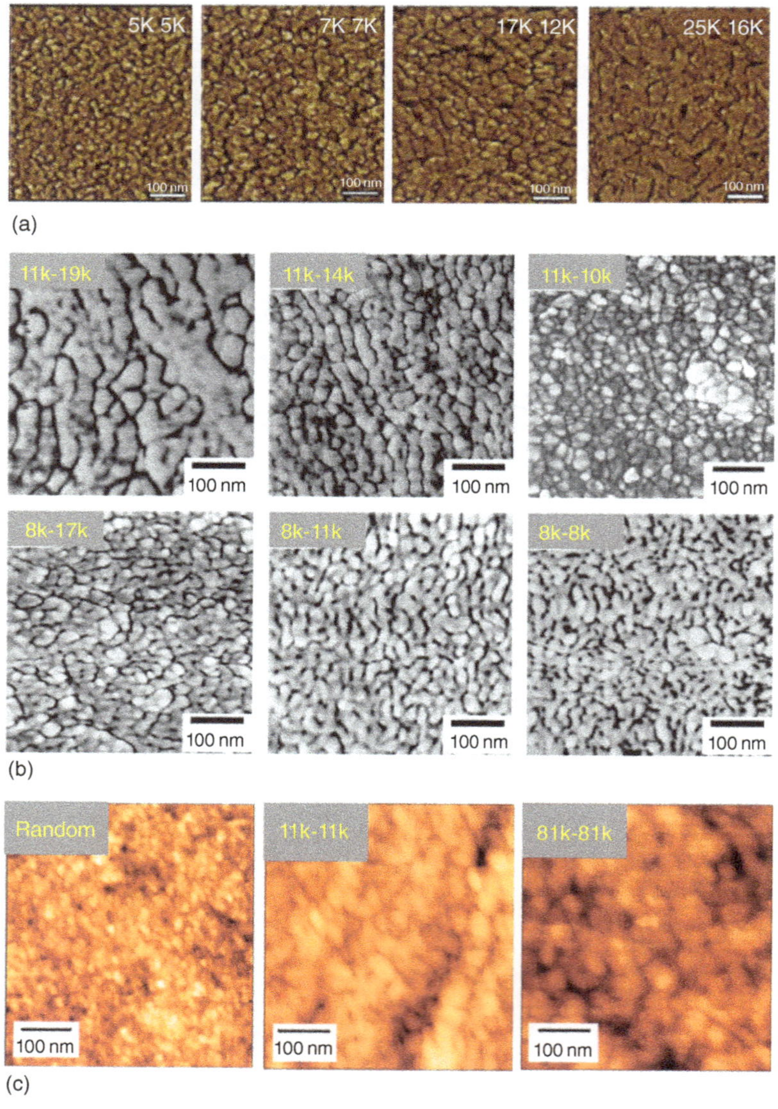

Figure 4.21 AFM images of block copolymers. (a) 14a (IEC = 1.29–1.32 meq g^{-1}). Source: Reproduced with permission from Yu et al. [58], © 2009/John Wiley & Sons, Inc., (b) 14b (IEC = 1.02–1.67 meq g^{-1}). Source: Reproduced with permission from Takamuku and Jannasch [59], © 2012/John Wiley & Sons, Inc., and (c) 14c (IEC = 1.64–1.71 meq g^{-1}). Source: Reproduced with permission from Kins et al. [57], © 2014/American Chemical Society.

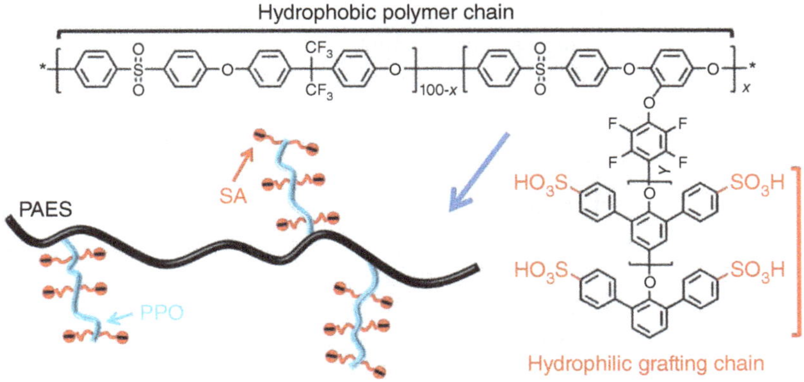

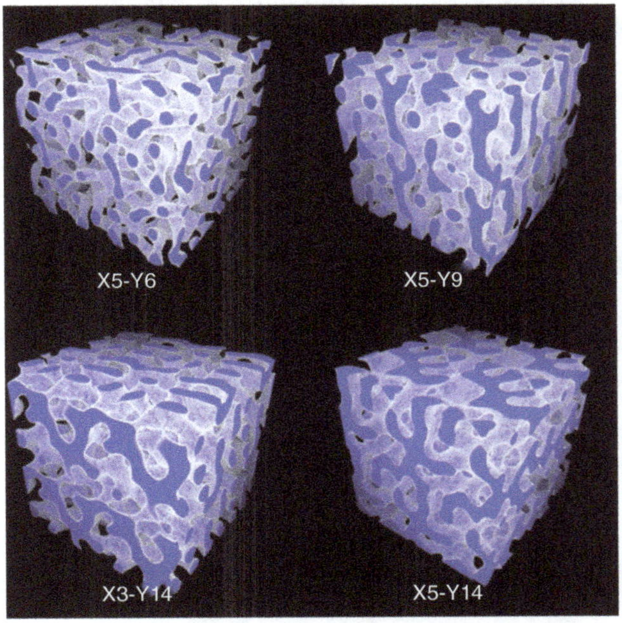

Figure 4.22 Chemical structure of comb-shaped SPAES and simplified illustration of fully aromatic comb-shaped SPAES copolymers. SA is sulfonic acid. Bottom panel illustrates the phase diagram of four membranes (hydrophilic side chain (blue) surrounded by isodensity surface (white). Source: Reproduced from Li et al. [60]/John Wiley & Sons.

We reported the synthesis of an SPAE-based copolymer with sulfonated diphenyl ketone (SK) and sulfonated PPO (SPPO) side chains, as shown in Figure 4.22. These comb-shaped copolymers containing sulfonic acid groups in the side chain showed well-defined phase-separated hydrophilic and hydrophobic domains, which resulted from the hydrophobic SPAE main chain and the hydrophilic SK and SPPO side chains. In particular, the hydrophilic domain of this comb-shaped copolymer consisted of ionic clusters interconnected with an ionic channel-like morphology similar to PFSAs. The SPAE–SK membrane had isolated ionic clusters at the 50–60 nm scale, and the SPAE–SPPO membranes represented connected ionic channels at the 3–20 nm scale. The continuous hydrophilic domains facilitated proton transport, leading to high proton conductivity over a wide range of relative humidity conditions. In addition, the proton conductivity was stably maintained due to the nature of the chemical structure with pendent sulfophenyl groups. As a result, the SPAE–SPPO showed excellent proton conductivity (20 mS cm^{-1}) at 90 °C and 30% RH.

Cross-linking is an effective and simple method for improving the chemical and mechanical stabilities of polymer membranes. Moreover, polymer morphology can be transformed during the cross-linking process because polymer chains are rearranged after cross-linking. However, cross-linked membranes are insoluble in organic solvents, so cross-linking must be performed during or after membrane formation. The cross-linker structure affects the morphology of cross-linked membrane. Oligomers can play a role as a cross-linker to provide graft cross-linked copolymers. Wang and coworkers reported graft cross-linked copolymers based on poly(arylene ether ketone)-*gc*-sulfonated poly(arylene ether sulfone) (PAEK-*gc*-SPAES). The phase-separation behavior of such graft cross-linked membranes depended on the molecular weights of each sequence. The introduction of cross-linkable groups onto polymer chains is also an attractive approach, instead of adding a cross-linker separately, because of the possibility of incompatible cross-linker solubility in polymer solutions. Acetylene groups, which trimerize to aromatic rings (see Figure 4.23), were induced at the ends of random copolymer

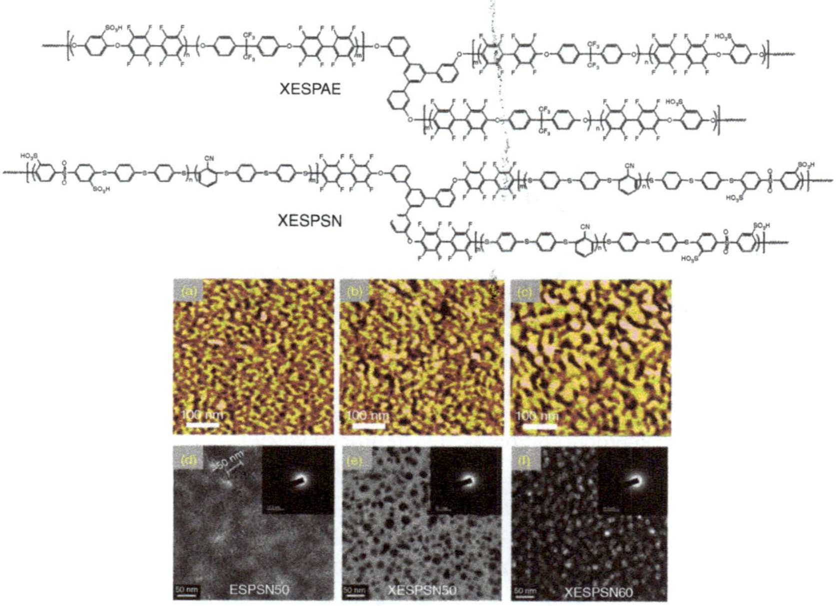

Figure 4.23 Chemical structures and morphology images of (a) XESPAE85 (IEC = 1.68 meq g^{-1}), (b) XESPAE90 (IEC = 1.79 meq g^{-1}), (c) XESPAE95 (IEC = 1.93 meq g^{-1}), (d) ESPSN50 (IEC = 1.96 meq g^{-1}), (e) XESPSN50 (IEC = 2.02 meq g^{-1}), and (f) XESPSN60 (IEC = 2.28 meq g^{-1}). Source: (a)–(c) Reproduced with permission from Lee et al. [61], © 2010/American Chemical Society. (d)–(f) Reproduced with permission from Lee et al. [62], © 2012/Royal Society of Chemistry.

chains to avoid excessive cross-linking. End-group cross-linked polymer membranes exhibited a well-ordered morphology. In particular, the hydrophilic domain size was enlarged with increasing DS, as described in Figure 4.23. The end group cross-linked membrane with a high DS showed higher proton conductivity and fuel cell performance than non-cross-linked membrane and Nafion, as shown in Figure 4.24.

Compared with random copolymers, block copolymers induce more ordered micro-phase separated morphology because of the different characteristics and lengths of each block-induced phase separation. The chemical structure of hydrophobic sequences is one of the critical factors in leading phase separation. Fluorine atoms, which have an extremely hydrophobic characteristics, magnify the contrast between hydrophilic and hydrophobic domains, resulting in well-ordered phase-separated morphology. Furthermore, the lengths of oligomer blocks influence morphological features and inter-domain distances. Longer block lengths induced clearer phase separation and increased domain sizes.

Pendent or side-chain grafted copolymers also showed developed hydrophilic domains. In these grafted copolymers, the flexibility and chain length of the pendent or side chain are important for morphology control. Long aliphatic side chains form much more connected ionic domains than short aromatic pendent groups because they can easily aggregate. Different characteristics between hydrophobic main chain and hydrophilic side chain (or hydrophilic main chain and hydrophobic side chain) induced a well-ordered morphology similar to block copolymers.

Furthermore, hydrophilic–hydrophobic phase separation behavior of polymers is enhanced by introducing densely clustered ion-conducting functional groups (e.g. sulfonic acid, quaternary ammonium [QA], guanidine, and imidazolium [IM]) into hydrophobic units. Densely clustered functional groups shorten the ion-conducting pathway by self-aggregation, improving the connectivity of hydrophilic domains to a greater extent compared with those of conventional copolymers.

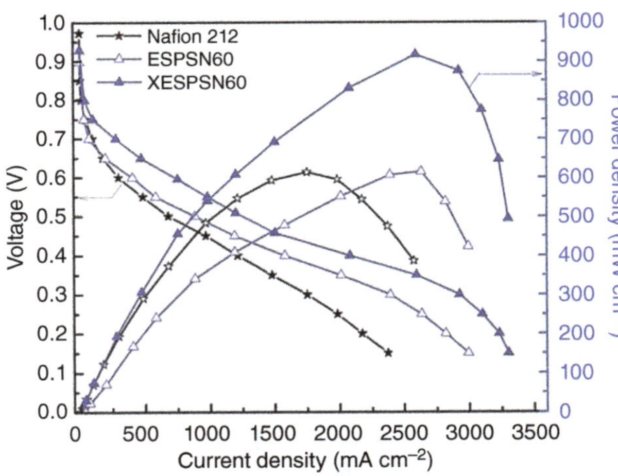

Figure 4.24 Fuel cell performance of pristine and end group cross-linked membranes at 80 °C-50% RH. Source: Adapted from Lee et al. [62].

Modification methods, such as composite, blend, and cross-linking, also influence polymer morphology as well as the chemical architecture. Inorganic fillers can act as a seed for ionic clusters, leading to a well-defined morphology. In addition, phase contrast is intensified by blending various homo- or copolymers with different characteristics. Application of pressing, E-fields, and shear fields to the membrane changed the morphology (connectivity and alignment) of ionic domains compared with the as-cast membrane that influenced the ionic conductivity. Surface plasma treatment on sulfonated hydrocarbon polymer membranes produced nano-cracks while maintaining the bulk morphology, which acted as nano-valves to retard water desorption and to maintain ion conductivity in the membrane on dehumidification. These membranes with surface nano-crack coating, operated at intermediate temperatures, showed improved electrochemical performance (Figure 4.25) [63]. Polymer chain rearrangement during cross-linking causes morphological transformations. For this reason, cross-linked membranes showed well-defined morphology, whereas non-cross-linked membranes exhibited homogeneous phase images.

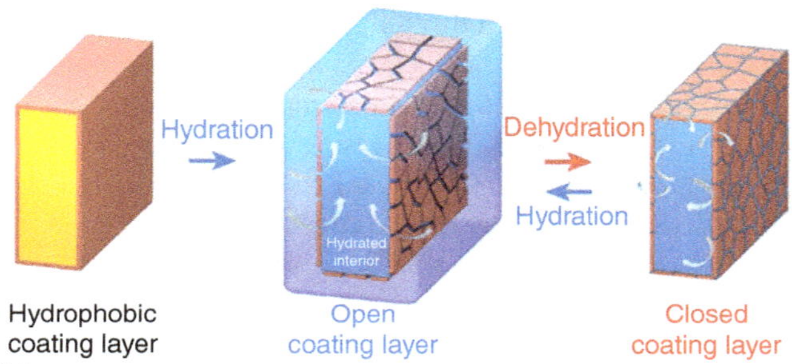

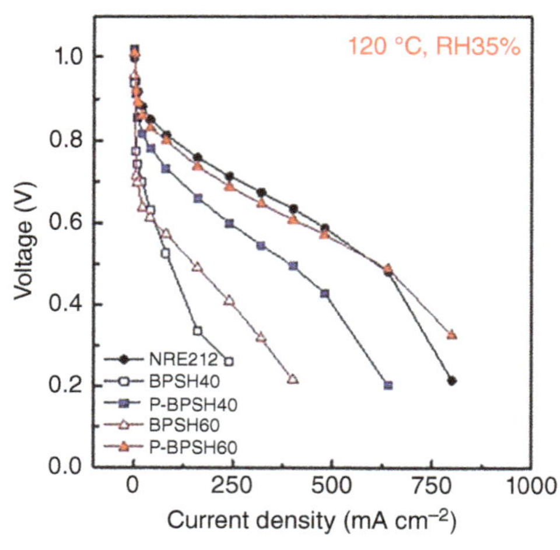

Figure 4.25 (a) Schematic drawing of nano-cracks on polymer membrane and (b) PEMFC performances of pristine membranes (BPSH40 and BPSH60), plasma-treated membranes (P-BPSH40 and P-BPSH60). Source: Adapted from Park et al. [63].

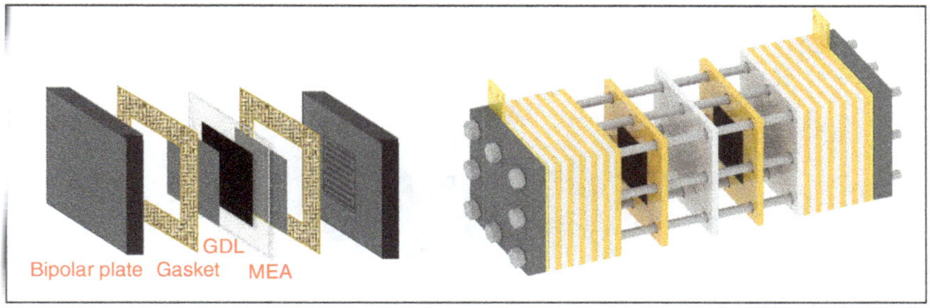

Figure 4.26 Schematic of single-cell composed of membrane, catalyst layer, and gas diffusion layer. Single cell is called membrane electrode assembly (MEA).

4.3.7 Fuel Cell Stacks-MEA

PEMFCs generate high energy density and operate with simplicity. Therefore, PEMFCs are widely used in various applications, including power units of fuel cell vehicles, military vehicles, mobile devices, and spaceships. The PEMFC has an important component called the MEA that consists of an anode and cathode located on both sides of a PEM (Figure 4.26). A catalyst (usually platinum) is uniformly coated on the polyelectrolyte membrane surface, and a GDL is present over the CL. A separator or PEM is present over the GDL. Hydrogen is fed into the system and reacts with the catalyst to produce protons (H^+) and electrons (e^-).

4.4 Membranes in Direct Methanol Fuel Cell

Direct methanol fuel cells (DMFCs) have developed interests regarding portable electronic devices and offer 10 times higher power densities than lithium-ion (Li-ion) rechargeable batteries [64]. Although there are several different types of fuel cells, the DMFC offers the most promising alternative for portable power applications because it is a low-temperature device, environmentally benign and portable, and inexpensive methanol fuel. Therefore, considerable efforts are focused on miniaturizing and improving the efficiency of the DMFC. In general, a fuel cell works by converting chemical fuel into electrical energy. In the DMFC, energy is converted when liquid methanol is directly and catalytically oxidized at the anode, producing protons and electrons. Protons diffuse through a PEM to the cathode, where they combine with electrons to produce electrical energy, with water and carbon dioxide as byproducts (Figure 4.27). The anode, cathode, and overall cell reactions, respectively, are $2CH_3OH + 2H_2O \rightarrow 412H^+ + 12e^- + 2CO_2$, $3O_2 + 12H^+ + 12e^- \rightarrow 46H_2O$, and $CH_3OH + 3/2O_2 \rightarrow 4CO_2 + 2H_2O$.

Although requirements may vary for different applications, common requirements for a PEM in DMFC include [65]: (i) operation at higher temperatures, (ii) low methanol crossover (MCO) ($<10^{-6}$ mol min^{-1} cm^{-1}) or low methanol diffusion coefficient in the membrane ($<5.6 \times 10^{-6}$ cm^2 s^{-1} at $T = 25\,°C$) [66], (iii) high proton conductivity (>80 mS cm^{-1}), (iv) high chemical and mechanical

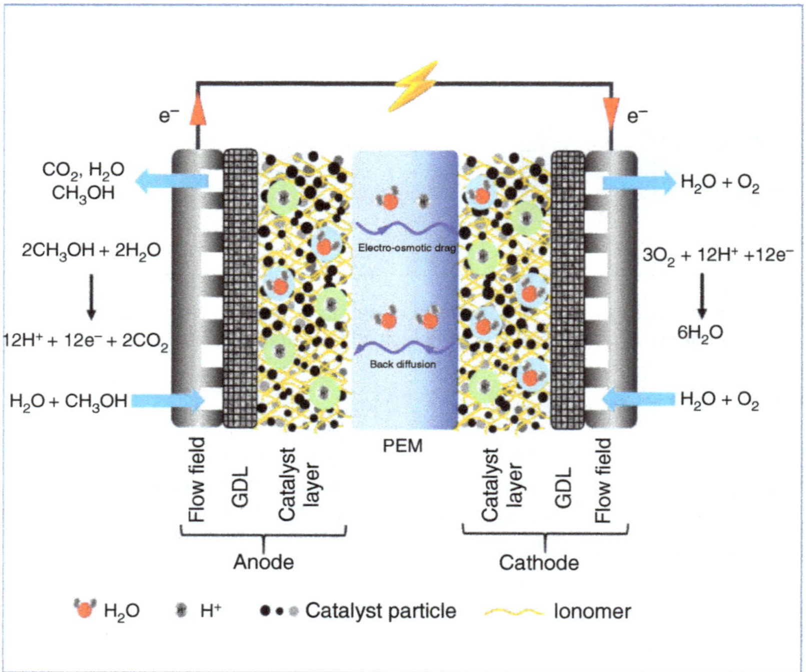

Figure 4.27 Schematic of a direct methanol fuel cell.

durability, especially at $T > 80\,°C$ (for increased CO tolerance), (v) low ruthenium crossover, and (vi) low cost (<\$10 kW^{-1} based on a PEMFC).

Although current DMFCs report power densities double that of current LIBs with an overall efficiency of ~20–25%, there are critical factors that hinder the ability of the DMFCs to reach their maximum efficiency (~100%); these include slow reaction kinetics at the anode and MCO [67]. In addition, MCO (high methanol permeability in the PEM) contributes to decrease in overall cell efficiency and lifetime. The reaction of methanol at the cathode causes a loss of fuel and cathode voltage referred to as a mixed potential. Typically, low methanol concentrations (~1–2 M; 4–8 vol %) are used in the DMFC to combat with this issue. If MCO in PEMs was not a problem, a higher methanol concentration could be used, resulting in a significantly higher cell voltage. Cell operating temperature, anode and cathode flow rates can all affect DMFC performance. However, methanol concentration, in particular, has a significant impact, where a 66% reduction in voltage at an optimal operating current density and a decrease in OCV are observed when the methanol concentration in the anode feed increases from 2 to 6 M [67].

Nafion has been typically used in DMFCs. Nafion shows high proton conductivity at optimal water contents. However, Nafion suffers from high MCO. Therefore, there are numerous studies that report on the development and evaluation of new PEMs (Nafion replacements) that are methanol resistant.

The proton conductivity and methanol permeability showed a linear relationship with several sulfonic acid-containing PEMs (Nafion and sulfonated polysulfone [PSF]) [68]. All the membranes have a similar selectivity regardless of ion content, water content, polymer chemistry, architecture, or morphology. This indicates that protons and methanol have similar molecular transport mechanisms in sulfonic acid-containing PEMs, which makes it difficult to improve selectivity for the DMFC applications.

Although Nafion 112 is the most commonly used membrane for PEMFC, Nafion 117 is preferred for DMFC applications despite higher ionic resistance due to the significant reduction in MCO [65]. It has been shown that the concentration of crossed methanol decreases with increasing thickness and equivalent [65]. A number of different approaches have been taken to mitigate the effects of MCO in DMFCs, including changes in operational parameters, the use of different cathode catalysts and cell designs, and the development of polymer membranes less permeable to methanol. Wang et al. [69] have used PBI membranes doped with phosphoric acid. Their MCO rates are about an order of magnitude less than those for Nafion membranes. Other membranes with low methanol permeability relative to Nafion include electrochemically polymerized sulfonated poly(oxyphenylene) [70] and various polymer blends containing sulfonated PSF (Udel) or sulfonated poly(ether ether ketone) [71]. Nafion can be impregnated with polymers by *in situ* polymerization of 1-methylpyrrole that decreased methanol permeability by more than 90% [72].

Scott et al. [46] reported on the DMFC performance using radiation-grafted PE and ETFE membranes. They concluded that these membranes exhibited low methanol diffusion coefficients, and MEAs from these membranes were as good as Nafion in DMFCs for short durations (<100 hours) tested under identical operating conditions. However, the stability of contact between the membrane and the CL requires improvement.

Aricò [73] grafted a 100 μm ETFE film (NEM 20 by Solvay) with an electron beam at a dose of 80 kGy in air. The grafting of the membranes was performed by maintaining the irradiated films for six hours at 75 °C in N_2-purged styrene, in which DVB was added as a cross-linker. The grafted membranes were then sulfonated with chlorosulfonic acid to obtain ion-conducting membranes. These membranes showed a power density of about 160 mW cm^{-2} at 130 °C of operation using 1 M methanol as a feed. The MCO was less than 10 mol min^{-1} cm^{-2} at 300 mA cm^{-2} test conditions, indicating comparable performance with Nafion 117. MCO of sulfonated ETFE membranes was less than 10% of Nafion 115. The PPD achieved with the ETFE-based membranes was about 40–65% lower than the corresponding values of the Nafion 115 membrane because of the lower conductivity and noticeably higher IR losses.

ETFE-g-polystyrene and subsequent sulfonation membranes were cross-linked with DVB and bis(*p,p*-vinyl phenyl)ethane (BVPE) by radiation cross-linking [74] for DMFC purposes. The membrane cross-linked by 2.5% DVB + 17.5% BVPE cross-linkers and post-cross-linked by γ-ray irradiation of the corresponding grafted film possessed the highest chemical stability among the prepared membranes, a significantly lower methanol permeability (0.63×10^{-6} cm^3 s^{-1}) compared with the Nafion membrane and better DMFC performance for high methanol feed concentration.

Saarinen et al. [75] later reported the sulfonated ETFE films (30 μm) for DMFC testing. They first irradiated ETFE film with a proton beam with a total dose of 500 kGy and a dose rate of 2.7 kGy s^{-1} to activate the film for functionalization. During irradiation, the films were kept in high vacuum (10^{-6} to 10^{-5} Pa) and at temperatures below −70 °C. After irradiation, the films were stored in liquid N_2 and sulfonated later. Later, they characterized the membranes for DMFC [76] and found that methanol permeability of ETFE–SA was less than 2% of that of Nafion membranes.

Recently, Li et al. [77] reported on the high-temperature fuel cell based on ETFE grafted with neutral acrylic monomers such as 2-hydroxyethyl methacrylate (HEMA) and 2-hydroxyethyl acrylate (HEA) with a maximum degree of grafting of 193%. Phosphoric acid-doped graft ETFE membrane showed a power density of 108 mW cm^{-2} and a current density of 200 mA cm^{-2} at 120 °C in H_2/O_2 fuel cell testing.

An initial study on the operation of DMFC with a commercial AEM (Morgane® ADP membrane, Solvay, Belgium) has been conducted by Yu and Scott; maximum power densities of around 11 mW cm^{-2} and limiting current densities of around 70 mA cm^{-2} were obtained when Pt/C electrodes were used at the anode and cathode of alkaline membrane DMFCs with air at the cathode [78]. The best performances were obtained when the GDL was omitted from the anode and with nontefluonized carbon paper; humidification of the cathode air stream further improved performances (expected as water is essential at the cathode in order to form the OH^- conducting species). Morgane ADP membrane exhibited a resistance across the membrane that was 600% of the value observed with Nafion, a factor responsible for the limited fuel cell performances observed. However, the Morgane ADP membrane discolored when immersed in aqueous NaOH (1 mol dm^{-3}), a symptom of the presence of some chemical instability under such conditions [78].

4.5 Membranes in Anion Exchange Membrane Fuel Cell

Although there has been a significant advancement in the development of AEMs, AEMFCs have displayed lower performances than the more well-established PEMFCs. One of the reasons for this is associated with the lower hydroxide (OH^- conductivity of many AEMs compared with H^+ conduction in PEMs; this often leads to employing AEMs with higher IECs, leading to the risk of additional mechanical instabilities [79]. AEMs possess lower ion conductivity at similar IECs due to the large volume of anions (such as OH^-) compared with protons in PEMs such as Nafion. This poor AEMFC performance is also due to the unavailability of suitable anion exchange ionomer (AEI) candidates, which are necessary to provide OH^- conductivity in the CLs of the electrodes. Hydrogen is supplied into and reacts with a catalyst to produce water and electrons (e^-) in the anode ($2H_2 + 4OH^- \rightarrow 44H_2O + 4e^-$). Oxygen, water, and electrons are consumed to

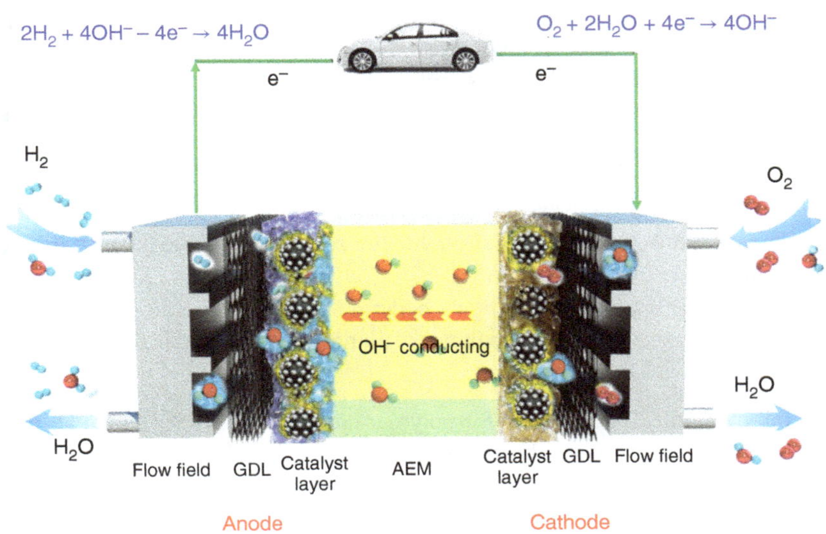

Figure 4.28 Schematic of anion exchange membrane fuel cell. Source: Chen and Lee [80]/with permission of Elsevier.

produce hydroxide ion (OH^-) in the cathode ($O_2 + 2H_2O + 4e^- \rightarrow 44OH^-$). The overall reaction therefore is $2H_2 + O_2 \rightarrow 42H_2O$. Note that electroosmotic drag in AEMFC is opposite to the PEMFC case (see Figure 4.28) [80]. Four water molecules are produced in the anode while, only two water molecules are consumed in the cathode, resulting in cathode dry-out. Therefore, water diffusion from anode to cathode or water back diffusion is very important to obtain the optimum electrochemical performance.

The chemical, thermal, and mechanical stabilities of AEMs strongly depend on the nature of the anion-conducting functional groups and also the polymer backbone chain. Other than temperature, the main reason for the decomposition of anion-conducting groups and polymer chains is the alkalinity of the medium (high pH), a problem that is especially serious in materials having lower hydration numbers (λ). The hydration number, λ, represents the number of water molecules surrounding each OH-ion. Low λ values represent the high alkalinity of corresponding alkaline conditions. AEMs are made of a variety of backbones, including non-fluorinated, partially fluorinated, or fully fluorinated polymers, such as polyethylene, PSFs, poly(phenylene oxide)s (PPO), PPs, PBI, and modified poly(ethylene-co-tetraflyoroethylene) (ETFE). The covalently bonded side-chain-type QA cationic groups are responsible for conducting the hydroxide anions (OH^-).

The potential advantages over PEMFCs and DMFCs when using AEMs are summarized below:

(1) Lower fuel (hydrogen and methanol) crossover because the OH^- ions conduct from the cathode to the anode (opposite direction to H^+ transport in PEMFC);

this, and electro-osmotic drag, oppose the direction of hydrogen flux through the membrane. Reduced fuel crossover allows the use of thinner membranes, improving fuel cell performances significantly.

(2) Electrocatalysis of each hydrogen oxidation reaction (HOR), and ORRs in alkaline media (high pH region) is more facile, permitting the use of cheaper non-noble metal catalysts, e.g. nickel and silver [81, 82].

(3) This wider range of catalysts should allow the choice of a cathode catalyst that is inactive toward hydrogen oxidation, further reducing performance losses due to crossover.

(4) Lowering the flooding issue at the cathode in DMFCs and better overall water management cause a simplified balance of plants (essential with passive DMFCs for portable devices).

The oxidative radical mechanism for polymer degradation is suppressed at high pH [48], and therefore, it will be possible to use hydrocarbon-only membranes (C—H backbone) with enhanced green credentials (manufacturing and disposal) when compared with inherently expensive fluorinated polymers.

4.5.1 Ammonium Groups and Their Stability

Figure 4.29 depicts the degradation mechanism of QA groups, where the Hofmann degradation is mainly triggered by the elimination of β-hydrogen via an anti-coplanar intermediate. Thus, the degradation rate is mainly determined by the reactivity of dehydrogenation and the rate of formation of the intermediate. On the other hand, S_N2 degradation is conducted by a direct nucleophile attack from behind. Therefore, the reaction kinetics are mainly determined by the steric hindrance of QA groups. Hence, the synthesis of specific QA without a β-hydrogen is a typical way to avoid the Hofmann degradation and improve the alkaline stability.

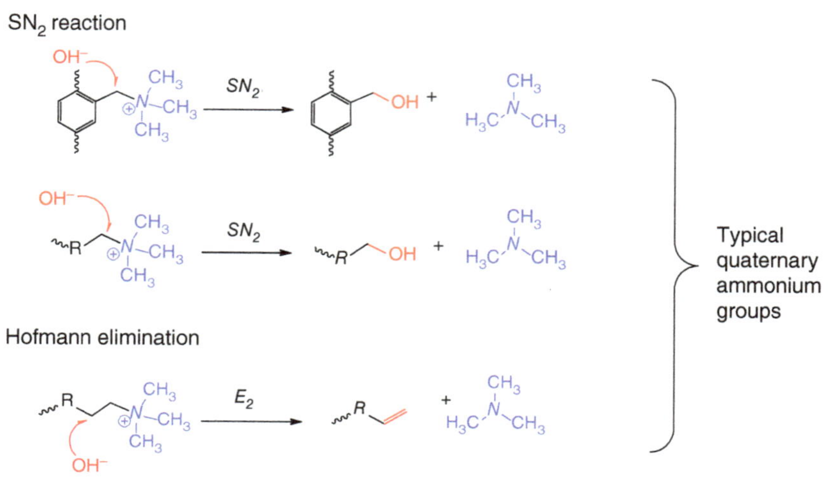

Figure 4.29 Degradation mechanism of typical quaternary ammonium groups.

Tetramethyl ammonium (TMA) group was the most stable QA group early on due to the absence of the β-hydrogens. Because of the difficulties in adding TMA directly into the polymer backbone, benzyltrimethylammonium (BTMA)-based anion exchange polymers (AEPs) have been extensively developed. For example, Danks et al. [83] prepared a series of BTMA-functionalized PVDF and FEP membranes for DMFCs via a radiation-grafting method, where these AEMs exhibited ion conductivities of ~20 mS cm^{-1} at room temperature. BTMA groups have been regarded as more stable than other alkyl benzyl ammonium (BA), such as benzyl triethylammonium (BTEA), under alkaline conditions since BTMA does not contain any β-hydrogens. However, because of the withdrawing effect of phenyl group on benzyl hydrogen (α-H), BTMA-based AEPs exhibit accelerated S_N2 degradation at the benzyl position, especially at elevated temperatures.

Durability and ion conductivity of AEMs are governed by their cationic groups in the side chain. Consequently, numerous cationic groups have been employed in AEMs to enhance the alkaline stability and ion conductivity over the past decades, such as ammonium, IM, phosphonium, sulfonium, and organic-metal cations, and other recent new groups such as carbazolium and triazatriangulenium (see Figure 4.30). However, a majority of cationic groups exhibit insufficient alkaline stability under alkaline conditions and undergo different degradation pathways like the previously mentioned nucleophilic substitution (S_N2), Hofmann degradation (E_2), ring-opening reaction, and Ylide reaction. Studies have revealed that QA and IM groups are the most efficient cationic groups in current AEMs, displaying high alkaline stability and ion conductivity.

Figure 4.31a depicts the $t_{1/2}$ of different QA groups under $\lambda = 4.8$ conditions in $NaOD/D_2O/CD_3OD$ at 80 °C. The $t_{1/2}$ is the half-life time of the cationic group, which is the specific time when a cationic group or an AEM raises a 50% degradation in its chemical structure under a specific alkaline condition [85]. ^1H NMR detects the change of chemical structures.

Three strategies have been suggested to increase the alkaline stability of cationic groups. (i) *Discovering new types of cationic groups*. To avoid the S_N2 and E_2 degradations of QA groups, several research groups [86–88] have proposed using organometallic groups for alkaline electrolyte polymers as new types of cationic groups. However, most organometallic groups easily take off ligands in an alkaline solution, and only a few examples display interesting alkaline stability, such as cobaltocenium [88] and crown ether cations [89, 90] (Figure 4.30).

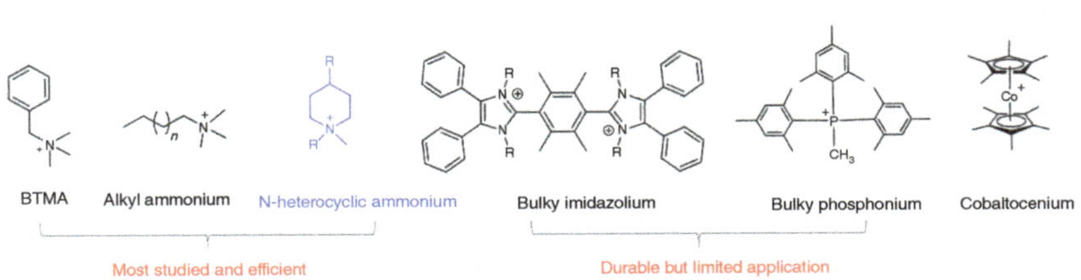

Figure 4.30 Most studied and efficient cationic groups.

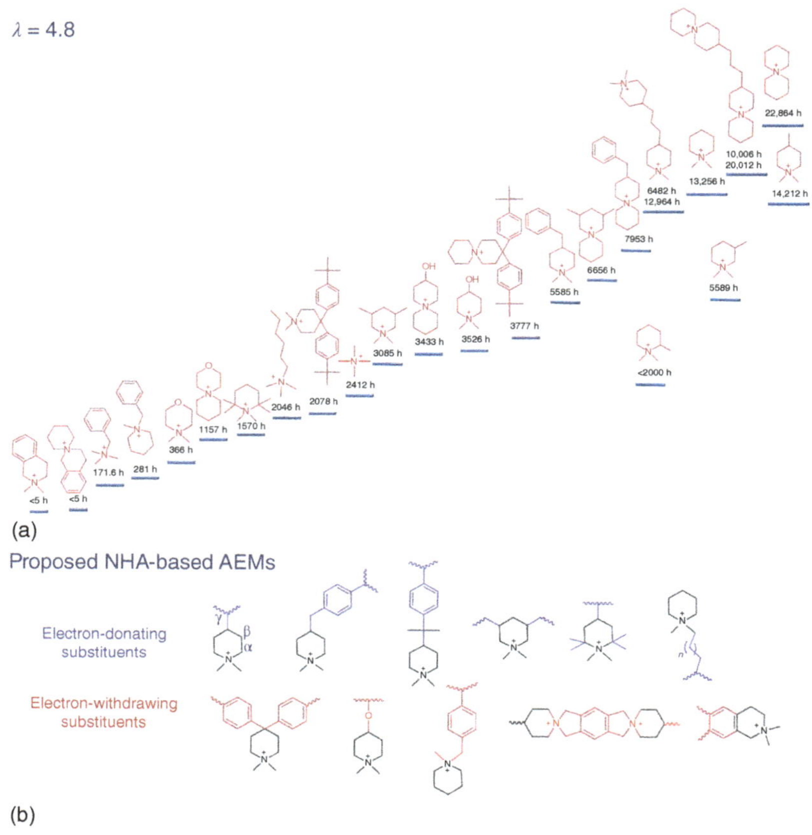

Figure 4.31 (a) The $t_{1/2}$ of different QA groups under $\lambda = 4.8$ conditions in NaOD/D_2O/CD_3OD at 80 °C, (b) the possible chemical structure of NHA-based polymers for AEMs. ^1H NMR spectra monitored total degradation ratios of NHAs. The degradation ratio of Bis-DMP and Bis-ASU with diammonium groups is twice that of the other NHA groups containing a single ammonium group. Therefore, the actual half-life time of Bis-DMP and Bis-ASU can be predicted by multiplying the first half-life by 2 [84].

Most organometallic-based AEPs exhibit poor AEMFC performance. (ii) *Enhancing the steric hindrance to address the degradations.* Many bulky IM groups display outstanding alkaline stability under different hydration number (λ, number of water molecules per OH^-) conditions. Holdcroft et al. [91, 92] and Coates and coworkers [93–95] have developed a series of bulky aryl IM groups that can remain stable in 5 M KOH ($\lambda = 11.1$) at 80 °C over 720 hours. Unfortunately, excessively high hindrance impedes the ion conductivity of bulky IM-based AEMs and makes the synthesis difficult. (iii) *Introducing ring hindrance and geometry to cationic groups.* Recently, *N*-heterocyclic ammonium (NHA) groups (Figure 4.31a) have seen great success in AEMs, which significantly improved the alkaline stability of AEMs due to special ring hindrance and geometric conformations [80, 96–100]. In particular,

Marino and Kreuer [101] demonstrated that 6-azonia-spiro[5.5]undecane (ASU) and dimethyl piperidinium (DMP) ammonium groups possess outstanding alkaline stability among QA groups. The half-life time of ASU and DMP groups even exceeded the well-acknowledged TMA in 6 M KOH at 160 °C resulting from cyclohexane conformations.

Chen et al. [84] explored 24 representative QA groups in $NaOD/D_2O/CD_3OD$ udder different λ conditions (i.e. 4.5, 7.0, and 10.0), and presented several durable and promising NHA groups for the future design of AEMs for energy devices. The electronic effect of substituents on NHA groups and degradation mechanisms of NHA groups are systematically investigated by ^{1}H nuclear magnetic resonance (^{1}H NMR) and density functional theory (DFT) calculations using poly(aryl-co-aryl piperidinium) (c-PAP) AEMs and ionomers as examples to explain the *ex situ* and *in situ* durability under different λ values. Figure 4.31a shows 24 model compounds, and the number under the chemical structure is the half-life time $t_{1/2}$ measured at $\lambda = 4.8$ conditions. The original ASU and DMP groups show outstanding alkaline stability due to their highly symmetric structure and special ring strain. The $t_{1/2}$ is greater than 20,000 hours compared with 13,256 hours for DMP, which is comparable to bulky IMs and is higher than bulky QA compounds [102–104].

NHAs with electron-withdrawing groups, such as phenyls or heteroatoms, in the α- or β-position show much lower half-life time than those of the pristine DMP or ASU. The electron-withdrawing groups activate α-carbon and β-hydrogens in the NHA ring and, therefore, accelerate S_N2 and Hofmann elimination reactions, respectively, which decreases the alkaline stability of NHAs. DMA or AUS groups with electron-withdrawing groups in the γ-position present significantly improved $t_{1/2}$ compared to the aforementioned NHAs with electron-withdrawing substituents in the α- or β-position due to the weakening electron-withdrawing effect. The DMP or ASU series displays much higher $t_{1/2}$ than that of BTMA, which is consistent with previous findings [101].

However, $t_{1/2}$ of NHA groups containing electron-donating substituents (such as $-CH_2-$, $-CH_3-$, benzyl, or aliphatic chain) surpasses that of the TMA benchmark, which is much higher than that of NHAs with electron-withdrawing substituents. The electron-donating groups contribute to an increase in steric hindrance and electron density of *N*-heterocyclic ring, thus reducing the possibility of S_N2 and Hofmann degradation reactions. Moreover, NHAs with electron-donating substituents in the γ-position exhibited much higher alkaline stability than NHAs with electron-donating substituents in the α- or β-position due to their higher geometric symmetry. In fact, Pham et al. [105] presented benzyl-DMP and benzyl-AUS-based PP AEMs with electron-donating substituents in the γ-position. They found that the alkaline stability of these AEMs significantly increased compared with PP-based AEMs with electron-withdrawing substituents (phenyl groups) in the γ-position. NHAs with a symmetric structure (such as ASU, DMP, Bis-ASU, and Bis-DMP) seem to possess higher alkaline stability than that of asymmetric NHAs, such as benzyl-DMP and benzyl-ASU. In summary, linking NHA substituents with the polymer backbone *via* the γ-position could be a rational design strategy for highly stable AEMs, particularly using electron-donating substituents.

Regarding the degradation mechanism of NHA groups and NHA-based AEMs, two interesting conclusions have been found from their study [84]. (i) Based on ^1H NMR study and analysis, the S_N2 reaction instead of E_2 reaction dominates the degradation pathway of most NHA substituents in $NaOD/D_2O/CD_3OD$ under different λ conditions. (ii) Under lower λ conditions (4.8 or 7), ASU and substituted ASU groups showed higher alkaline stability than DMP or substituted DMPs. However, under higher λ conditions ($\lambda = 10$), DMP displays comparable alkaline stability to ASU or substituted ASUs.

Figure 4.31b shows the possible chemical structure of NHA-based polymers for AEMs. ^1H NMR spectra monitored total degradation ratios of NHAs. The degradation ratio of Bis-DMP and Bis-ASU with diammonium groups is twice that of the other NHA groups containing a single ammonium group. Therefore, the actual half-life time of Bis-DMP and Bis-ASU can be predicted by multiplying the first half-life by 2.

Note that the cationic groups should be grafted onto the polymer backbone in AEMs. Hence, research on the alkaline stability of original NHA groups without substituents may not clearly indicate the durability of corresponding NHA-based AEMs. From the studies of the alkaline stability of 24 representative NHA groups with different substituents under different λ conditions and AEM studies of poly(aryl piperidinium) copolymers, Chen et al. found that (i) NHAs with electron-donating substituents exhibited outstanding alkaline stability (half-life time >100,000 hours under $\lambda = 10$ conditions), whereas electron-withdrawing substituents had a detrimental effect on the stability of NHAs. (ii) γ-Substituted NHAs possess much higher alkaline stability than α- or β-substituted NHAs. (iii) Moreover, the alkaline stability of NHAs or NHA-based AEMs is not always the same but is determined by the specific degradation pathways of NHAs. That is, the predominant degradation pathway of NHAs and NHA-based AEMs could be different under *ex situ* or *in situ* conditions, resulting in different alkaline stability behaviors.

In conclusion, NHA groups with electron-donating substituents possess outstanding and promising alkaline stability among current cationic groups. Linking NHAs at the γ position to the polymer backbone could be a rational design for NHA-based AEMs [55]. On the other hand, previous reports [55–57] indicated that the alkaline stability of cationic groups could decrease orders of magnitude under low hydration levels (such as $\lambda = 0$–4). Therefore, identifying the real alkalinity of AEMFCs during operation is crucial for future AEMFC research.

Dang et al. [106, 107] prepared a series of PPO membranes tethered with flexible cationic alkyl side chains. They found that placing QA groups on flexible spacer units tethered to the PPO backbone facilitated alkaline stability and ion conductivity. Kim et al. also systematically investigated the alkaline stability of BTMA-functionalized polyaromatic membranes by computational modeling and experiments. The computational modeling study suggests that the aryl-ether backbone is easily cleaved under alkaline conditions due to the electron-withdrawing effect of BTMA groups. Their experimental methods showed that aryl-ether groups degraded after only two hours of treatment in 0.5 M NaOH at 80 °C, while BTMA-type PP membrane was much stable in some cases. Based on these results, BTMA groups in aryl-ether backbones have been regarded as unstable for AEPs.

Hibbs [108] presented a series of PP AEMs with five different cationic head groups and evaluated their alkaline stability in 4 M KOH at 90 °C. He found that BTMA-type PP exhibited over 30% loss in ion conductivity, while side-chain-type PP with a six-carbon spacer showed only 5% ion conductivity loss. After that, many side-chain-type AEPs have been developed. Park et al. also investigated the alkaline stability of side-chain-type PP membranes under various stability testing conditions. They found that side-chain-type PP membranes exhibited a slight degradation in 0.5 M NaOH at 80 °C after 11,000 hours.

In 2015, Kreuer and coworker [101] systematically investigated the alkaline stability of 26 different QA groups in 6 M NaOH at 160 °C. The results demonstrated that ASU and DMP exhibited higher alkaline stabilities among these QA groups. The half-lives of ASU (110 hours) and DMP (87.3 hours) exceeded the TMA (61.9 hours) benchmark. Consequently, the appearance of ASU and DMP groups opened up newly recognized QA research since there are four β-hydrogens in DMP and eight in ASU (Figure 4.31a). They pointed out that the outstanding alkaline stability of ASU and DMP groups was attributed to the geometric conformation of the ring that constrained the transition state of S_N2 or E_2 reactions.

4.5.2 Stable Polymer Backbones

4.5.2.1 Aryl-Ether Polymers

Early studies on AEMs focused on the poly(aryl ether)s, such as PPO (Figure 4.32a), poly(ether ether ketone) (PEEK) (Figure 4.32b), PSF (Figure 4.32c) due to their

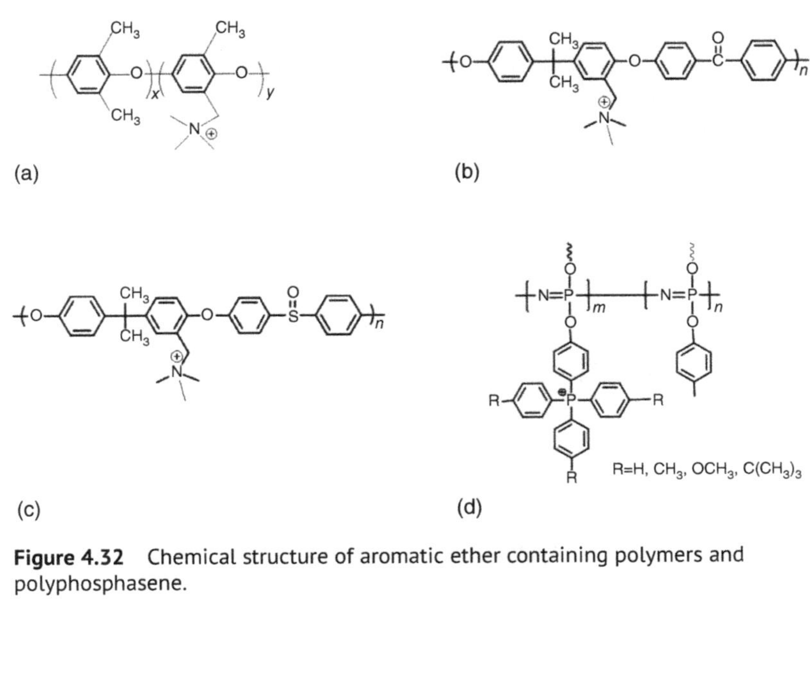

Figure 4.32 Chemical structure of aromatic ether containing polymers and polyphosphasene.

ease of synthesis, relatively low cost, and good mechanical properties. Most of these polymers are commercially available, which is beneficial for realizing the near-term commercialization of AEMs. Generally, the popular way to prepare these poly(aromatic ether)-based AEMs is either the chloromethylation of phenyl groups or the radical halogenation of benzyl groups in the aryl-ether polymer backbone followed by quaternarization. However, aryl-ether cleavage reactions in aryl-ether polymers were first identified under alkaline conditions and were a serious concern for use as AEM. In 2016, Mohanty et al. [109] investigated the chemical stability of different polymers. They found that withdrawing groups, such as carbonyl groups and sulfuryl groups, in the polymer backbones could decrease the chemical stability of the polymer backbone, and aryl-ether-free polymers, such as styrene–ethylene–butylene–styrene polymer (SEBS) and poly(biphenyl alkylene)s exhibited much higher chemical stability than that of the aryl-ether polymers. This work highlighted the importance of the chemical stability of the polymer backbone.

Han et al. [110] prepared a series of tetraphenylphosphonium-functionalized polyphosphazene (PPMPP) membranes (Figure 4.32d) that exhibited OH^- conductivities of 22.3 mS cm^{-1} at 80 °C and 9.8% degradation after alkaline treatment in 1 M KOH at 80 °C for 240 hours. Unfortunately, phosphonium groups have not been widely utilized in AEMs due to their complicated and low-efficiency utilization process.

4.5.2.2 Polybenzimidazole and SEBS

PBI contains IM group in the polymer backbone. IM-based AEMs are still popular in current research. Henkensmeier et al. [111] reported a *meta*-PBI-based AEMs, as shown in Figure 4.33a. This *meta*-PBI membrane exhibited an OH^- conductivity of 58 mS cm^{-1} at 60 °C, while the membrane quickly degraded under alkaline conditions. The expected degradation mechanisms of the PBI membrane were assumed to be S_N2 and hydrolysis, triggered by hydroxide attack at the methyl groups and the C2 position, respectively.

SEBS-type AEPs received much attention because of their aryl-ether-free backbone. Gao et al. [112] prepared BTMA-type SEBS (QASEBS, Figure 4.33b) that exhibited an OH^- conductivity of 100 mS cm^{-1} at 80 °C and was stable in 1 M KOH at 60 °C for 1700 hours. Mohanty et al. [113] also prepared a series of BA-type SEBS

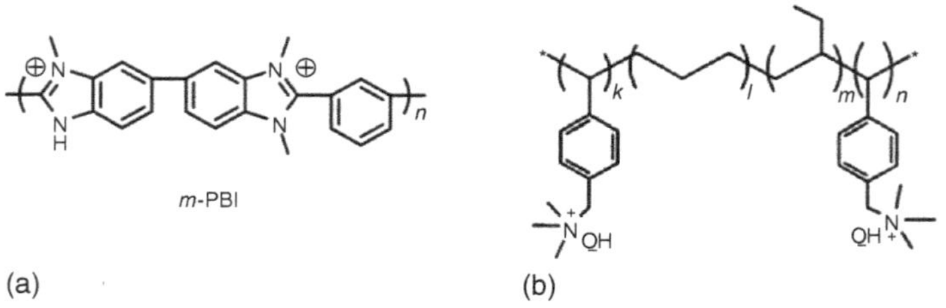

Figure 4.33 Chemical structure of polybenzimidazole and poly(styrene-ehtylene-butyrene-styrene) anion exchange membranes.

membranes. They indicated that BTMA-type SEBS membranes exhibited higher ion conductivity (over 100 $mS\,cm^{-1}$ at 60 °C) and water uptake than other large-volume BA-type SEBS. Jeon et al. [114] developed a series of side-chain-type and cross-linked SEBS (SEBS-Cn-TMA-x) membranes that displayed acceptable OH^- conductivities at 80 °C (50–92 $mS\,cm^{-1}$) and exhibited a PPD of 520 $mW\,cm^{-2}$ at 60 °C.

4.5.2.3 Poly(norbonene)

Clark et al. [115] proposed a ROMP method to fabricate aryl ether-free polymers by constructing alkyl–alkyl bonds in 2009. They reported a series of QA-based polynorbornene (PNB) membranes via a facile ROMP reaction (Figure 4.34a) and that were further crosslinked (Figure 4.34b) for stable AEMs. These membranes exhibited a high OH^- conductivity of 111 $mS\,cm^{-1}$ at 50 °C and acceptable mechanical properties. Recently, PNB-based AEMs have attracted wide attention due to their incredible power density, as shown in Figure 4.35a. Kohl and coworkers [117–119] have reported a series of side-chain-type PNB copolymers for AEMs by the ROMP method. The OH^- conductivity of these PNB membranes reached 198 $mS\,cm^{-2}$ at 80 °C. Subsequently, Huang et al. [120] prepared a series of composite PNB membranes with a PTFE reinforcement (GT64). The PPD of H_2–O_2 AEMFCs was dramatically improved to 3.37 $W\,cm^{-2}$ at 80 °C along with a 545 hours *in situ* durability under 0.6 $A\,cm^{-2}$ at 80 °C. They pointed out that high power density was associated with lower water uptake and high mechanical properties. Meanwhile, they reduced the thickness of PTFE-reinforced PNB membranes to 10 μm, and the PPD further

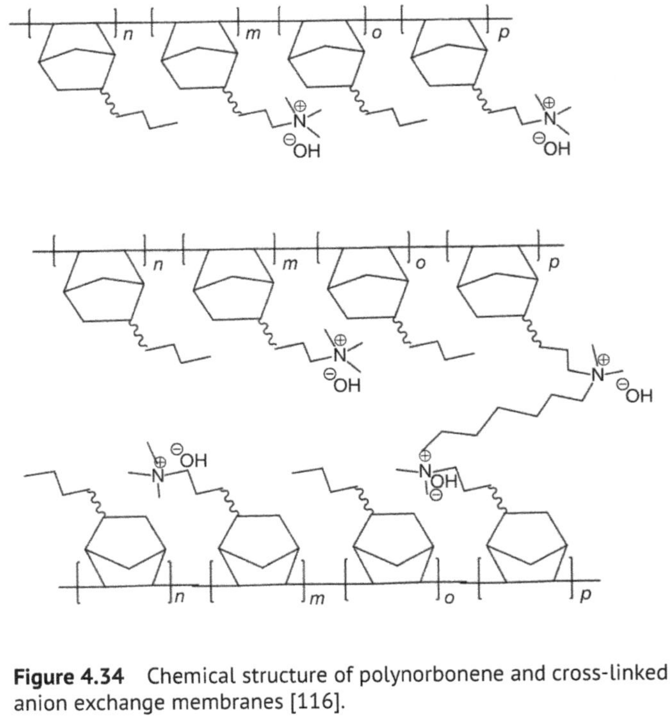

Figure 4.34 Chemical structure of polynorbonene and cross-linked polynorbonene for anion exchange membranes [116].

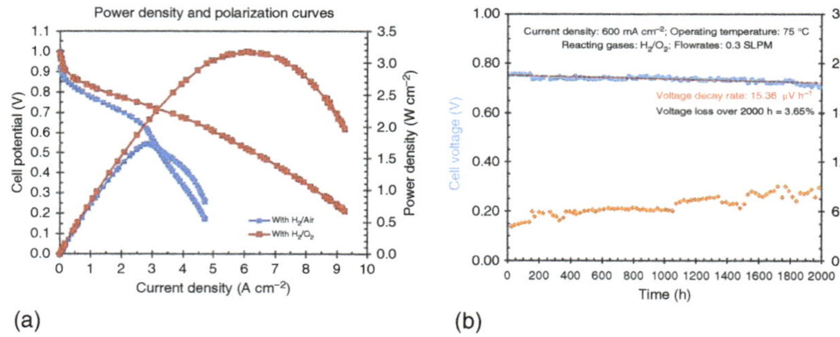

Figure 4.35 (a) Polarization and power density curves for AEMFCs with a Type 2 GT78 GDE in the anode and Type 2 GT32 GDE in the cathode operating on both H_2/O_2 and H_2/CO_2-free air feeds, at 80 °C, (b) cell voltage and HFR during a long-term durability test with the same GDE in (a) measured at 75 °C, operating at 600 mA cm^{-2}, and flow rate of 0.3 l min^{-1} on both electrodes. Source: Ul Hassan et al. [116]/John Wiley & Sons.

increased to 3.5 W cm^{-2} under 9.7 A cm^{-2} current density, along with a low ASR of 29 mΩ cm^2. Ul Hassan et al. achieved 2000-hours stability in AEMFC by manipulating ionomer and electrode optimization [116]. Three PNB tetrablock copolymer ionomers with a range of physical properties are synthesized and incorporated into AEMFC anode and cathode electrodes. Systematic electrode engineering with these ionomers allowed the PPD to be increased by 100% (1.6 W cm^{-2} → 43.2 W cm^{-2}) and the current density at 0.2 V to be increased by 59% (5.9 A cm^{-2} → 49.4 A cm^{-2}). Moreover, the best-performing electrode configuration was tested in an operating AEMFC at 0.6 A cm^{-2} for 2000 hours, showing a low voltage decay rate of 15.36 mV h^{-1}, which is only 3.65% over 2000 hours (see Figure 4.35b).

4.5.2.4 Diels-Alder Polymer – Polyphenylene

The Diels-Alder (D-A) polymerization has been employed to construct aryl–aryl polymers. Hibbs et al. [108, 121] presented a BTMA-poly(phenylene) membrane (BTMA-PP) for AEMs by D-A polymerization, as shown in Figure 4.36. They indicated that D-A polymerization is conducted by heating equimolar amounts of bis(cyclopentadienone)s and 1,4-diethynylbenzene with the loss of one carbon monoxide molecule for each D-A reaction accompanying with the formation of a highly stable aromatic ring that makes the polymerization irreversible. Subsequently, the aforementioned side-chain-type PP membranes have been widely

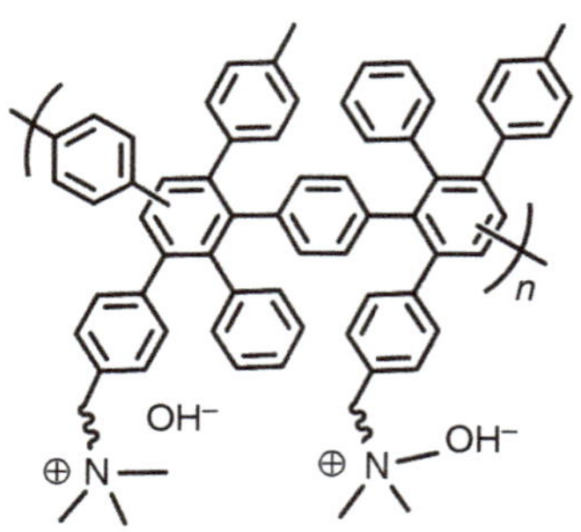

Figure 4.36 Chemical structure of polyphenylene prepared by ROMP polymerization.

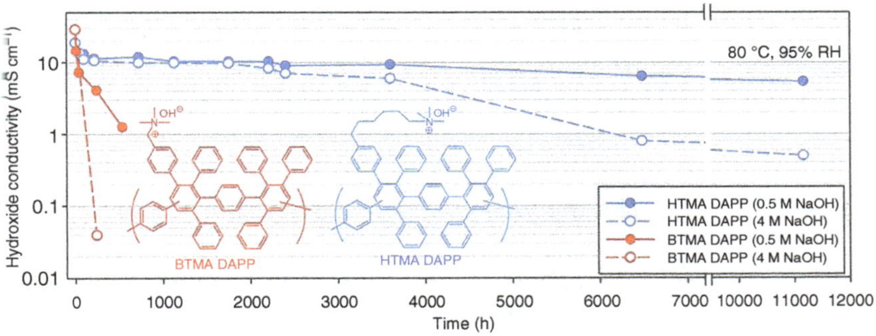

Figure 4.37 Hydroxide conductivity change of quaternized Diels-Alder polyphenylenes (DAPPs) during the alkaline stability test at 80 °C in 0.5 M NaOH solutions. DAPPs with benzyltrimethyl ammonium (BTMA) and hexyltrimethylammonium (HTMA) were tested. The hydroxide conductivity of the AEMs was measured at 80 °C, 95% relative humidity. Source: Park et al. [122]/John Wiley & Sons.

developed and exhibited excellent alkaline stability. Park et al. [122, 123] highlighted the advantages of D-A polymerization, including (i) metal-free conditions to avoid potential contamination of metals in the resulting materials, (ii) irreversible high molecular weight growth, and (iii) good solubility, and thus good processability of the resulting polymer for sequential quaternization. However, the synthesis of these D-A PPs is complicated and requires multiple reactions (six steps) to fabricate hydroxide-conducting polymers.

They investigated the chemical stability of D-A polyphenylenes (DAPPs) with different chain lengths in the trimethylammonium group, as illustrated in Figure 4.37 [122]. They found that the degradation pathways of quaternized DAPPs vary depending on the temperature, alkaline concentration, duration, and RH levels. They confirmed that HTMA DAPPs showed much higher alkaline stability compared with the BTM DAPPs. The degradation mechanism of HTMA DAPP is β-elimination reaction of cations, whereas BTMA DAPPs decomposed mainly by S_N2 reaction.

Polyfluorenes can be prepared by metal-catalyzed coupling reactions. Akiyama et al. polymerized a fluorene monomer containing hexamethyl ammonium side chains using Ni catalysts [124]. Yamaguchi and coworkers prepared a series of BTMA-spirofluorene ionomers for AEMFCs and water lectrolysis by Suzuki–Miyaura coupling reactions [125]. They suggested that typical polyfluorenes often suffer from low solubility and processability because of the strong p–p stacking interactions of the polymer backbone and that spirobifluorene polymers could improve the solubility and mechanical properties of AEMs. These membranes showed a hydroxide conductivity of 86.2 mS cm^{-1} at 70 °C, and were stable in 1 M NaOH solution at 80 °C for one week.

4.5.2.5 Poly(aryl piperidinium)s

Super acid-catalyzed polycondensation is another class of reactions to build aryl–alkyl polymers, and this reaction is primarily based on aromatic compounds and specific ketones, as shown in Figure 4.38a–c. Acid-catalyzed polycondensation

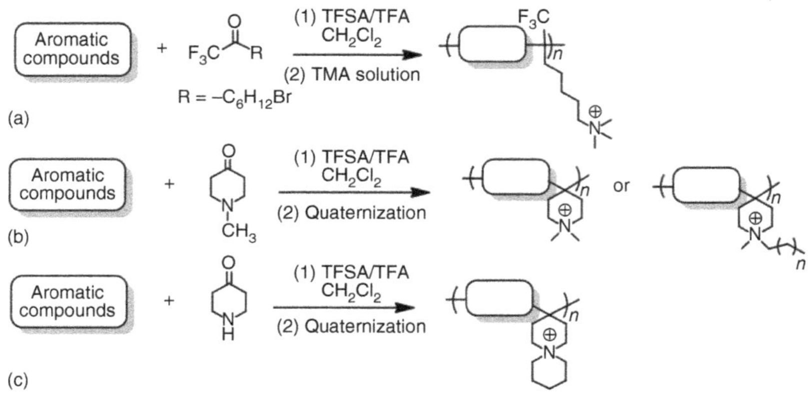

Figure 4.38 Preparation of acid-catalyzed polymer with aryl-ether-free piperidinium using (a) alkyl trifluoro ketone, (b) N-methyl piperidone, and (c) piperidone as comonomers.

provides several advantages, such as oxygen insensitivity, high molecular weight, metal-free reaction, and extensive species options. Typically, the aforementioned PAPs and some side-chain-type PPs were produced by this method.

Olsson et al. [100] reported a series of PAP-based AEMs via a super-acid polycondensation, effectively incorporating DMP in an aryl-ether-free polymer backbone, as shown in Figure 4.39a. Poly(terphenyl piperidinium) (PTP) membranes were stable in 2 M NaOH at 90 °C for 15 days and exhibited a OH^- conductivity of 89 mS cm^{-1} at 80 °C. Peng et al. also reported the same structure of PTP membrane (QAPPT) that possessed the higher OH^- conductivity of 137 mS cm^{-1} at 80 °C and was almost

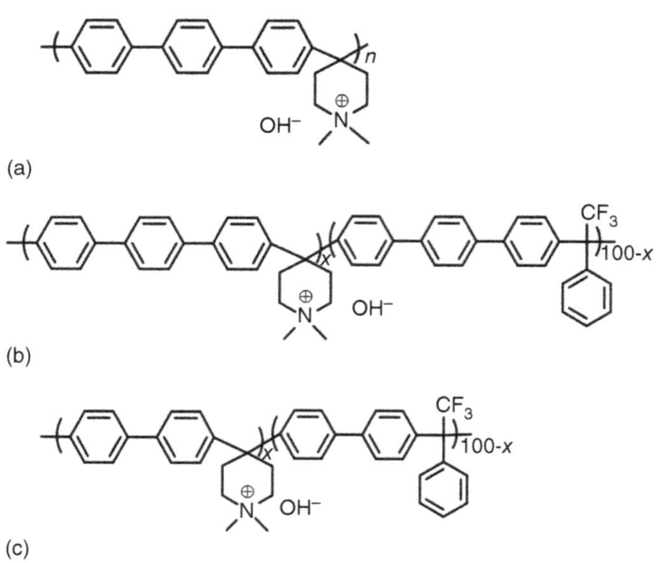

Figure 4.39 Examples of ether-free poly(aryl piperidinium) for anion exchange membranes and ionomers.

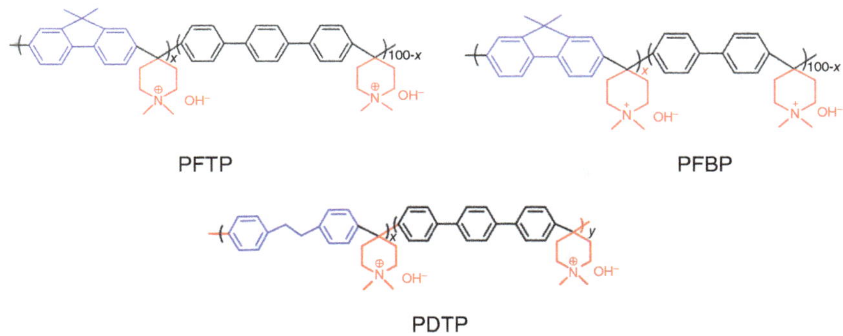

Figure 4.40 Chemical structure of poly(fluorenyl terphenyl piperidinium) (PFTP) for membranes and poly(fluorenyl biphenyl piperidinium) (PFBP) as ionomers. Poly(diphenylethanyl fluorenyl piperidinium (PDTP) can be used as membrane and ionomers.

stable in 1 M NaOH at 80 °C for 210 days, as shown in Figure 4.39b [126]. QAPPT membrane (thickness: 30 ± 5 μm)-based fuel cells reached a PPD of 1.5 W cm^{-2} at 80 °C under H_2–O_2 conditions. Wang and coworkers [97] presented a series of PAP copolymers (PAP-TP-x) which had the same chemical structure with Figure 4.39b by introducing a 2,2,2-trifluoroacetophenone block to adjust IEC values and using the ionomer having the chemical structure shown in Figure 4.39c. They pointed out that the inherent viscosity of PAP-TP-x polymers was about 12 times higher than Olsson et al.'s PTP (4.71 dl g^{-1} vs. 0.39 dl g^{-1}). These PAP-TP-x membranes (thickness: 25 μm) showed a high OH^- conductivity of 193 mS cm^{-1} at 95 °C, and were stable in 1 M KOH at 100 °C for 2000 hours. PAP-TP-x-based fuel cells achieved a PPD of 0.92 W cm^{-2} with an Ag cathode under H_2–air conditions.

However, the production of these PAP-based AEMs highly relies on an excess of trifloromethanesulfonic acid (TFSA), which is one of the strongest known acids (pK_a [TFSA]: −14.6 vs. pK_a[pure H_2SO_4]: −11.93).

Chen et al. reported on poly(fluorenyl aryl piperidinium) (PFAP) membranes and ionomers for AEMFCs [98] (Figure 4.40). By adding the fluorene group into terphenyl (PFTP), membranes showed high OH- conductivity of 208 mS cm^{-1} at 80 °C, low H_2 permeability, excellent mechanical properties (84.5 MPa), and 2000 hours *ex situ* durability in 1 M NaOH at 80 °C, while the ionomers (PFBP) have high water vapor permeability and low phenyl adsorption. Based on the design of PFTP and PDTP membranes and ionomers, they demonstrated AELFC performance of 2.34 W cm^{-1} in H_2–O_2 and 1.38 W cm^{-2} in H_2–air (CO_2-free) at 80 °C (Figures 4.41 and 4.42).

4.5.2.6 Radiation-Grafted AEM

A series of radiation-grafted AEMs with degrees of grafting (by mass) in the range of 20–25% were first produced and characterized by Slade et al. [83, 127, 128]. The radiation grafting of vinylbenzyl chloride (VBC) into FEP (forming intermediate FEP-g-PVBC) with subsequent amination with trimethylamine and ion exchange with aqueous KOH yields alkaline, as opposed to acidic (proton conducting)

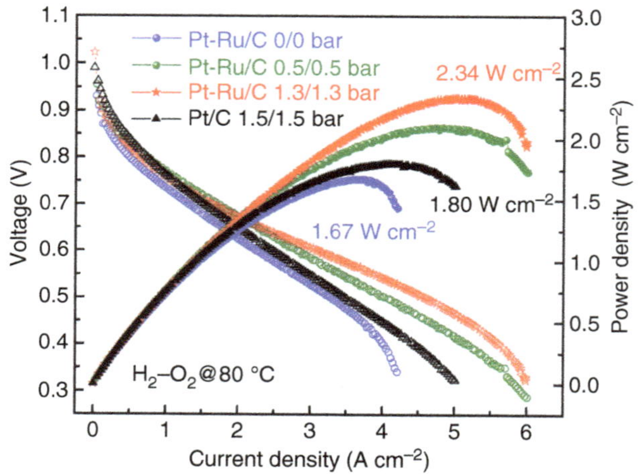

Figure 4.41 The power density of AEMFCs based on Pt-Ru/C anode with back pressure: PFBP 14 ionomers and PFTP-13 membranes (20 μm) at 80 °C, 75%/100% A/C RH, 1000/1000 ml min^{-1} H_2-O_2 flow rate, different back pressures [98].

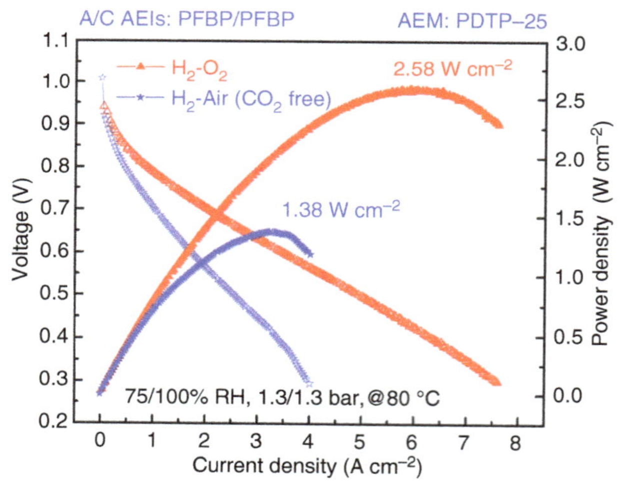

Figure 4.42 Polarization curves of PDTP-based AEMFC with back pressure. Source: Taken from Chen et al. [96].

membranes (FEP-g-PVBTMAOH). Radiation grafting of VBC into PVDF are inappropriate due to the chemical degradation caused by dehydrofluorination. FEP-g-PVBC were exposed to a radiation dose of 4–7 MRad and subsequent grafting in monomer solution for up to 166 hours at around 50 °C, yielding a degree of grafting of 20–26%. IEC ranged around 1–1.2 meq g^{-1}, depending on the degree of grafting. Ion conductivities ranged from 10 to 40 mS cm^{-1} at temperatures ranging from 25 to 100 °C. They concluded that FEP was better than PVDF-based AEM application in terms of long-term stability.

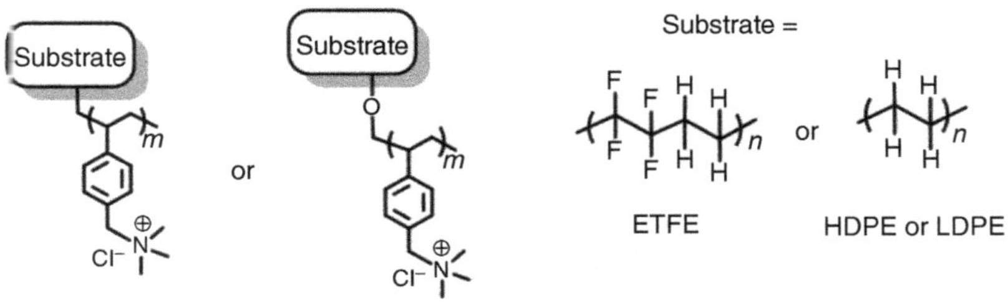

Figure 4.43 Radiation-grafted anion exchange membranes on ETFE or PE-based substrate.

Varcoe et al. first prepared an ETFE-grafted AEM [129] for fuel cell (Figure 4.43). They irradiated ETFE films followed by a treatment of VBC and aqueousTMA and KOH solution to prepare ETFE-based radiation-grafted alkaline anion exchange membrane (ETFE–AAEM). An H_2/O_2 fuel cell PPD of 130 mW cm^{-2} was obtained for AEMFC, while a PPD of 8.5 mW cm^{-2} was obtained in a metal-cation-free methanol/O_2 fuel cell. Mass transport phenomena in the electrodes and inadequate interface between the membrane and electrodes were identified as the major limits of power performance; the need for electrode architecture to be optimized for alkaline membrane fuel cells was identified. They later improved the chemical stability and electrochemical performance of ETFE–AAEMs in both H_2/O_2 and methanol-fed fuel cells [130].

Although BTMA-type AEPs have been previously demonstrated to be unstable under alkaline conditions, BTMA-type AEPs have currently received much attention due to the emergence of radiation-grafting AEPs. Typically, in 2014, Varcoe and coworkers reported the use of a BTMA-functionalized ethylene-tetrafluoroethylene (BTMA-ETFE) as AEIs by radiation grafting, and the BTMA–ETFE ionomer displayed a remarkable fuel cell performance. Subsequently, Wang et al. [131] developed a BTMA-functionalized low-density polyethylene (BTMA–LDPE) membrane (40 ± 5 μm) for AEMs. Combined with BTMA–EFTE ionomer, BTMA–LDPE membrane-based fuel cells exhibited distinguished PPD over 2 W cm^{-2} at 80 °C. Moreover, they found that switching from LDPE to high-density polyethylene (HDPE) could lead to remarkable enhancement in fuel cell performance (Figure 4.44a). The PPD of H_2–O_2 AEMFCs has been advanced to 2.55 W cm^{-2} with BTMA–HDPE membrane and BTMA–ETFE ionomers. BTMA–LDPE and BTMA–HDPE membranes exhibited high OH^- conductivity up to 200 mS cm^{-1} at 80 °C along with excellent mechanical properties. The key to the success of these BTMA-type AEPs is that these membranes are modified from commercial preforming HDPE (10 μm) and LDPE (15 μm) films that endow them with excellent performance. Although the chemical structure of BTMA–LDPE and BTMA–HDPE membranes is very similar, the *in situ* durability is totally different. HDPE-based cells operated stably at a current density of 0.6 A cm^{-2} for 440 hours at 70 °C, along with stable area-specific resistance (ASR), while LDPE-based cells displayed a rapid degradation at an initial 100 hours with an increase in ASR (38–52 mΩ cm^{-2}) (Figure 4.44b). The improved *in situ* durability has been hypothesized to be due to

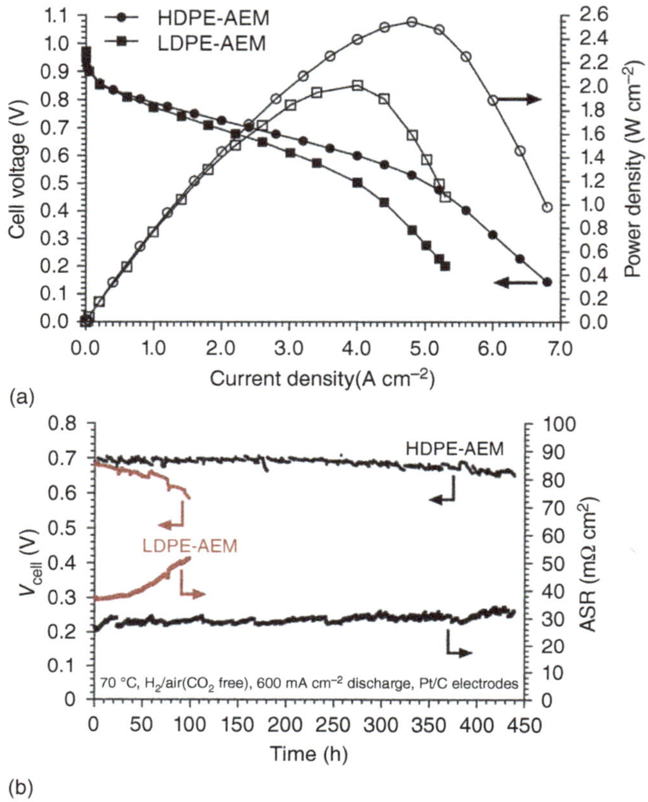

Figure 4.44 (a) H_2/O_2 AEMFC performance at 80 °C for the radiation-grafted AEMs from LDPE and HPDE support. PtRu/C anodes with a Pt loading of 0.4 mg cm^{-2}, (b) H_2/air(CO_2 free) AEMFC stability test at 70 °C and 600 mA cm^{-2}. Pt/C anode with a Pt loading of 0.6 mg cm^{-2}; Pt/C cathode with a PT loading of 0.6 mg cm^{-2}. Source: Wang et al. [131]/Royal Society of Chemistry/CCBY 3.0.

enhanced water transport characteristics caused by the change in the nanomorphology/microstructure of the precursor HDPE film. On the other hand, the alkaline stability issues associated with BTMA–ETFE ionomer still have not been well addressed. BTMA–ETFE membrane became brittle and showed severe breakage after being soaked in aqueous NaOH at 80 °C for 28 days, accompanied by a color change from light to dark yellow.

Fang et al. [132] developed a series of cross-linked AEM for fuel cells. They first irradiated ETFE film with VBC, followed by quaternization and cross-linking with 1,4-diazabicyclo(2,2,2)octane (DABCO), alkylation with p-xylenedichloride (DCX), and quaternization again with TMA. OH^- conductivity of these membranes is as high as 39 mS cm^{-1} at 30 °C in DI water. The membrane is stable after being treated in 10 M KOH solution at 60 °C for 120 hours. The fuel cell performance with the final AEM in H_2/O_2 single cell at 40 °C was 48 mW cm^{-1} at a current density of 69 mA cm^{-2}.

A study on ETFE-based radiation-grafted AEMs [133] containing three different cationic head groups: BTMA, benzyl-*N*-methylpyrrolidinium (MPY), or benzyl-*N*-methylpiperidinium (MPRD), among others. Among these, the chemistries involving TMA-, MPY-, and MPRD showed that MPRD-containing (and MPY-containing) materials have higher alkaline stabilities (when hydrated) compared to TMA-based materials. Among these, ETFE-g-TMA and ETFE-g-MPY showed OH^- conductivity of 138 and 159 mS cm^{-1} at 80 °C in a 95% RH condition. ETFE-g-TMA showed the highest performance when tested in H_2/O_2 AEMFC at 60 °C with PtRu/C anodes, Pt/C cathodes, and a polysulfone ionomer. The thickness of the base ETFE membrane and the resulting graft membrane was critical for the PPD of the AEM; the 100 μm thick variant (synthesized from 50 μm thick ETFE) produced PPD of 800 and 630 mW cm^{-2} (with and without 0.1 MPa back pressurization, respectively), while a 52 μm thick variant (synthesized from 25 μm thick ETFE) yielded 980 and 800 mW cm^{-2} under the same conditions.

In continuation of radiation-grafted ETFE, they used three cationic head groups (TMA, MPY, or MPRD) for both AEM and AEI [133] to investigate the efficacy of the same chemistry in AEMFC performances. The resulting AEMFC performance of three head group chemistries yielded a PPD of >1 W cm^{-2} at 60 °C in the H_2/O_2 fuel cells. In particular, the MPRD-based ETFE membranes showed notable characteristics related to the power outputs of the AEMFCs.

AEMFC with a PPD of 1.4 W cm^{-2} at 60 °C was accomplished by taking a more precise look at balancing necessary membrane hydration while preventing electrode flooding, which somewhat surprisingly can occur both at the anode and the cathode.

Mandal et al. [134] reported on the PTFE-reinforced PNB-based AEM for H_2/O_2 and H_2/air (CO_2 free) feed gases entering the cell at a volumetric flow rate of 1 l min^{-1}, and achieved a PPD of 3.5 W cm^{-2} and a maximum current density of 9.7 A cm^{-2} at 80 °C (Figure 4.45). The high performance was due to a combination of factors: (i) very high ionic conductivity (i.e. low internal resistance loss) and (ii) very high water transport through the AEM from the hydrogen anode to the oxygen cathode during the operation. They concluded that a critical current-limiting factor in AEMFC is the rate of water transport from the water-producing negative electrode to the water-consuming positive electrode. They also demonstrated that the values for the optimized dew points of the feed gases were different for the 10 and 20 μm thick membranes. The performance for all cells at optimized dew points was 75 and 77 °C, respectively, for the anode and cathode of the 10 μm AEM. The optimized dew points for the 20 μm membrane were 70 °C for the anode and 73 °C for the cathode. The membranes with very high OH^- conductivities (e.g. >200 mS cm^{-1} at 80 °C) are favored. Ultra-thin membranes facilitate high flux rates for water and minimize the parasitic voltage drop across the membrane. Very light cross-linking in the AEMs provided enough control over water uptake so that very high IEC values could be used without mechanical distortion of the membranes.

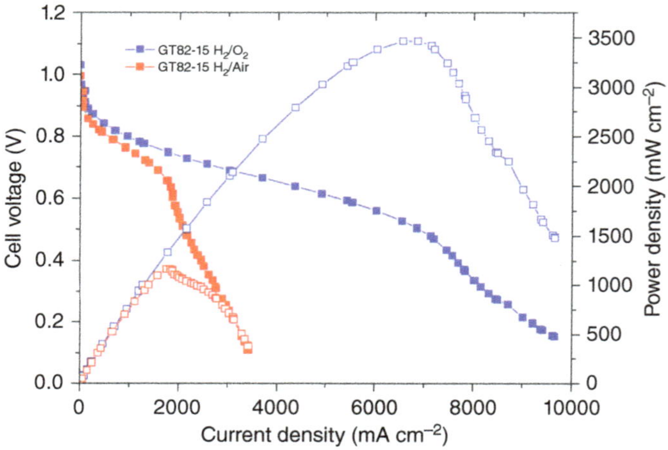

Figure 4.45 Current–voltage (filled) and current-power density (empty) curves of poly(norbonene) copolymer-based AEM with IEC of 3.88 meq g^{-1} and reinforced with PTFE (10 μm GT82-15) operating with H_2/O_2 (blue) and H_2/air (red) feeds at 1000 ml min^{-1} gas flow rate at 80 °C. The anode was loaded with 0.7 mg PtRu/C, and cathode was 0.6 mg Pt/C with back pressure of 0.05 and 0.1 MPa at the anode and cathode, respectively. Source: Adapted from Mandal et al. [134].

4.5.3 Water Management

In AEMFC operation, water is produced at the anode with the H_2 feed gas and consumed at the cathode with the O_2 gas feed. For every four electrons transferred, four water molecules are formed at the anode, and only two water molecules react at the cathode. This creates a hydration difference between the two electrodes. During the cell operation, water is moved from the cathode to the anode through the AEM by an electro-osmotic drag as OH^- is being produced in addition to a back diffusion of water from the anode to the cathode. These two processes can lead to imbalanced water transport and, thus, anode water flooding and cathode dry-out.

In fact, Eriksson et al. [135] reported that using a Tokuyama A201 membrane, the water fluxes are shown to be approximately three times larger for a membrane in contact with liquid water compared to vapors. The flux during fuel cell operation showed that the transport rate of water in the membrane is affected by an applied current. High current produces fast production of water. Through modeling, an apparent water drag coefficient is determined to be −0.64, indicating that the current-induced transport of water occurs in the opposite direction to the transport of OH^- ions. These results implicate that water flooding on one or both electrodes is a larger concern than dry-out in an AEMFC.

4.5.4 Outlook

Nowadays, BTMA-type, DMP-type, and side-chain type (five or six alkyl spacers) AEPs have seen great progress in current research. Although some other cationic groups also possessed high alkaline stability, the corresponding AEPs, such as ASU, large-hindrance IM, and cobaltocenium groups, exhibited limited ion conductivity and fuel cell performance. Besides, most high-performance AEIs are based on BTMA and alkyl ammonium groups, while other cationic functional groups in the AEIs are rare. Therefore, QA group is so far the best cationic group. Overall, DMP-type and side-chain-type AEPs show higher *ex situ* durability than BTMA-type AEMs. Meanwhile, the alkaline stability of BTMA-type AEMs shows a big difference depending on the commercial pre-forming films. Typically, BTMA–HDPE membrane is more stable than BTMA–ETFE and BTMA–LDPE membranes.

Moreover, the *ex situ* durability and *in situ* durability of AEPs have been demonstrated to be mismatching. Some of AEPs exhibited excellent *ex situ* durability over 5000 hours, while no cells to date have been reported to show sufficient *in situ* durability (~2000 hours) at DOE-relevant current densities (0.6 A cm^{-2}), particularly at high temperature and in low humidity conditions. Although DMP-type AEPs exhibited outstanding *ex situ* durability, their long-term *in situ* durability has reached below 300 hours at 0.6 A cm^{-2}. On the contrary, BTMA–HDPE membrane and BTMA–ETFE ionomers-based AEMFCs achieved a 440 hours *in situ* durability under 0.6 A cm^{-2}. As for side-chain-type AEPs, the *in situ* durability is also limited without reinforcement, while it is dramatically improved to 545 hours (under 0.6 A cm^{-2}) after PTFE reinforcement. Therefore, the mechanical degradation of AEMs should be elaborately investigated during *in situ* durability testing, and the mechanical properties of AEMs should be seriously taken into account when the AEMs are used for *in situ* durability testing, particularly in OH^- form and at 100% RH. Since the PPD of AEMFCs has reached to equal level to PEMFCs, the *in situ* durability should be emphasized in future research to promote the commercialization of AEMFCs.

4.6 Anion Exchange Ionomers

Nafion has been used as a binder for catalyst slurry in PEMFC as Nafion has been used as PEM. Extensive studies have been conducted to optimize the structure and composition of cathodes for good oxygen transport and proton transport in the CL. Several researches have focused on the effects of Nation loading on the performance of cathodes. Raistrick [136] first introduced Nafion into the CL by impregnating it with a Nafion solution. The loading of Pt required was greatly reduced by this method without compromising the cell performance. Wilson and Gottesfeld [137] further improved the performance by mixing the catalyst with a Nafion solution. They showed that the optimum Nafion loading for cathodes prepared by mixing the catalyst with Nafion solution is 33 mass %. Li and Pickup [138] showed that the optimum Nafion loading of 30 mass % was determined to balance the improved proton conductivity of the CL with the loss of oxygen permeability. Loss of electronic conductivity in the CL is not a concern at high Nafion loading, but a resistance between the CL and the carbon paper support appears to be a problem if insufficient Nafion is used.

Later, researchers realized that Nafion binder causes carbon corrosion, which is a serious durability issue that occurs in an H_2/air fuel cell when H_2–air mixture is present in the anode during start-up. The limited wettability of the catalyst carbon surface is critically important for increasing its corrosion resistance in MEA. The addition of hydrophobic PVDF into the CL slows carbon corrosion rates. However, the use of PVDF electrode binder is challenging because it does not conduct protons, and its oxygen permeability is low. PVDF has been employed with some success as the electrode binder in PBI-based H_2–air fuel cell electrodes [139]. Blending of Nafion and PVDF has been tried as well. Landis and Moore [140] showed that Nafion and PVDF are incompatible and immiscible polymers, leading to phase separation when the solution is cast into thin membranes. Pintauro et al. developed well-mixed Nafion/PVDF blends with nanometer-sized domains prepared by electrospinning Nafion/PVDF mixtures [141]. They compared the PVDF binder or Nafion/PVDF blend binder to the carbon corrosion durability. A nanofiber MEA with a 33/67 Nafion/PVDF cathode binder generated the highest peak power. They also showed that MEAs with an electrospun particle/polymer cathode generated high beginning-of-life power and also exhibited excellent durability. In their case, Pt/C particles and a binder of Nafion + PAA produced 900 mW cm^{-2} at maximum power in an H_2/air cell at 80 °C, 100% RH, and high feed gas flow rates at 2 atm backpressure. Nanofiber cathode mats were prepared with a metal–organic framework (MOF)-derived Fe–N–C catalyst and a blended binder of Nafion/PVDF [142]. The introduction of PVDF into the electrode binder of nanofiber and sprayed cathodes produced a stable power output for 300 hours, whereas the sprayer-electrode MEA with neat Nafion binder exhibited a 63% power loss. They later replaced PAA with PEO that showed a higher maximum power of 822 mW cm^{-2} at 40% RH, 80 °C and 2 bar backpressure, whereas PAA-based ink electrode MEAs produced 527 mW cm^{-2}. Changes in PEO molecular weight had no notable impact on fuel cell performance [143].

AEI plays an essential role in enhancing mass transport and lowering internal resistance that further leads to superior output. Therefore, AEI plays an important role in the CL, with active participation in the ion transport to and from the surfaces of the electrocatalysts. In addition, well-dispersed electrocatalysts resulted in a sufficient three-phase boundary, beneficial for improving the cell performance. The AEIs are often supplied in solution or dispersion form, combining the binder and OH^- ion conductor role in the CL. Similar to the AEMs, the main requirements for use in gas diffusion electrodes (GDE) are the combination of high ion conductivity and stability in alkaline medium, especially in the cathode where dry-out leads to reduced conductivity and durability. The development of AEIs is necessary because it is difficult to formulate ionomer solutions without deterioration of their chemical properties [133]. AEI concepts based on both cross-linked and uncross-linked polymers have been proposed, but their performances in AEMFCs are still poor compared to the PFSA ionomers used in PEMFCs. Advances in AEI development are desperately required.

Ünlü et al. [144] investigated the effect of thin-film and ionomer impregnation methods for AEM electrodes on AEM/PEM hybrid fuel cells. They found that the

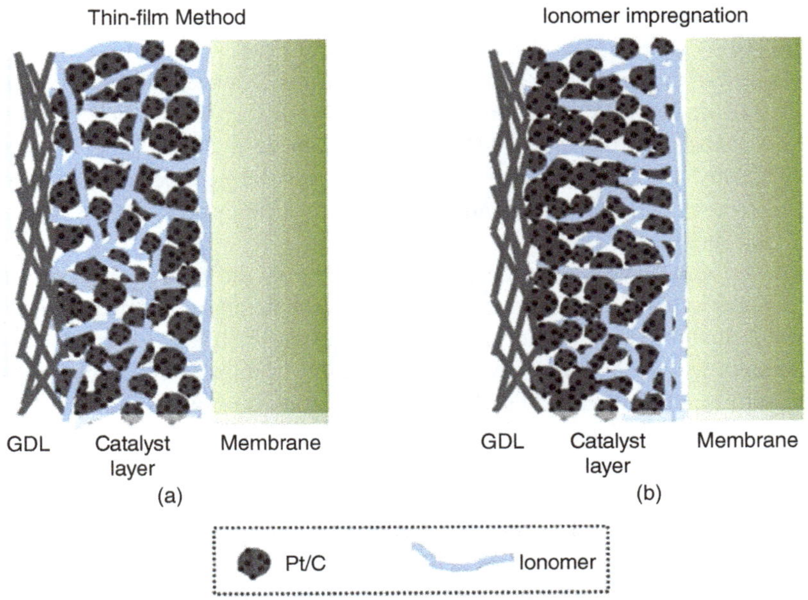

Figure 4.46 Schematic diagram of the electrode structures made by (a) thin film and (b) ionomer impregnation methods. The performance of hybrid fuel cell comprising the AEM cathode made by impregnation with 9% PTFE relative to the catalyst layer performed better than thin film method, as PTFE protects the water from the cathode, whereas ionomers are much more concentrated along the membrane to prevent the dry-out in the cathode. Source: Adapted from Ünlü et al. [144].

ionomer impregnation method led to superior fuel cell performance compared to the thin-film method, as seen in Figure 4.46. The improvement was attributed to the non uniform distribution of the AEM ionomer in the CL, creating free volumes near the GDL. These domains increase the catalyst surface area in contact with gas reactants. The use of a hydrophobic PTFE additive in the CL also improved the catalyst utilization by decreasing the loss of catalytic sites due to excessive water uptake. Improvements in electrode performance were examined with electrochemical impedance spectroscopy. This study showed that the physical properties of polymer electrolytes (e.g. water uptake and ionic conductivity) require different electrode architectures and fabrication protocols to obtain high performances.

In 2014, Poynton et al. [40] first reported the use of a powdered AEI in AEMFCs. The AEI powder was prepared by the radiation-grafting of VBC monomer onto ETFE powders, followed by functionalization with TMA to form the QA forms, which can be directly used within the CLs. Subsequently, this AEI powder has been used in combination with optimized AEMs, leading to impressive AEMFC performances; this demonstrated that ionomer powders can be effectively incorporated into CLs, opening up new opportunities for high-performance AEMFCs. They irradiated ETFE powders in the air to a total dose of 70 kGy, submerged in a premixed solution of VBC and propan-2-ol, purged with N_2 for 2 hours, and then sealed and heated for 72 hours at 60 °C. The resultant ETFE-g-poly(VBC)-grafted powder was recovered by filtration, washed with toluene, and dried in a vacuum oven at 50 °C, followed by immersion in aqueous TMA solution. An average AEI particle size of 25 ± 10 μm led to a poor, uneven distribution of the AEI particles across the surface of the electrodes. The 20% mass AEI showed a PPD = 240 mW cm^{-2} measured at 50 °C. Considering that a large proportion of the electrocatalyst surface was not in contact with an ionic-conducting phase, the high performances were regarded as surprising.

In 2018, the same group reported [133] the synthesis and characterization of electron-beam irradiated ETFE-based AEI powders and membranes functionalized with three different cationic head groups: TMA, MPY, or MPRD (Figure 4.46a). MPRD-containing ETFE–AEMs have higher alkaline stabilities when hydrated compared to TMA-based materials. A subsequent study on both AEM and AEI with the same chemistry confirmed the alkaline stability of MPRD-based radiation-grafted AEMs and AEIs. When MPRD-based AEM and AEI were used for fuel cell testing at 60 °C, they obtained a PPD of ~1.4 W cm^{-2} in H_2/O_2 fuel cell.

In fact, the same group reported on polyethylene-based AEM membranes [131]. They concluded that replacing the ETFE-based AEMs with LDPE–AEMs leads to more mechanically robust AEMs. It is not the mechanical weakness of the ETFE-based AEMs that limits AEMFC testing at 60 °C. They expect the more robust LDPE-based AEMs are expected to be tested at 80 °C for *in situ* durability (not achievable with ETFE-based AEMs).

When HDPE was irradiated and grafted with VBC with six hours grafting at 40 °C for LDPE and four hours at 50 °C for HDPE. The changes in OH^- conductivity of the two radiation-grafted AEMs maintained very stable conductivity up to 500 hours and showed a PPD of up to 2.6 W cm^{-2} for HDPE–AEM and about 2 W cm^{-2} in H_2/O_2 AEMFC performance test at 80 °C where 0.4 mg cm^{-2} of PtRu/C in the anode and Pt/C in the cathode was loaded and gases were supplied at 1 l min^{-1} at RH = 92% with no back pressure applied (Figure 4.44).

4.7 Fuel Cell Vehicle Market

The need to conserve natural resources and protect the environment is driving many vehicle manufacturers to take the lead in developing new, more efficient, and cleaner powertrains. Fuel cell vehicles are driven by the electricity generated onboard through fuel cell stack by passing hydrogen gas. Fuel cell vehicles include passenger vehicles, buses, and trucks. According to The Fuel Cell Industry Review (2019), the global fuel cell vehicle market size was valued at $446.7 million in 2018 and is projected to reach $6731.4 million by 2026, exhibiting a compound annual growth rate (CAGR) of 29.7% during the forecast period according to Fortune Business Insights. Among automakers, Toyota Motor Corporation and Hyundai Motor Company are leading the market, followed by Honda, Nissan Motor Corporation,

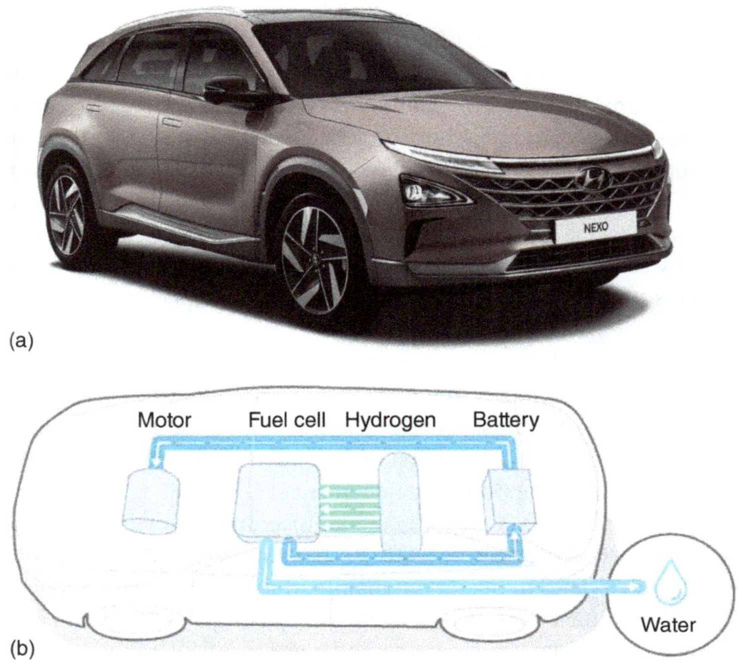

Figure 4.47 (a) Fuel cell vehicle Nexo produced by Hyundai Motor Company and (b) the schematic of powertrain and the principles of proton exchange membrane fuel cells. High-pressure hydrogen storage tank is stored at the bottom of the powertrain, whereas the fuel cell stack is placed in the front of the train. Source: Hyundai Motor America /https:// www.hyundaiusa.com/us/en/vehicles/nexo / last accessed September 25, 2023.

Daimler AG, AUDI AG, and BMW AG. Figure 4.47a shows the newly developed fuel-cell passenger cars and the schematic of the power train (Figure 4.47b). More than 8000 Hyundai NEXO vehicles have been deployed in 2020. Fuel cell electric vehicle (FCEV) revolution is underway. Leading FCEV plans are from European countries. Europe's FCEV plans are ambitious; 111 FCEVs are in operation in 2020, and 350 are planned for 2021. This could expand to 8000–17,000 vehicles by 2030. FCEV in the United States is also growing, though at a slower pace. In 2020, 64 FCEVs were active, and another 30 planned for 2021 and beyond. Many are present in California. The BMW Group maintains its plans for a few X5s to have fuel cell powertrains by 2022. China presents its hydrogen strategy plan, aiming for 10 FCEV models, 30,000 shipments, and a 10% market share by 2025. Nikola plans to sell trucks in collaboration with GM. China has the world's largest bus and coach market and has already delivered 3600 buses. In 2020, Nanjing announced the plan to replace its electric battery buses with FCEVs with a total of 7000 vehicles. While Korea's FCEV activity is smaller than China's, it is ambitious. The national roadmap includes 2000 FCEBs by 2022 and 40,000 by 2040, with Hyundai leading the charge, using its NEXO fuel cell module. Japan's FCEB bus activities are dominated by Toyota. Japan plans to produce 1200 FCEBs expected on the roads by 2030.

Heavy-duty trucks are an important in the fuel cell sector. GM will supply FC powertrains for Nicola's North American heavy-duty trucks. Hyundai Motor shipped the first 10 units of its Xcient fuel cell heavy-duty truck in 2020, the world's first to come off the production line, to logistics operators in Switzerland. Plans include shipping 1600 units by 2025. Hyundai plans to produce 2000 trucks per year by 2021 and to deploy 25,000 trucks in Europe by 2030. Hyundai also plans for North America to sell hundreds of trucks by 2022 and 12,000 by 2030; and for China, targeting 27,000 units by 2030. Nikola plans to manufacture 35,000 trucks annually, expecting to launch initial sale in 2023.

Hydrogen fueling stations (HRS) need to be installed for FCVs and FCEVs. New high-capacity stations can fill vehicles simultaneously at 700 bar, dispensing up to 1000 kg d^{-1} (enough for up to 250 passenger cars), and around 25 m^2 are more compact than before. France has installed the first of 20 stations in the Zero Emission Valley. Two HRS have opened to support the deployment of Hyundai's first 50 Xcient commercial fuel cell trucks. China is planning to manufacture HRS in the next few years. Germany has a goal of 100 stations with 700 bar capacity by 2021. In 2020, European countries had 78 HRS in operation, mainly in France, UK, and Norway. In the United States, 42 stations were opened in 2020 with a target of 200 stations by 2025. Japan leads with 133 stations in operation and plans to install 320 stations by 2025. Korea is set to expand its current 43 stations to 310 in 2022. China had 37 HRS operating in 2020.

4.8 Conclusions and Future Perspectives

Fuel cell technology has advanced significantly over the past half-century, leading to the realization of commercial hydrogen fuel cell vehicles in recent decades, thanks to the success of PEMs and catalysts. After successful invention of PFSA polymer, first represented as Nafion by Du Pont, as membranes and ionomers for PEMFCs, PEM fuel cell has made a significant progress. Other than Nafion membrane, different side-chain-type PFSAs and radiation-grafted fluoropolymers have been investigated extensively for PEMFC during the 1990s. Although PEMFC has been well-established and commercialized, high cost of PFSA and Pt catalysts has still been an issue. There are concerns over the fluorinated polymers such as PFSA as they are pointed out as a source of environmental pollutants inducing world climate changes, particularly in European Union. In this regard, AEMFC system receives much attention because of potential advantages of AEM over PEM; that is, (i) reduced fuel crossover, (ii) facile electrocatalysis of HOR and ORR in alkaline media, (iii) wide range of catalysts other than expensive Pt, and (iv) reduction in water flooding in cathode.

AEMFCs have been utilized for their facile modification and ease of synthesis for a wide variety of hydrocarbon polymers, including fluoropolymers. Fluoropolymers, such as PVDF, ETFE, and FEP have been extensively researched for AEM, particularly through radiation grafting techniques. For example, BTMA-g-ETFE has recently been used as both AEM and AEM and showed 1.4 W cm^{-2} at 60 °C and H_2–O_2 conditions. ETFE-based HDPE leads to more mechanically robust AEMs, resulting in higher operation temperatures at 80 °C. It is expected to see significant advances in the *in situ* durability of these fluoropolymer-based AEMFCs in the near future. Aryl-ether-free PP-based polymers have exhibited significant advancements in their electrochemical performance over the past decade. The ongoing development of these AEMs and ionomers, particularly for long-term durability, is expected to advance the commercialization of AEMFCs in the future.

References

1 Grove, W.R. (1839). XXIV. On voltaic series and the combination of gases by platinum. *The London, Edinburgh, and Dublin Philosophical Magazine and Journal of Science* 14: 127–130.

2 Chen, E. (2003). History, Chapter 2. In: *Fuel Cell Technology Handbook* (ed. G. Hoogers), 2-1–2-35. CRC Press.

3 Barbir, F. (2012). *PEM Fuel Cells: Theory and Practice*. Academic Press.

4 Banerjee, S. and Curtin, D.E. (2004). Nafion® perfluorinated membranes in fuel cells. *Journal of Fluorine Chemistry* 125: 1211–1216.

5 Johns, K. and Stead, G. (2000). Fluoroproducts—the extremophiles. *Journal of Fluorine Chemistry* 104: 5–18.

6 Cele, N. and Ray, S.S. (2009). Recent progress on Nafion-based nanocomposite membranes for fuel cell applications. *Macromolecular Materials and Engineering* 294: 719–738.

7 Hsu, W.Y. and Gierke, T.D. (1983). Ion transport and clustering in Nafion perfluorinated membranes. *Journal of Membrane Science* 13: 307–326.

8 Larmine, J. and Dicks, A. (2018). Efficiency and open-circuit voltage. In: *Fuel Cell Systems Explained* (ed. A. Dicks and D.A.J. Rand), 27–41. Wiley.

9 Ciureanu, M. (2004). Effects of Nafion® dehydration in PEM fuel cells. *Journal of Applied Electrochemistry* 34: 705–714.

10 Song, D., Wang, Q., Liu, Z. et al. (2005). A method for optimizing distributions of Nafion and Pt in cathode catalyst layers of PEM fuel cells. *Electrochimica Acta* 50: 3347–3358.

11 Haubold, H.-G., Vad, T., Jungbluth, H., and Hiller, P. (2001). Nano structure of NAFION: a SAXS study. *Electrochimica Acta* 46: 1559–1563.

12 Mauritz, K.A. and Moore, R.B. (2004). State of understanding of Nafion. *Chemical Reviews* 104: 4535–4586.

13 Xie, G. and Okada, T. (1995). Water transport behavior in Nafion 117 membranes. *Journal of The Electrochemical Society* 142: 3057–3062.

14 Sone, Y., Ekdunge, P., and Simonsson, D. (1996). Proton conductivity of Nafion 117 as measured by a four-electrode AC impedance method. *Journal of the Electrochemical Society* 143: 1254–1259.

15 Okada, T., Xie, G., Gorseth, O. et al. (1998). Ion and water transport characteristics of Nafion membranes as electrolytes. *Electrochimica Acta* 43: 3741–3747.

16 Motupally, S., Becker, A.J., and Weidner, J.W. (2000). Diffusion of water in Nafion 115 membranes. *Journal of the Electrochemical Society* 147: 3171–3177.

17 Divisek, J., Eikerling, M., Mazin, V. et al. (1998). A study of capillary porous structure and sorption properties of Nafion proton-exchange membranes swollen in water. *Journal of the Electrochemical Society* 145: 2677–2683.

18 Chiou, J.S. and Paul, D.R. (1988). Gas permeation in a dry Nafion membrane. *Industrial & Engineering Chemistry Research* 27: 2161–2164.

19 Broka, K. and Ekdunge, P. (1997). Oxygen and hydrogen permeation properties and water uptake of Nafion® 117 membrane and recast film for PEM fuel cell. *Journal of Applied Electrochemistry* 27: 117–123.

20 Morris, D.R. and Sun, X. (1993). Water-sorption and transport properties of Nafion 117 H. *Journal of Applied Polymer Science* 50: 1445–1452.

21 Majsztrik, P.W., Satterfield, M.B., Bocarsly, A.B., and Benziger, J.B. (2007). Water sorption, desorption and transport in Nafion membranes. *Journal of Membrane Science* 301: 93–106.

22 Zhao, Q., Majsztrik, P., and Benziger, J. (2011). Diffusion and interfacial transport of water in Nafion. *The Journal of Physical Chemistry B* 115: 2717–2727.

23 Mittal, V.O., Kunz, H.R., and Fenton, J.M. (2006). Is H_2O_2 involved in the membrane degradation mechanism in PEMFC? *Electrochemical and Solid-State Letters* 9: A299–A302.

24 Tang, H., Peikang, S., Jiang, S.P. et al. (2007). A degradation study of Nafion proton exchange membrane of PEM fuel cells. *Journal of Power Sources* 170: 85–92.

25 Yu, T.H., Sha, Y., Liu, W.G. et al. (2011). Mechanism for degradation of Nafion in PEM fuel cells from quantum mechanics calculations. *Journal of the American Chemical Society* 133: 19857–19863.

26 Wu, B., Zhao, M., Shi, W. et al. (2014). The degradation study of Nafion/PTFE composite membrane in PEM fuel cell under accelerated stress tests. *International Journal of Hydrogen Energy* 39: 14381–14390.

27 Satterfield, M.B., Majsztrik, P.W., Ota, H. et al. (2006). Mechanical properties of Nafion and Titania/Nafion composite membranes for polymer electrolyte membrane fuel cells. *Journal of Polymer Science Part B: Polymer Physics* 44: 2327–2345.

28 Kannan, R., Kakade, B.A., and Pillai, V.K. (2008). Polymer electrolyte fuel cells using Nafion-based composite membranes with functionalized carbon nanotubes. *Angewandte Chemie International Edition in English* 47: 2653–2656.

29 Tominaga, Y., Hong, I.-C., Asai, S., and Sumita, M. (2007). Proton conduction in Nafion composite membranes filled with mesoporous silica. *Journal of Power Sources* 171: 530–534.

30 Zhai, Y., Zhang, H., Hu, J., and Yi, B. (2006). Preparation and characterization of sulfated zirconia (SO_4^{2-}/ZrO_2)/Nafion composite membranes for PEMFC

operation at high temperature/low humidity. *Journal of Membrane Science* 280: 148–155.

31 Zhai, Y., Zhang, H., Zhang, Y., and Xing, D. (2007). A novel H_3PO_4/Nafion–PBI composite membrane for enhanced durability of high temperature PEM fuel cells. *Journal of Power Sources* 169: 259–264.

32 Liu, F., Yi, B., Xing, D. et al. (2003). Nafion/PTFE composite membranes for fuel cell applications. *Journal of Membrane Science* 212: 213–223.

33 Amjadi, M., Rowshanzamir, S., Peighambardoust, S.J. et al. (2010). Investigation of physical properties and cell performance of Nafion/TiO_2 nanocomposite membranes for high temperature PEM fuel cells. *International Journal of Hydrogen Energy* 35: 9252–9260.

34 Adjemian, K.T., Dominey, R., Krishnan, L. et al. (2006). Function and characterization of metal oxide–Nafion composite membranes for elevated-temperature H_2/O_2 PEM fuel cells. *Chemistry of Materials* 18: 2238–2248.

35 Jalani, N.H., Dunn, K., and Datta, R. (2005). Synthesis and characterization of Nafion®-MO_2 (M=Zr, Si, Ti) nanocomposite membranes for higher temperature PEM fuel cells. *Electrochimica Acta* 51: 553–560.

36 Cui, Z., Drioli, E., and Lee, Y.M. (2014). Recent progress in fluoropolymers for membranes. *Progress in Polymer Science* 39: 164–198.

37 Souzy, R. and Ameduri, B. (2005). Functional fluoropolymers for fuel cell membranes. *Progress in Polymer Science* 30: 644–687.

38 Nasef, M. (2004). Preparation and applications of ion exchange membranes by radiation-induced graft copolymerization of polar monomers onto non-polar films. *Progress in Polymer Science* 29: 499–561.

39 Ponce-González, J., Whelligan, D.K., Wang, L. et al. (2016). High performance aliphatic-heterocyclic benzyl-quaternary ammonium radiation-grafted anion-exchange membranes. *Energy & Environmental Science* 9: 3724–3735.

40 Poynton, S.D., Slade, R.C.T., Omasta, T.J. et al. (2014). Preparation of radiation-grafted powders for use as anion exchange ionomers in alkaline polymer electrolyte fuel cells. *Journal of Materials Chemistry A* 2: 5124–5130.

41 Ponce-González, J., Ouachan, I., Varcoe, J.R., and Whelligan, D.K. (2018). Radiation-induced grafting of a butyl-spacer styrenic monomer onto ETFE: the synthesis of the most alkali stable radiation-grafted anion-exchange membrane to date. *Journal of Materials Chemistry A* 6: 823–827.

42 Rouilly, M., Kötz, E., Haas, O. et al. (1993). Proton exchange membranes prepared by simultaneous radiation grafting of styrene onto Teflon-FEP films. Synthesis and characterization. *Journal of Membrane Science* 81: 89–95.

43 Büchi, F.N., Gupta, B., Haas, O., and Scherer, G.G. (1995). Study of radiation-grafted FEP-G-polystyrene membranes as polymer electrolytes in fuel cells. *Electrochimica Acta* 40: 345–353.

44 Gupta, B., Buchi, F.N., Staub, M. et al. (1996). Cation exchange membranes by pre-irradiation grafting of styrene into FEP films. II. Properties of copolymer membranes. *Journal of Polymer Science Part A: Polymer Chemistry* 34: 1873–1880.

45 Nasef, M.M., Saidi, H., Nor, H.M., and Foo, O.M. (2000). Proton exchange membranes prepared by simultaneous radiation grafting of styrene onto poly(tetrafluoroethylene-co-hexafluoropropylene) films. II. Properties of sulfonated membranes. *Journal of Applied Polymer Science* 78: 2443–2453.

46 Scott, K., Taama, W., and Argyropoulos, P. (2000). Performance of the direct methanol fuel cell with radiation-grafted polymer membranes. *Journal of Membrane Science* 171: 119–130.

47 Dargaville, T.R., George, G.A., Hill, D.J.T., and Whittaker, A.K. (2003). High energy radiation grafting of fluoropolymers. *Progress in Polymer Science* 28: 1355–1376.

48 Horsfall, J.A. and Lovell, H.A. (2002). Synthesis and characterisation of sulfonic acid-containing ion exchange membranes based on hydrocarbon and fluorocarbon polymers. *European Polymer Journal* 38: 1671–1682.

49 Brack, H.-P., Bührer, H.G., Bonorand, L., and Scherer, G.G. (2000). Grafting of pre-irradiated poly (ethylene-alt-tetrafluoroethylene) films with styrene: influence of base polymer film properties and processing parameters. *Journal of Materials Chemistry* 10: 1795–1803.

50 Horsfall, J. and Lovell, K. (2001). Fuel cell performance of radiation grafted sulphonic acid membranes. *Fuel Cells* 1: 186–191.

51 Chen, J., Asano, M., Maekawa, Y. et al. (2006). Preparation of ETFE-based fuel cell membranes using UV-induced photografting and electron beam-induced crosslinking techniques. *Journal of Membrane Science* 283: 373–379.

52 Chen, J., Asano, M., Yamaki, T., and Yoshida, M. (2006). Effect of crosslinkers on the preparation and properties of ETFE-based radiation-grafted polymer electrolyte membranes. *Journal of Applied Polymer Science* 100: 4565–4574.

53 Chen, J., Asano, M., Maekawa, Y., and Yoshida, M. (2006). Suitability of some fluoropolymers used as base films for preparation of polymer electrolyte fuel cell membranes. *Journal of Membrane Science* 277: 249–257.

54 Gubler, L., youcef, H.B., Gürsel, S.A. et al. (2008). Cross-linker effect in ETFE-based radiation-grafted proton-conducting membranes. *Journal of The Electrochemical Society* 155: B921.

55 Ben youcef, H., Gubler, L., Yamaki, T. et al. (2009). Cross-linker effect in ETFE-based radiation-grafted proton-conducting membranes. *Journal of The Electrochemical Society* 156: B532.

56 Sohn, J.-Y., Sung, H.-J., Song, J.-M. et al. (2012). Radiation-grafted proton exchange membranes based on co-grafting from binary monomer mixtures into poly(ethylene-co-tetrafluoroethylene) (ETFE) film. *Radiation Physics and Chemistry* 81: 923–926.

57 Kins, C.F., Sengupta, E., Kaltbeitzel, A. et al. (2014). Morphological anisotropy and proton conduction in multiblock copolyimide electrolyte membranes. *Macromolecules* 47: 2645–2658.

58 Yu, X., Roy, A., Dunn, S. et al. (2009). Synthesis and characterization of sulfonated-fluorinated, hydrophilic-hydrophobic multiblock copolymers for proton exchange membranes. *Journal of Polymer Science Part A: Polymer Chemistry* 47: 1038–1051.

59 Takamuku, S. and Jannasch, P. (2012). Multiblock copolymers with highly sulfonated blocks containing di- and tetrasulfonated arylene sulfone segments for proton exchange membrane fuel cell applications. *Advanced Energy Materials* 2: 129–140.

60 Li, N., Wang, C., Lee, S.Y. et al. (2011). Enhancement of proton transport by nanochannels in comb-shaped copoly(arylene ether sulfone)s. *Angewandte Chemie International Edition in English* 50: 9158–9161.

61 Lee, K.-S., Jeong, M.-H., Lee, J.-P. et al. (2010). Synthesis and characterization of highly fluorinated cross-linked aromatic polyethers for polymer electrolytes. *Chemistry of Materials* 22: 5500–5511.

62 Lee, S.Y., Kang, N.R., Shin, D.W. et al. (2012). Morphological transformation during cross-linking of a highly sulfonated poly(phenylene sulfide nitrile) random copolymer. *Energy & Environmental Science* 5: 9795–9802.

63 Park, C.H., Lee, S.Y., Hwang, D.S. et al. (2016). Nanocrack-regulated self-humidifying membranes. *Nature* 532: 480–483.

64 DeLuca, N.W. and Elabd, Y.A. (2006). Polymer electrolyte membranes for the direct methanol fuel cell: a review. *Journal of Polymer Science Part B: Polymer Physics* 44: 2201–2225.

65 Neburchilov, V., Martin, J., Wang, H., and Zhang, J. (2007). A review of polymer electrolyte membranes for direct methanol fuel cells. *Journal of Power Sources* 169: 221–238.

66 Arico, A., Srinivasan, S., and Antonucci, V. (2001). DMFCs: from fundamental aspects to technology development. *Fuel Cells* 1: 133–161.

67 Ge, J. and Liu, H. (2005). Experimental studies of a direct methanol fuel cell. *Journal of Power Sources* 142: 56–69.

68 Elabd, Y.A., Napadensky, E., Walker, C.W., and Winey, K.I. (2006). Transport properties of sulfonated poly (styrene-b-isobutylene-b-styrene) triblock copolymers at high ion-exchange capacities. *Macromolecules* 39: 399–407.

69 Wang, J.-T., Wainright, J., Savinell, R., and Litt, M. (1996). A direct methanol fuel cell using acid-doped polybenzimidazole as polymer electrolyte. *Journal of Applied Electrochemistry* 26: 751–756.

70 Küver, A. and Potje-Kamloth, K. (1998). Comparative study of methanol crossover across electropolymerized and commercial proton exchange membrane electrolytes for the acid direct methanol fuel cell. *Electrochimica Acta* 43: 2527–2535.

71 Walker, M., Baumgärtner, K.M., Kaiser, M. et al. (1999). Proton-conducting polymers with reduced methanol permeation. *Journal of Applied Polymer Science* 74: 67–73.

72 Jia, N., Lefebvre, M.C., Halfyard, J. et al. (2000). Modification of Nafion proton exchange membranes to reduce methanol crossover in PEM fuel cells. *Electrochemical and Solid-State Letters* 3: 529–531.

73 Aricò, A. (2003). Investigation of grafted ETFE-based polymer membranes as alternative electrolyte for direct methanol fuel cells. *Journal of Power Sources* 123: 107–115.

74 Chen, J., Asano, M., Yamaki, T., and Yoshida, M. (2006). Chemical and radiation crosslinked polymer electrolyte membranes prepared from radiation-grafted ETFE films for DMFC applications. *Journal of Power Sources* 158: 69–77.

75 Saarinen, V., Kallio, T., Paronen, M. et al. (2005). New ETFE-based membrane for direct methanol fuel cell. *Electrochimica Acta* 50: 3453–3460.

76 Saarinen, V., Karesoja, M., Kallio, T. et al. (2006). Characterization of the novel ETFE-based membrane. *Journal of Membrane Science* 280: 20–28.

77 Li, X., Drache, M., Gohs, U., and Beuermann, S. (2015). Novel concept of polymer electrolyte membranes for high-temperature fuel cells based on ETFE grafted with neutral acrylic monomers. *Journal of Membrane Science* 495: 20–28.

78 Yu, E.H. and Scott, K. (2004). Direct methanol alkaline fuel cell with catalysed metal mesh anodes. *Electrochemistry Communications* 6: 361–365.

79 Chen, Z. (2020). Water balancing. *Nature Energy* 5: 12–13.

80 Chen, N. and Lee, Y.M. (2021). Anion exchange polyelectrolytes for membranes and ionomers. *Progress in Polymer Science* 113: 101345.

81 Demarconnay, L., Coutanceau, C., and Léger, J.M. (2004). Electroreduction of dioxygen (ORR) in alkaline medium on Ag/C and Pt/C nanostructured catalysts—effect of the presence of methanol. *Electrochimica Acta* 49: 4513–4521.

82 Abdel Rahim, M.A., Abdel Hameed, R.M., and Khalil, M.W. (2004). Nickel as a catalyst for the electro-oxidation of methanol in alkaline medium. *Journal of Power Sources* 134: 160–169.

83 Danks, T.N., Slade, R.C.T., and Varcoe, J.R. (2002). Comparison of PVDF- and FEP-based radiation-grafted alkaline anion-exchange membranes for use in low temperature portable DMFCs. *Journal of Materials Chemistry* 12: 3371–3373.

84 Chen, N., Jin, Y., Liu, H. et al. (2021). Insight into the alkaline stability of N-heterocyclic ammonium groups for anion-exchange polyelectrolytes. *Angewandte Chemie* 133: 19421–19429.

85 Chen, N. and Lee, Y.M. (2022). Anion-conducting polyelectrolytes for energy devices. *Trends in Chemistry* 4: 236–249.

86 Zhu, T., Sha, Y., Firouzjaie, H.A. et al. (2019). Rational synthesis of metallo-cations toward redox- and alkaline-stable metallo-polyelectrolytes. *Journal of the American Chemical Society* 142: 1083–1089.

87 Zhu, T., Xu, S., Rahman, A. et al. (2018). Cationic metallo-polyelectrolytes for robust alkaline anion-exchange membranes. *Angewandte Chemie International Edition in English* 57: 2388–2392.

88 Zha, Y., Disabb-Miller, M.L., Johnson, Z.D. et al. (2012). Metal-cation-based anion exchange membranes. *Journal of the American Chemical Society* 134: 4493–4496.

89 Chen, Y., Li, Z., Chen, N. et al. (2017). Preparation and characterization of cross-linked polyphosphazene-crown ether membranes for alkaline fuel cells. *Electrochimica Acta* 258: 311–321.

90 Yang, Q., Li, L., Gao, X.L. et al. (2019). Crown ether bridged anion exchange membranes with robust alkaline durability. *Journal of Membrane Science* 578: 230–238.

91 Holdcroft, S. and Fan, J. (2019). Sterically-encumbered ionenes as hydroxide ion-conducting polymer membranes. *Current Opinion in Electrochemistry* 18: 99–105.

92 Fan, J., Wright, A.G., Britton, B. et al. (2017). Cationic polyelectrolytes, stable in 10 M KOHaq at 100 °C. *ACS Macro Letters* 6: 1089–1093.

93 You, W., Padgett, E., MacMillan, S.N. et al. (2019). Highly conductive and chemically stable alkaline anion exchange membranes via ROMP of trans-cyclooctene derivatives. *Proceedings of the National Academy of Sciences of the United States of America* 116: 9729–9734.

94 Hugar, K.M., You, W., and Coates, G.W. (2019). Protocol for the quantitative assessment of organic cation stability for polymer electrolytes. *ACS Energy Letters* 4: 1681–1686.

95 You, W., Hugar, K.M., and Coates, G.W. (2018). Synthesis of alkaline anion exchange membranes with chemically stable imidazolium cations: unexpected cross-linked macrocycles from ring-fused ROMP monomers. *Macromolecules* 51: 3212–3218.

96 Chen, N., Hu, C., Wang, H.H. et al. (2021). Poly(alkyl-terphenyl piperidinium) ionomers and membranes with an outstanding alkaline-membrane fuel-cell performance of 2.58 W cm(-2). *Angewandte Chemie International Edition* 60: 7710–7718.

97 Chen, N., Hu, C., Wang, H.H. et al. (2021). Chemically & physically stable crosslinked poly(aryl-co-aryl piperidinium)s for anion exchange membrane fuel cells. *Journal of Membrane Science* 638: 119685.

98 Chen, N., Wang, H.H., Kim, S.P. et al. (2021). Poly(fluorenyl aryl piperidinium) membranes and ionomers for anion exchange membrane fuel cells. *Nature Communications* 12: 2367.

99 Wang, J., Zhao, Y., Setzler, B.P. et al. (2019). Poly(aryl piperidinium) membranes and ionomers for hydroxide exchange membrane fuel cells. *Nature Energy* 4: 392–398.

100 Olsson, J.S., Pham, T.H., and Jannasch, P. (2018). Poly(arylene piperidinium) hydroxide ion exchange membranes: synthesis, alkaline stability, and conductivity. *Advanced Functional Materials* 28: 1702758.

101 Marino, M.G. and Kreuer, K.D. (2015). Alkaline stability of quaternary ammonium cations for alkaline fuel cell membranes and ionic liquids. *ChemSusChem* 8: 513–523.

102 Fan, J., Willdorf-Cohen, S., Schibli, E.M. et al. (2019). Poly(bis-arylimidazoliums) possessing high hydroxide ion exchange capacity and high alkaline stability. *Nature Communications* 10: 1–10.

103 Zhang, B., Kaspar, R.B., Gu, S. et al. (2016). A new alkali-stable phosphonium cation based on fundamental understanding of degradation mechanisms. *ChemSusChem* 9: 2374–2379.

104 Hugar, K.M., Kostalik, H.A. IV, and Coates, G.W. (2015). Imidazolium cations with exceptional alkaline stability: a systematic study of structure–stability relationships. *Journal of the American Chemical Society* 137: 8730–8737.

105 Pham, T.H., Olsson, J.S., and Jannasch, P. (2019). Effects of the N-alicyclic cation and backbone structures on the performance of poly(terphenyl)-based hydroxide exchange membranes. *Journal of Materials Chemistry A* 7: 15895–15906.

106 Dang, H.-S. and Jannasch, P. (2015). Exploring different cationic alkyl side chain designs for enhanced alkaline stability and hydroxide ion conductivity of anion-exchange membranes. *Macromolecules* 48: 5742–5751.

107 Dang, H.-S. and Jannasch, P. (2016). Alkali-stable and highly anion conducting poly(phenylene oxide)s carrying quaternary piperidinium cations. *Journal of Materials Chemistry A* 4: 11924–11938.

108 Hibbs, M.R. (2013). Alkaline stability of poly(phenylene)-based anion exchange membranes with various cations. *Journal of Polymer Science Part B: Polymer Physics* 51: 1736–1742.

109 Mohanty, A.D., Tignor, S.E., Krause, J.A. et al. (2016). Systematic alkaline stability study of polymer backbones for anion exchange membrane applications. *Macromolecules* 49: 3361–3372.

110 Han, H., Ma, H., Yu, J. et al. (2019). Preparation and performance of novel tetraphenylphosphonium-functionalized polyphosphazene membranes for alkaline fuel cells. *European Polymer Journal* 114: 109–117.

111 Henkensmeier, D., Cho, H.-R., Kim, H.-J. et al. (2012). Polybenzimidazolium hydroxides – structure, stability and degradation. *Polymer Degradation and Stability* 97: 264–272.

112 Gao, X., Yu, H., Qin, B. et al. (2019). Enhanced water transport in AEMs based on poly(styrene–ethylene–butylene–styrene) triblock copolymer for high fuel cell performance. *Polymer Chemistry* 10: 1894–1903.

113 Mohanty, A.D., Ryu, C.Y., Kim, Y.S., and Bae, C. (2015). Stable elastomeric anion exchange membranes based on quaternary ammonium-tethered polystyrene-b-poly(ethylene-co-butylene)-b-polystyrene triblock copolymers. *Macromolecules* 48: 7085–7095.

114 Jeon, J.Y., Park, S., Han, J. et al. (2019). Synthesis of aromatic anion exchange membranes by Friedel–crafts Bromoalkylation and cross-linking of polystyrene block copolymers. *Macromolecules* 52: 2139–2147.

115 Clark, T.J., Robertson, N.J., Kostalik, H.A. IV et al. (2009). A ring-opening metathesis polymerization route to alkaline anion exchange membranes: development of hydroxide-conducting thin films from an ammonium-functionalized monomer. *Journal of the American Chemical Society* 131: 12888–12889.

116 Ul Hassan, N., Mandal, M., Huang, G. et al. (2020). Achieving high-performance and 2000 h stability in anion exchange membrane fuel cells by manipulating ionomer properties and electrode optimization. *Advanced Energy Materials* 10: 2001986.

117 Mandal, M., Huang, G., and Kohl, P.A. (2019). 38-Anionic multiblock copolymer membrane based on vinyl addition polymerization of norbornenes: applications in anion-exchange membrane fuel cells. *Journal of Membrane Science* 570–571: 394–402.

118 Mandal, M., Huang, G., and Kohl, P.A. (2019). 15-Highly conductive anion-exchange membranes based on cross-linked poly(norbornene): vinyl addition polymerization. *ACS Applied Energy Materials* 2: 2447–2457.

119 Mandal, M., Huang, G., Hassan, N.U. et al. (2019). 9-The importance of water transport in high conductivity and high-power alkaline fuel cells. *Journal of The Electrochemical Society* 167: 054501.

120 Huang, G., Mandal, M., Peng, X. et al. (2019). Composite poly(norbornene) anion conducting membranes for achieving durability, water management and high power (3.4 W/cm^2) in hydrogen/oxygen alkaline fuel cells. *Journal of The Electrochemical Society* 166: F637–F644.

121 Hibbs, M.R., Fujimoto, C.H., and Cornelius, C.J. (2009). Synthesis and characterization of poly(phenylene)-based anion exchange membranes for alkaline fuel cells. *Macromolecules* 42: 8316–8321.

122 Park, E.J., Maurya, S., Hibbs, M.R. et al. (2019). Alkaline stability of quaternized Diels–Alder polyphenylenes. *Macromolecules* 52: 5419–5428.

123 Park, E.J. and Kim, Y.S. (2018). 13-Quaternized aryl ether-free polyaromatics for alkaline membrane fuel cells: synthesis, properties, and performance – a topical review. *Journal of Materials Chemistry A* 6: 15456–15477.

124 Akiyama, R., Yokota, N., and Miyatake, K. (2019). Chemically stable, highly anion conductive polymers composed of quinquephenylene and pendant ammonium groups. *Macromolecules* 52: 2131–2138.

125 Miyanishi, S. and Yamaguchi, T. (2019). Highly durable spirobifluorene-based aromatic anion conducting polymer for a solid ionomer of alkaline fuel cells and water electrolysis cells. *Journal of Materials Chemistry A* 7: 2219–2224.

126 Peng, H., Li, Q., Hu, M. et al. (2018). Alkaline polymer electrolyte fuel cells stably working at 80 °C. *Journal of Power Sources* 390: 165–167.

127 Slade, R. and Varcoe, J. (2005). Investigations of conductivity in FEP-based radiation-grafted alkaline anion-exchange membranes. *Solid State Ionics* 176: 585–597.

128 Danks, T.N., Slade, R.C.T., and Varcoe, J.R. (2003). Alkaline anion-exchange radiation-grafted membranes for possible electrochemical application in fuel cells. *Journal of Materials Chemistry* 13: 712–721.

129 Varcoe, J. and Slade, R. (2006). An electron-beam-grafted ETFE alkaline anion-exchange membrane in metal-cation-free solid-state alkaline fuel cells. *Electrochemistry Communications* 8: 839–843.

130 Varcoe, J.R., Slade, R.C., Yee, E.L.H. et al. (2007). Poly(ethylene-co-tetrafluoroethylene)-derived radiation-grafted anion-exchange membrane with properties specifically tailored for application in metal-cation-free alkaline polymer electrolyte fuel cells. *Chemistry of Materials* 19: 2686–2693.

131 Wang, L., Peng, X., Mustain, W.E., and Varcoe, J.R. (2019). Radiation-grafted anion-exchange membranes: the switch from low- to high-density polyethylene leads to remarkably enhanced fuel cell performance. *Energy & Environmental Science* 12: 1575–1579.

132 Fang, J., Yang, Y., Lu, X. et al. (2012). Cross-linked, ETFE-derived and radiation grafted membranes for anion exchange membrane fuel cell applications. *International Journal of Hydrogen Energy* 37: 594–602.

133 Gonçalves Biancolli, A.L., Herranz, D., Wang, L. et al. (2018). ETFE-based anion-exchange membrane ionomer powders for alkaline membrane fuel cells: a first performance comparison of head-group chemistry. *Journal of Materials Chemistry A* 6: 24330–24341.

134 Mandal, M., Huang, G., Hassan, N.U. et al. (2020). The importance of water transport in high conductivity and high-power alkaline fuel cells. *Journal of The Electrochemical Society* 167: 054501.

135 Eriksson, B., Grimler, H., Carlson, A. et al. (2019). Quantifying water transport in anion exchange membrane fuel cells. *International Journal of Hydrogen Energy* 44: 4930–4939.

136 Raistrick, I.D. (1986). Modified gas diffusion electrode for proton exchange membrane fuel cells. In: *Proceedings of the Symposium on Diaphragms, Separators, and Ion-Exchange Membranes*, Boston, MA, USA, Volume 86-14, pp. 172–178.

137 Wilson, M.S. and Gottesfeld, S. (1992). High performance catalyzed membranes of ultra-low pt loadings for polymer electrolyte fuel cells. *Journal of the Electrochemical Society* 139: L28.

138 Li, G. and Pickup, P.G. (2003). Ionic conductivity of PEMFC electrodes: effect of Nafion loading. *Journal of the Electrochemical Society* 150: C745.

139 Su, H., Pasupathi, S., Bladergroen, B. et al. (2013). Optimization of gas diffusion electrode for polybenzimidazole-based high temperature proton exchange membrane fuel cell: evaluation of polymer binders in catalyst layer. *International Journal of Hydrogen Energy* 38: 11370–11378.

140 Landis, F.A. and Moore, R.B. (2000). Blends of a perfluorosulfonate ionomer with poly (vinylidene fluoride): effect of counterion type on phase separation and crystal morphology. *Macromolecules* 33: 6031–6041.

141 Brodt, M., Wycisk, R., Dale, N., and Pintauro, P. (2016). Power output and durability of electrospun fuel cell fiber cathodes with PVDF and Nafion/PVDF binders. *Journal of the Electrochemical Society* 163: F401.

142 Slack, J., Halevi, B., McCool, G. et al. (2018). Electrospun fiber mat cathode with platinum-group-metal-free catalyst powder and Nafion/PVDF binder. *ChemElectroChem* 5: 1537–1542.

143 Waldrop, K., Slack, J., Gumeci, C. et al. (2019). Electrospun particle/polymer fiber electrodes with a neat Nafion binder for hydrogen/air fuel cells. *ECS Transactions* 92: 595.

144 Ünlü, M., Zhou, J., Anestis-Richard, I. et al. (2011). Improved gas diffusion electrodes for hybrid polymer electrolyte fuel cells. *Electrochimica Acta* 56: 4439–4444.

5

Membranes in Energy Storage System

5.1 Introduction

As many recent advances in battery technology have been achieved, lithium-ion battery has been used in our daily life represented as mobile phones and electrical vehicles. Separators play a key role in batteries. Their main function is to keep the positive and negative electrodes electrically contacting each other and allow rapid ionic transport to complete the circuit for the passage of current in lithium-ion batteries. Separator is required to minimize the ionic resistance, to be mechanically and chemically stable, to be easily wetted by electrolytes and so on. They should be very good electronic insulators and have the capability of conducting ions by either intrinsic ionic conductor or by soaking electrolyte. Therefore, they play very important roles in lithium-ion batteries and may affect the electrochemical energy efficiency. Polyolefins have been mainly used as separator and fluoropolymer as a binder for electrodes. In this chapter, types of separators for lithium-ion batteries and next-generation batteries will be introduced.

5.1.1 Li-Ion Battery

Many recent progresses have been made in battery technology, both through the development and introduction of new battery chemistries and through continued improvement of electrochemical energy storage devices. However, there is still no one “ideal” battery that provides optimum electrochemical performance under all operating conditions. At the same time, there is no one separator that can be considered “ideal” for all battery chemistries and geometries [1]. Among many batteries, this section focuses on the lithium-ion batteries (LIBs) as they have become the main power sources of the consumer electronics and transportation market [2]. Since the first LIBs became commercially available in 1991, they are well-known due to high specific energy and high specific power density which are the advantages that most other electrochemical energy storage systems cannot offer and are the most powerful technology in consumer mobile electronics and mobile transportation vehicles. Usually, LIBs are required with high specific energy (about 220 $Wh\,kg^{-1}$), high energy density (about 550 $Wh\,l^{-1}$), long lifetime cycle (>1000 cycles), low self-discharge rate (<5%/month), and high operational voltage

Membranes for Energy Applications, First Edition. Young Moo Lee.

(2.5–4.2 V) [3]. They have been widely used in consumer electronics, such as notebook computers and cell phones, as well as in electric vehicles (EV). However, LIBs face severe problems, such as high cost and relatively low compaction density. Among them, safety is an urgent issue to be overcome in the large-scale application of LIBs, especially for the high-energy-density EVs.

Figure 5.1a describes a typical LIB, which consists of a positive electrode (cathode, $LiMn_{1.5}Ni_{0.5}O_4$ spinel; layered structure $LiMnO_2$; $LiMPO_4$ olivine, M = Mn, Fe and Co) and a negative electrode (anode, intercalated graphite; alloying materials Si and Sn), together with the electrolyte that allows Li-ion transport but prevents electrodes from electronic contact [4]. In a commercial LIB, a porous separator is soaked in an electrolyte ($LiPF_6$) which is dissolved in a mixture of organic solvents such as ethylene carbonate (EC), ethyl methyl carbonate (EMC), or diethyl carbonate (DEC). Both electrodes and electrolytes participate in the electrochemical reaction as described in Figure 5.1a in a battery cell, while the separator is an inactive component. Most of LIB use carbon as the active anode materials, although silicon, lithium titanate, tin dioxide, and some intermetallic have also been intensively studied. This active material is adhered to a copper current collector with a binder, typically poly(vinylidene fluoride) (PVDF), and a high surface area carbon is used in the electrode formulation in order to provide an electron flow path to the current collector. During the battery operation, lithium ion (Li+) is shuttled between the anode and cathode via the electrolyte from the anode to the cathode during discharging and in the reverse direction during the charging of the battery.

Separators are inactive materials in LIB and are sandwiched between the anode and cathode but play a key role in all batteries. The major role of the battery separator is to physically isolate the anode from the cathode while allowing mobile Li+ ions to transport back and forth to complete the circuit for the passage of current in LIBs. They should be very good electronic insulators and have the capability of conducting ions by either intrinsic ionic conductor or by soaking electrolyte. Therefore, they play very important roles in LIBs, and may affect the electrochemical energy efficiency.

Figure 5.1b describes a next-generation Li battery. In this case, solid polymer electrolytes replace liquid electrolytes in Li-ion to fabricate all solid Li battery for Li–S and Li–air batteries. Li–S and Li–air batteries both offer substantial increase in specific energy compared with LIB, delivering electric vehicles with a driving range of approaching the goal of ~500 km.

5.1.1.1 Battery Market, Separator Market

The battery market has undergone a drastic change over the past decade. For example, cell phone sales increased from less than 500 million units in 2000 to 2050 million units in 2016, and the portable PC market skyrocketed to 15 times during the same period [5, 6]. The global market for battery reached to 450 GWh in 2016 (see Figure 5.2a) [6]. Although lead acid batteries still have a 90% share, it is noteworthy that investment and demand for LIB is rapidly increasing due to their high energy density [9–11]. Reports predicted that the LIB market for electric vehicle (EV) receiving the most attention in recent years will increase from 123 GWh in 2020 to 1293 GWh in 2030 [12]. Li-ion cells and packs are expected

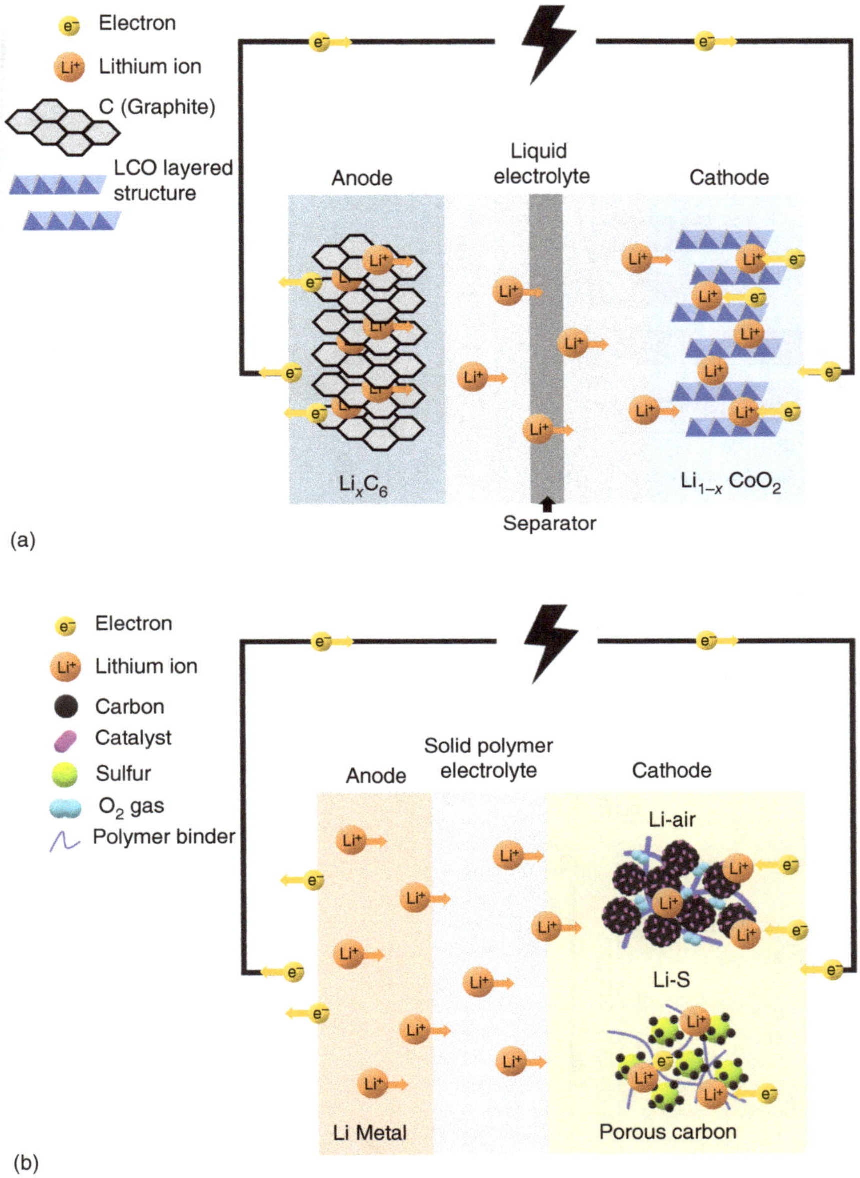

Figure 5.1 (a) Schematic of lithium-ion battery (LIB) consisting of the anode negative electrode (graphitic carbon), separator, liquid electrolyte, and positive cathode electrode (Li-intercalation compound) in conventional Li battery. (b) Schematic of next-generation Li battery; solid polymer electrolyte replaces liquid electrolyte to fabricate all solid Li battery for Li-S and Li-air batteries.

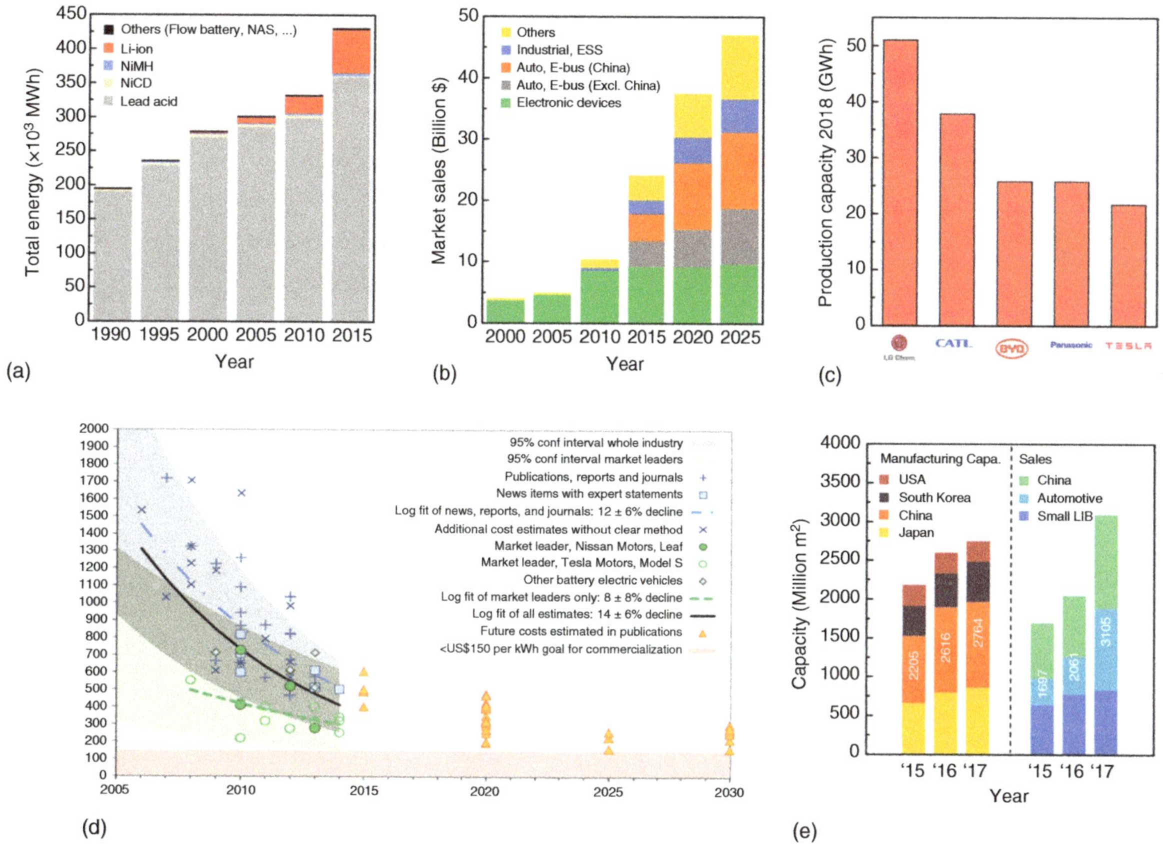

Figure 5.2 (a) The worldwide battery market from 1990 to 2015 [6]. (b) Market forecast for Li-ion packs from 2000 to 2025 [6]. (c) Top five LIB producers by capacity in 2019. Lithium-ion battery arms race, in, Obtenido de Benchmark Mineral Intelligence: https://www.benchmarkminerals.com/who-is-winning-the-global-lithium-ion-battery-arms-race/?source=content_type%3Areact%7Cfirst_level_url%3Aarticle%7Csection%3Amain_content%7Cbutton%3Abody_link, 2019 (d) Prediction of LIB prices from 2005 to 2030 [7]. (e) Manufacturing capacity and sales of separator depending on the major regions and the applications, respectively [8].

to show compound annual growth rates (CAGR; 2015–2025) of ~12% in the E-bus and energy storage system (ESS) fields, whereas the market for electronic devices exhibits 0% CAGR [6]. Considering that the market share of LIB has been concentrated on electronic devices so far, this prediction means that the market paradigm is rapidly shifting now.

According to the report, China is the largest LIB producer, meeting 51% of global capacity (76.3 GWh; 2015), followed by Korea (21%), Japan (16%), and USA (7%) [13]. In terms of individual companies, LG Chem (Korea) is leading the global market, showing LIB manufacturing capacity of 17 GWh in Q1 2017) [12]. Although other companies such as BYC (China) and Panasonic (Japan) are also spurring LIB production, the supply is still not enough considering that LIB demand for EV in 2030 is expected to be 1293 GWh. Another issue facing LIB manufacturers is the cost of LIBs. For example, the market share of EV in automobile market in 2020 only occupies 5% compared to internal combustion engines [8]. However, the proportion of EV in total automobile market will exceed 30% by 2030 and therefore the price of LIB becomes the most important factor to be considered in the automobile market [8]. Although LIB producers have reduced manufacturing costs by more than 70% over the past decade (273 \$ kWh^{-1}; 2016), intense price competition is leading manufactures to develop new materials and chemical processes to cut down the production cost [12]. Bloomberg New Energy Finance forecasts that LIB cost will continue to fall to as little as 73 \$ kWh^{-1} in a Korean manufacturing plant by 2030 [12].

Batteries are made of many components such as cathode, anode, electrolyte, and separator. Since the supply chain of each component operates in a complicated way, it is difficult to increase the supply to meet the soaring demand for LIB. Separators serve as a bottleneck in the LIB supply chains because producers did not have to produce separators at the expense of complicated production processes and low profits. As represented in Figure 5.2e,f, the increase of manufacturing capacity of separator was 6% from 2016 to 2017, but sales of separator increased by 51% [12]. This situation will be intensified as the EV market in China expands, and some report predicted that the demand for separator would triple from 2015 to 2025 with 12% CAGR [14]. For this reason, LIB manufacturers need special efforts to secure the supply chain of the separators.

Separator production in 2013 was estimated to be 1018 billion m^2 of which, 848 million m^2 was produced by wet processing and the rest by dry processing. (chinamarketresearchreports.com/114844.htlm) In 2015, Japan produced 48%, 17% was from China, 10% was from Korea, and so on. Separator production in 2025 is estimated to be 2.7 billion m^2 with CAGR of 12%. Semcorp from China had an annual production capacity of 2.0 billion m^2 in 2020, 900 million m^2 from Asahi Kasei, and 870 million m^2 from SK IE Technology. LG Chem and Toray coats Semcorp's bare film with nano-ceramics. Semcorp is offering bare films at US\$1 per m^2. Coated separators prices range from US\$1.5–2 per m^2. Separator accounts for around 20% of battery production costs, which makes price and safety of them a priority for battery cell manufacturers. Among the application of separators, electrical vehicles, and portable electronics occupy more than half of a total sales volume of separator in LIBs.

5.2 Requirements of Li-Ion Battery Separators

Although separators do not participate in the electrochemical reactions in a LIB, they perform the critical functions of physically separating the positive and negative electrodes while permitting the free flow of lithium ions through the liquid electrolyte that fill in their open porous structure. Therefore, the separator is a critical part in rechargeable LIBs and prevent internal short while serving as the electrolyte reservoir to enable ionic transport [15]. The ideal separator should have large electrolyte uptake to lower the cell internal resistance, have very thin thickness with strong mechanical strength, be electrochemically and structurally stable, and have a highly porous structure with great tortuosity to prevent the growth of dendritic lithium. In addition, the separator should shut the battery down when overheating occurs in order for the safety of battery. However, commercial battery separators cannot simultaneously possess all these ideal properties, and it becomes extremely important to balance different separator properties to achieve high-performance batteries (Figure 5.3).

A number of factors must be considered in selecting the best separator for a particular battery and application [15]. The following characteristics of each available separator must be considered against the requirements and one selected that best fulfils these needs as a wide variety of properties are required of separators used in batteries.

- Electronic insulator.
- Minimal electrolyte (ionic) resistance.
- Mechanical and dimensional stability.
- Sufficient physical strength to allow easy handling.
- Chemical resistance to degradation by electrolyte, impurities, and electrode reactants and products.

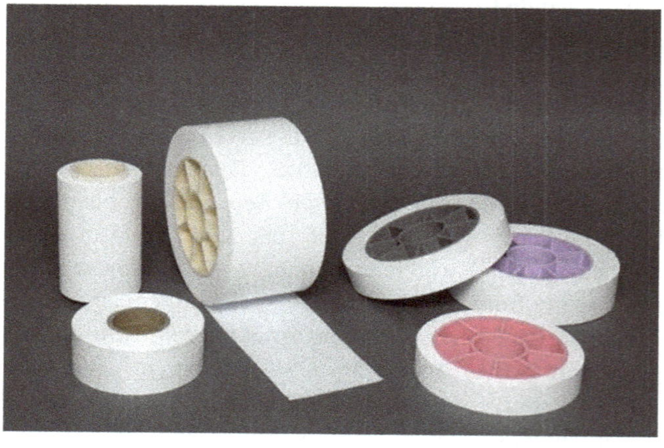

Figure 5.3 Battery separator. Source: THE ELEC/ http://www.thelec.net/news/articleView.html?idxno=2436/last accessed 21 August, 2023.

- Effective in preventing migration of particles or colloidal or soluble species between the electrodes.
- Readily wetted by electrolyte.
- Uniform in thickness and other properties.

Table 5.1 lists the ideal values of typical parameters for LIB separators. Thickness, porosity, pore size and distribution, electrical resistance, air permeability, mechanical strength, electrolyte uptakes as well as thermal shrinkage are very important physical properties of separators.

Separator should have an appropriate **porosity** to retain adequate liquid electrolyte so that the LIBs have sufficient ionic conductivity. If the porosity is too low, the internal resistance will be high because of insufficient liquid electrolyte between the two electrodes. However, if the porosity is too high, it adversely affects the battery safety because it could cause low mechanical strength. In addition, separators with too high porosities tend to shrink when the temperature increases. The porosity is defined by the ratio of void volume to apparent geometric volume, and it can be calculated by the following equation:

$$\text{Porosity}\,(\%) = \left(1 - \frac{\rho_M}{\rho_P}\right) \times 100 \tag{5.1}$$

where ρ_M is the apparent density of the separator and ρ_P is the density of the polymer. Often, the porosity is measured by the weights of the separator before and after the absorption of a liquid as the following equation according to ASTM D2873:

$$\text{Porosity}\,(\%) = \left(\frac{W - W_n}{\rho_L V_o}\right) \times 100 \tag{5.2}$$

where W and W_n represent the weights of the separator before and after immersing a liquid, respectively, ρ_L is the density of the liquid, and V_o is the geometric volume of the separator. In this equation, it is assumed that the volume occupied by the liquid is equal to the porous volume of the separator. The porosity of around 40–60% is desirable for normal LIB separators.

Separators should have small mean pore sizes and uniform pore size distributions. Small mean pore sizes are preferred as they can effectively prevent shorts caused by dendrite between battery electrodes. Uniform pore size distributions can avoid the non-uniform current densities across the electrode–separator interfaces. A non-uniform current accelerates the formation of Li dendrites in a battery cell.

Pore dimension and **pore size distribution** can be estimated from porosimetry techniques based on the Laplace equation as shown below in Eq. (5.3):

$$d = -\frac{4\gamma \cos\theta}{\Delta P} \tag{5.3}$$

where d is the pore diameter, ΔP is the pressure difference across the pore, γ is the surface tension of the liquid, and θ is the contact angle. Based on this equation, different types of porometers have been developed to characterize complex porous structures. For a more qualitative approach, cross section and surface pore morphologies can be examined simply from scanning electron microscopy images.

Table 5.1 Ideal value of typical parameters for lithium-ion battery separator.

Parameter	Ideal value	Relevance	Test method	References
Thickness	<15 μm	• The thickness determines the mechanical strength of the separator. • Ideal value for the thickness has decreased to increase battery performance and capacity.	ASTM D5947, ASTM D2103	[16–18]
Porosity Pore size	40–60% <1 μm	• Porosity and pore size affect the conduction of Li^+ ion between the electrodes. • Excessive porosity and pore size not only hinder the pore-close in overheated battery, but induce the short due to the dendritic Li growth on the anode.	ASTM E128-99	[19, 20]
Electrical resistance (MacMullin no.)	<8	• MacMullin no. represents the ratio of the resistance of the electrolyte-filled separator to that of electrolyte alone.		[21, 22]
Air permeability (Gurley)	<20 s/10 cc	• Gurley no. represents the time required for a specific amount of air to pass through the separator under a specified area and pressure. • It reflects the tortuosity of the pores, when the porosity and thickness of the separator are fixed.	ASTM D726	[1, 23, 24]
Mechanical strength	<2% offset at 1000 psi	• The separator has to withstand the mechanical stress applied during battery manufacturing. • The separator should be safe from external shocks.	ASTM D882	[3, 25]
Electrolyte uptake (Wettability)	>150%	• The separator must be completely wet with a specific electrolyte.		[26–28]
Electrochemical stability	>4.5 V in electrolyte	• The separator material should be electrochemically stable against the electrolyte and electrode under the strongly reactive environments.		[29–31]
Handling (Skew)	<0.2 mm m^{-1}	• In the manufacture of batteries, the separator must be straight and not skewed.		[22]
Shutdown or thermal runaway temperature	130 °C	• When thermal runaway occurs, the membrane should offer the ability to shut down, while retaining its mechanical and dimensional properties.		[32]
Thermal shrinkage	<5% at 150 °C	• Typically, when manufacturing batteries, the area of the separator is 5% larger than that of the electrode. • The separator remains stable over a wide temperature range without dimensional deformation.	ASTM D1204	[1, 22]

The **tortuosity factor** indicated the ratio of the mean actual path in comparison with the direct distance. Tortuosity τ can be estimated from the following equation shown below:

$$\tau = \sqrt{\varepsilon \frac{R_S}{R_0}} \tag{5.4}$$

where ε is the porosity ratio. A high tortuosity hinders the growth of Li dendrites but increases the ionic resistance of the inter-electrode medium.

Permeability of separators are described by the MacMullin number, which is the ratio of the resistance of the separator immersed with liquid electrolyte to the resistance if the electrolyte alone. The MacMullin number is proportional to air permeability, and it is often expressed by the Gurley value. The permeability can be evaluated from the permeability coefficient P according to Darcy's law, from Eq. (5.5) shown below. Here, l is the thickness of a separator, μ is the fluid viscosity, v is the fluid velocity, and ΔP is the pressure drop across the separator.

$$P = \frac{l\mu v}{\Delta P} \tag{5.5}$$

The Gurley value measure the time required for air to pass through the unit area of the separator under a fixed pressure. If the porosity and thickness of the separator are fixed, the Gurley value is determined by the tortuosity of the pores. If the Gurley value is low, the separator has high porosity and low tortuosity. In addition, the uniform permeability of a separator is critical for achieving a long cycle life because non-uniform permeability leads to uneven distribution of current density and the dendrite formation on the electrodes. Typical Gurley value s for LIB separators is less than 2 s cm^{-3} measured by ASTM D726 and D737.

Thermal shrinkage of a separator is generally less than 5% after exposure for 60 min at 90 °C. Separator should not shrink or wrinkle when the temperature rises. The thermal shrinkage during the drying process for battery assembly should be minimized.

When thermal runaway occurs, the membrane should offer the ability to shut down, while retaining its mechanical and dimensional properties. **Shut down** function is the ability to shut the batteries down when overheating or short circuit occurs and is a desirable safety function of separators. The shutdown function should not be achieved by sacrificing the mechanical integrity, which could cause the direct contact of electrodes. Some separators can close the pores and block ionic flows when the thermal runaway temperature is reached. These separators typically have a multilayer design, in which at least one-layer melts to close the pores near the thermal runaway temperature and the other layers provide mechanical strength. When these separators are used, the electrode reactions are stopped in the batteries before an explosion occurs. Currently, thermal runaway temperatures of PE or PP separators are typically 130 or 165 °C, which is close to their melting temperature.

Acceptable **ionic conductivities** in batteries are in the range of 10^{-3} S cm^{-1} to 10^{-1} S cm^{-1} at room temperature. Most separators are not ionically conductive and ion transport occurs only after they are filled with the liquid electrolyte. For cellular phones, ionic conductivity of 10^{-3} S cm^{-1} is sufficient in low power applications.

In high-power automobile applications, higher conductivities are needed. For long-term use, electrolyte retention is also critical.

Electrolyte uptake or wettability. The separator should quickly absorb and retain the electrolyte during battery operation. Poor wettability limits the performance of a cell by increasing the internal ionic resistance. Good wettability shortens the electrolyte filling time during the cell assembly and extends the life cycle of a battery under normal operating conditions. There is no standard method for the wettability test although measuring the contact angle between the liquid electrolyte and the separator is a simple indicator. An alternative method to determine wettability is by hanging a separator vertically in a liquid electrolyte and calculating the wicking (capillary rise) rate or height.

Mechanical properties. A certain mechanical strength is required for separator handling, battery assemble, and abuse tolerance consideration. The most important mechanical properties are the tensile strength and puncture strength. Tensile properties can be measured by the conventional stretching process according to ASTM D882. The strength is strongly influenced by the post-extrusion stretching. Porous membranes usually show a high tensile strength in the stretched direction. The puncture strength is the load required for a needle to puncture a separator and can be tested according to ASTM F1306. High puncture strength helps to deter dendrites or particles from penetrating the battery separator.

5.3 Fabrication of Separator

5.3.1 Type of Polymers

At present, polyolefin membranes, such as polyethylene (PE) and polypropylene (PP), are the most widely used battery separators in LIBs. Melting points of PE and PP are 135 and 160 °C, respectively. Because of their relatively low melting points, thermal stability of separators is a serious concern for the safety of a separator. When the operating temperature exceeds the melting points of polyolefin membranes, the mechanical properties of separators sharply drop and the separators shrink to cause the contact of the anode and the cathode.

To solve the thermal stability issue of polyolefin membranes, high-performance polymers have been reported such as polyimide, polyacrylonitrile (PAN), poly (vinylidene fluoride) (PVDF), and its copolymers, and poly(ethylene terephthalate) (PET). In fact, these polymers have been used for nonwoven nanofiber membranes that show high porosity and excellent thermal stability. However, the weak physical interaction between the fibers brings about the poor mechanical strength of nonwoven membranes, and lithium dendrites easily penetrate those membranes and thus cause internal shot circuits.

5.3.2 Type and Fabrication of Separator

5.3.2.1 Type of Separator

LIB separators can be divided into different types, depending on their physical and chemical properties. Separators are microporous in nature and can be classified into, microporous membrane, nonwoven membrane, and composite membrane. Composite membrane can be prepared by adding inorganic particles into polymer solution and cast the polymer/inorganic particle mixture to form microporous films. Alternatively, inorganic particles can be coated on the microporous membranes from either side (or both sides) of membranes.

Separators are categorized into layers: monolayer, bilayer, and multilayer. For example, most microporous membrane separators are made of PE, PP, and their combinations such as PE/PP and PP/PE/PP, PVDF and its copolymers.

5.3.2.2 Fabrication of Separator

Microporous membranes can be prepared mainly by two different methods[15]: the wet process and dry process.

Figure 5.4 describes the dry process and wet process of manufacturing separators.

Both methods basically involve an extrusion step to prepare polymer thin films and a stretching step to form porous membrane structures. Wet process involves the solvent extraction process after casting thin films and before stretching process. Microporous membranes prepared by the dry process have slit-like pores, while those from the wet process show interconnected and elliptical pores as can be seen in Figure 5.5 [1]. Slit-like and anisotropic pores in Celgard 2400 PP single-layer membrane are uniformly distributed whereas the larger, round-shaped, and isotropic pore structures in PE membranes from wet process are quite different from those made by dry process. The membranes formed by the dry process are more appropriate for high-power density batteries because of their open and straight porous structure. On the other hand, the membranes prepared by the wet process are more suitable in long cycle life batteries because interconnected pores and the tortuous pore structure are beneficial in preventing the growth of dendrites during charging and discharging.

Dry Process The **dry process** generally involves heating and melting of polyolefin (PE or PP) resin, extruding it into a film, thermally annealing it to increase the

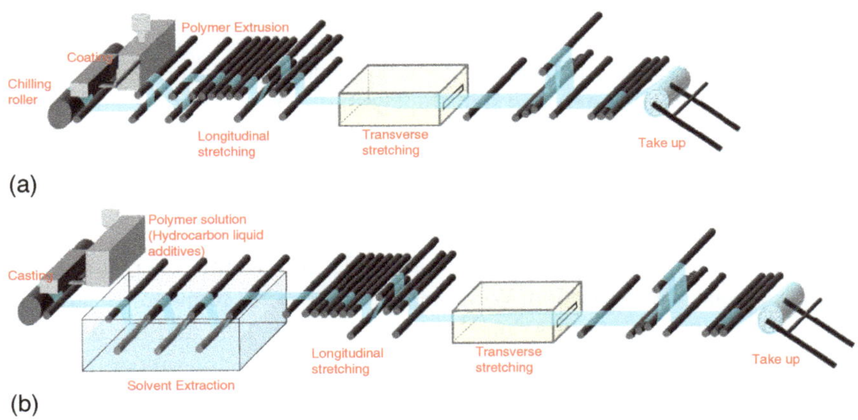

Figure 5.4 Schematic of fabrication process of microporous membranes from (a) dry and (b) wet process.

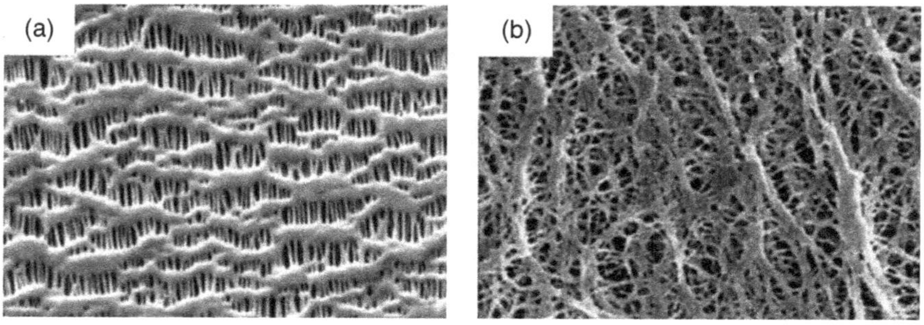

Figure 5.5 SEM images of microporous membrane separators prepared by (a) PP single layer Celgard 2400 manufactured from dry process and (b) UHMWPE membrane from wet process. Source: Reproduced with permission from Arora and Zhang [1]/American Chemical Society.

size and amount of lamella crystallites, and precisely stretching it to form tightly ordered micropores [1, 2]. After the heating process and extruding, the resulting polymer film has a nonporous structure with lamellae aligned in the machine direction. The annealing step is necessary to increase the crystallinity of the polymer membrane, thus facilitating the formation of a porous structure in the stretching step. The annealed membrane is then stretched along the machine direction to form micropores. This microporous structure is continuous throughout the bulk interior of the membrane.

PE and PP microporous films obtained by this method are available from Celgard and Ube. The dry process is technologically convenient because no solvents are required. However, only a uniaxial stretching method has been successful to date, and as a result, the pores are slit-like in shape and the mechanical properties of membranes are anisotropic. The tensile strength in the lateral direction is relatively low.

Wet Process The **wet process** involves the solvent mixing of hydrocarbon liquid or some other low-molecular weight substance generally with a polyolefin resin and extrusion process. This process generally involves in four steps. The first step is to mix and heat the polymer, hydrocarbon liquid, and other additives to form a homogeneous solution. The second step employs the extrusion of the solution into a nonporous film. The third step is to extract the liquid and other additives with a volatile solvent to form the microporous film structure. The fourth step is to stretch the film either uniaxially or biaxially, conducted before or after the extraction of liquid and additives to obtain desirable porosity and pore size.

Separators made by the wet process are available from Asahi Kasei, Tonen and Mitsui Chemicals, SK Innovation, W-Scope, Entek etc. First commercialization of microporous PE membrane was done by Asahi Kasei in 1998 and the separator "Hi-Pore" has been widely used in LIBs so far. The structure and properties of the membranes can be changed depending on the composition of the solutions and the evaporation or extraction of solvents in the gelation and solidification processes. The separators made by wet process use ultrahigh molecular weight polyethylene

(UHMWPE). The use of UHMWPE provides good mechanical properties as well as some degree of melt integrity.

Uniaxially oriented separator has generally high strength in only machine direction whereas biaxially stretched separator is relatively strong in both machine and transverse direction. One can expect biaxially oriented separator to be preferred over uniaxially oriented separators, but in practice, biaxial orientation provides no performance advantage. In fact, biaxial orientation tends to introduce TD shrinkage. This shrinkage, at elevated temperatures, can allow electrodes to contact each other. The separator should have sufficient strength in the MD direction because it should not decrease in width or break under the winding stress. In other words, strength in TD direction is not as important as that in MD direction during the manufacturing process of spirally wound batteries.

Efforts have been made to prepare biaxial orientation of a non-porous PP film with high β-crystal content [33]. Based on this technique, two Chinese manufacturers produce biaxial oriented PP separators. The porosity of these films range between 30% and 40%, with an average pore size of approximately 0.05 μm. The pores on the surface were almost circular compared to fluids with good mechanical properties and almost circular pore shape with narrow pore size distribution [34, 35].

The PE/PP bilayers and PP/PE/PP trilayer separators were developed by Celgard and Ube. Multilayer separators have advantages by combining the lower melting temperature of PE with the high-temperature strength of PP. Nitto Denko patented a single layer separator made from a blend of PE/PP by the dry stretching process [1] (see Figure 5.6).

Phase inversion is extensively investigated to manufacture microporous membranes and separators. Phase inversion process involves the preparation of a polymer solution in a good solvent and the precipitation of the polymer in a non-solvent to form the microporous structure by solvent exchange [36]. When the phase inversion method is used, an asymmetrical porous structure is formed, with a highly porous structure on the top surface, but a less porous structure on the bottom. The asymmetric membrane structure prepared by the phase inversion process is influenced by the polymer type and concentration, the solvent, the type of

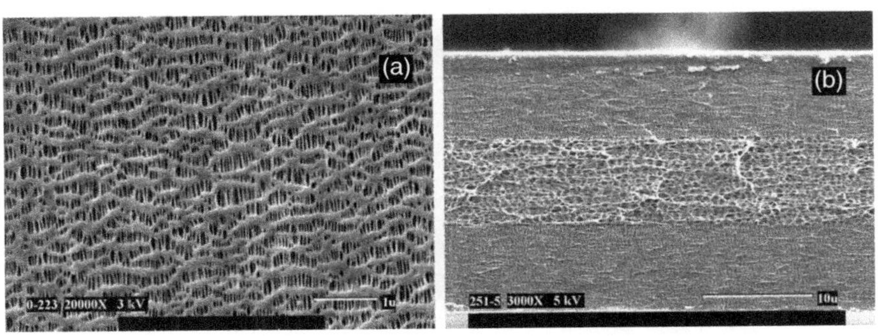

Figure 5.6 Scanning electron microscopes of Celgard 2325 (PP/PE/PP) separator used in LIBs: (a) surface SEM and (b) cross-sectional SEM images. Source: Reproduced from Pankaj Arora et.al. [1]/American Chemical Society.

additive and its concentration, membrane thickness, and processing temperature and time. Currently, microporous polyolefin membranes are the most commonly used separator for rechargeable LIBs. Several microporous multilayer membranes have been used in commercial LIBs, addressing the issues raised from the drawbacks of microporous monolayer membranes.

Modification **Microporous membranes** have been modified by using various techniques such as plasma [37], UV irradiation, gamma ray [38], electron beam irradiation, and polymer coating [39]. Modified microporous membranes impart the improved properties including mechanical strength, thermal stability, wettability, and ionic conductivity.

Nonwoven Membranes **Nonwoven membranes**, especially electrospun nonwoven mats, have been used as the separators in LIBs because of many advantages as battery separators, including high porosity, small pore size, large surface area, interconnected open structure, and high permeability. However, electrospun mats are relatively weak in mechanical strength compared to microporous polyolefin and fluoropolymer membranes and modified microporous membranes. Therefore, polymeric materials with high mechanical strength is required for electrospun mats. In fact, nonwoven mats have very limited applications in LIB separators. Thin nonwoven mat separators can thus be produced with nanofibers (see Figure 5.7), which provide the potential for LIB applications. A nonwoven mat is usually produced through a dry laid process, a wet laid process, a melt-blown process, or an electrospinning process. These processes allow fiber adhesion by sprayed resin bonding, thermoplastic fiber bonding, fiber self-bonding, or mechanical bonding [40].

Composite Membrane **Composite membrane separators** coupled with multilayer membranes and inorganic particles could be promising for high-performance LIBs with high thermal stability, and advanced physical and electrochemical properties [41], especially for EVs or HEVs. This can be achieved either by using high

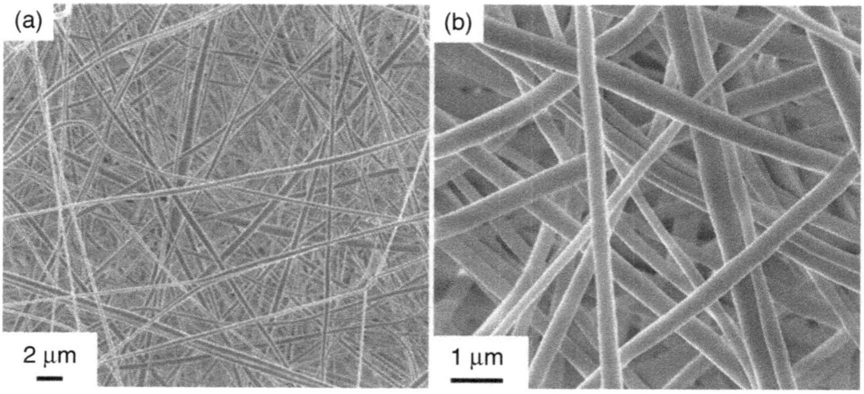

Figure 5.7 Sem images of the electrospun membrane coextruded with PVDF-CTFE in DMF and Sb_2O_3 nanoparticles. Source: Wang et al. [40]/with permission from Elsevier.

melting temperature polymers such as polyamide or polyimide, or by incorporating inorganic fillers into the porous or nonwoven membranes. The latter are called composite or ceramic-incorporated separators. In general, composite separators have excellent wettability by the non-aqueous electrolyte solvents, even when the temperature is low [42]. Composite separators show excellent thermal stability, which is critical to ensure satisfactory battery abuse tolerance at high temperature. In addition, inorganic ceramic particles exhibit relatively high thermal conductivity, dissipating a heat at a faster rate. Because of these advantages, composite separators have attracted much attention recently. However, one of the disadvantages is its low tensile strength compared with the plain separators. Other issues of ceramic composite separator are an increased separator weight due to the incorporation of heavy ceramic particles, unknown electrochemical stability of the ceramic, and the ceramic binder in the highly oxidizing and reducing atmosphere encountered in LIBs.

Ceramic composite separators can be produced by coating micron or submicron ceramic particles with polymeric binders onto either side (or both sides) of the separator [43] (see Figure 5.8). By casting a slurry of inorganic particles on the porous layer of polyolefin, ceramic-coated separators can be produced. This conceptual idea has been already attempted by battery industries (Samsung SDI, LG Chem., etc.) in order to strengthen the mechanical-thermal properties, as well as to improve the electrolyte wettability of polyolefin separators. By adopting ceramic-coated separator, LIBs have delivered improved electrochemical performances even at elevated temperatures. In addition, the thermal stability and mechanical strength of ceramic-coated separators are dramatically enhanced in comparison to the pristine PE separator (Figure 5.8).

Inorganic particles for coating polyolefin separators are nitrides, oxides, carbon etc. A number of research groups have demonstrated the effect of a nitride layer comprising either boron nitride (BN) or AlN on managing the localized heat generated inside the cell and homogenizing Li+ flux [42]. With these nitride coating, the formation of dendritic Li was noticeable suppressed, and the efficiency

Figure 5.8 Ceramic coated microporous separator. Source: From Mohamed Alamgir et al. [44]/IEEE.

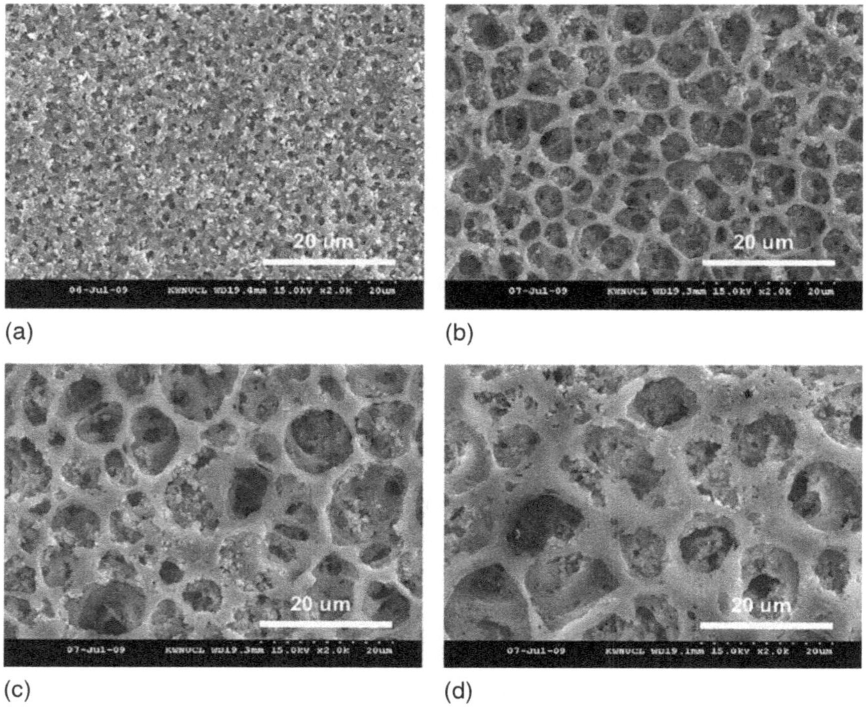

Figure 5.9 SEM images of Al_2O_3/PVDF-HFP binder (50/50, w.w in acetone as a solvent) coated PE membrane as a function of water content (as a non-solvent) in coating solution: (a) 1.89 wt%, (b) 3.77 wt%, (c) 5.66 wt%, and (d) 7.55 wt%. Source: Reproduced with permission from Lee et al. [36]/Elsevier.

of Li plating/stripping was dramatically improved when it is formed on the polymer separators. Oxides are known to be effective in improving the surface wettability with liquid electrolytes and the mechanical/thermal stability against high temperature. For example, Al_2O_3 or $Mg(OH)_2$ coated on the PE separator has exhibited improved thermal/mechanical stability and electrolyte wettability [45] (see Figure 5.9).

Ceramic composite separators can be fabricated by adding ceramic particles into the material formulation for making separators using the wet process. A mixture of PE resin, alumina particles and/or inorganic fibers, and mineral oil was kneaded into a sheet at the melting point of the polyolefin resin. A porous membrane was formed through the extraction of the mineral oils. However, it is still difficult to combine inorganic particles with organic materials effectively. For successful development, the novel composite membrane separators should be tried by understanding the relationship between organic and inorganic materials.

When used in LIBs, microporous membrane, nonwoven mat, and composite membrane separators must be filled with liquid electrolyte in order to transport Li ions between the anode and cathode [46]. Electrolyte membranes, on the other hand, are ionically conductive and can act as both separator and electrolyte between the two electrodes. They are suitable for flexible batteries and/or solid-state batteries with high safety.

In the future, LIB separators would demand unique properties including basic requirements as there has been a continued demand for advanced batteries. Unique techniques such as self-assembly and atomic layer deposition can also be applied to prepare advanced separators used in LIBs [23].

5.4 Gel Polymer Electrolytes

As previously mentioned, LIBs have been widely used in mobile applications due to its high energy density and long cycle life. However, as the energy storage capacity required in the market increases, the conventional LIBs with liquid electrolytes are challenged in terms of their stability (e.g. explosion due to liquid leakage). In particular, the formation of Li-dendrites in conventional LIBs not only induces a short circuit of the cell electrode, but also detrimental to the energy capacity [70]. Among the several methods to inhibit Li-dendrites and enhance cell safety, solid electrolytes have been spotlighted with their superior stability [71, 72]. Solid electrolytes can be classified into ceramic electrolyte and polymer electrolytes. Compared to polymer electrolyte, ceramic electrolytes represented high ion conductivity, but their processability was a concern due to brittle characteristics of the materials [73–75]. Mass production of LIBs should be possible based on excellent processability in the current battery market, thus the research on flexible, lightweight, and electrochemically stable polymer electrolytes are attracting increased attention compared to ceramic electrolytes [76–78].

Polymer electrolytes can be defined as an ionic membrane that is capable of transporting ions comparable to that of common liquid ionic solutions [79]. The advantages of polymer electrolytes over their liquid electrolytes include such as no internal shorting, leakage of electrolytes, and non-combustible reaction products at the electrode surface existing in the liquid electrolytes. A polymer electrolyte battery can be formed by sandwiching the electrolyte between a lithium metal anode and a composite cathode. The polymer electrolyte should allow good cycle lives at low temperatures and must withstand with the internal pressure build up during the battery operation.

The development of polymer electrolytes includes three stages [79]; (i) dry solid-state polymer, (ii) gel/plasticized polymer electrolyte systems, and (iii) polymer composites. The first example of "dry solid" polymer electrolytes is the poly(ethylene oxide) (PEO)-based systems that show low ambient temperature conductivities of the order of 10^{-8} S cm^{-1}. The cycling performance of this PEO system with metal electrodes was not satisfactory and was used as low as 200–300 cycles, due to the low ionic conductivity of the electrolytes. Gel polymer electrolytes or plasticized polymer electrolyte is neither liquid nor solid or conversely both liquid and solid. Gel polymer electrolyte is a two-component system, that is, a polymer matrix swollen with a liquid electrolyte. The gel polymer electrolyte approach to the LIB technology combines the positive attributes of both the liquid (high conductivity) and solid polymer electrolytes (elimination of leakage problems). Gels possess both cohesive properties of solids and the diffusive property liquids. Composite electrolytes are a subset of polymer electrolytes with an idea of incorporating electrochemically inert fillers into polymer matrices. Although several polymers, such as PEO, poly(vinyl chloride), poly(methyl methacrylate), poly(acrylonitrile), can be made for polymer electrolytes, fluoropolymers such as PVDF, and PVDF-HFP have been used for polymer electrolytes.

5.5 Polymers for Separators and Polymer Electrolytes

5.5.1 Polyolefin

PE and PP are the representative polymers for LIB separators. Melting points of PE and PP are 135 and 165 °C, respectively. For PE, HDPE has been used instead of LDPE due to high density and therefore high mechanical strength. In many cases, UHMWPE is used or modified with HDPE at least for better mechanical strength.

For gel polymer electrolytes, PEO has been widely used as discussed in Section 5.4.

5.5.2 PVDF

Because of its attractive properties, PVDF has been widely selected as a polymer for LIB applications. PVDF-based polymer electrolytes are highly anodically stable due to the presence of strong electron-withdrawing group (–C–F) and the polymer itself has a dielectric contact ($\varepsilon = 8.4$) which helps for greater dissolution of lithium salts and subsequently supports high concentration of charge carriers [79].

Table 5.2 includes the summary of LIB separators made of polyolefin and fluoropolymers (such as PVDF and its copolymers) according to the type of separators and polymer type. Depending on the number of layers in a separator, monolayer separator and multilayer separator can be produced. **Monolayer separator** can be prepared by phase inversion of PVDF or PVDF copolymers (typically PVDF-*co*-HFP) or their blends with PAN or PMMA in solvents such as NMP or DMSO [50–55]. The thickness of separators is in the range of 30–50 μm but in some cases, it ranges over 100 μm. Ionic conductivities of these separators range 0.3–5 mS cm^{-1}. Although monolayer separators exhibit good electrical stability and long-term stability, **multilayer separators** show enhanced C-rate performance and good cycle life capacity [58–60]. Separators can be modified with grafting of ethylene glycol containing acrylic monomers using gamma ray or UV grafting methods [61, 62]. **Nonwoven separators** are usually made from electrospinning of PVDF or PVDF copolymers or blends can be prepared using NMP or DMF solvents [40, 63, 64]. Because of high porosity of nonwoven separators, its ionic conductivities are relatively high. These nonwoven separators are usually exhibiting stable charge and discharge curves,

Table 5.2 Summary of polyolefin and fluoropolymer-based separators for lithium-ion battery.

Materials	Solvents	Processing method	Separator thickness (μm)	Electrolyte	Ion conductivity ($mS\,cm^{-1}$)	Performance	References
Monolayer separator							
PE, PP		Dry and wet process	13–57	$LiPF_6$-EC/DEC	0.4–2.1	Thin thickness, high porosity, and large pore size	[47]
PE, PP		Dry and wet process	25–50	$LiAsF_6$-cyclic ester/ether		Preventing thermal runaway	[48]
PE	Paraffin oil	Wet process	25	$LiPF_6$-EC/DEC			[49]
PVDF-HFP	NMP/Acetone	Phase separation	N/V	$LiPF_6$-EC/DEC, $LiClO_4$-EC/PC	0.31–1.6	Good electrochemical stability	[50]
PVDF	Acetone/Ethanol	Phase separation	35–50	$LiPF_6$-EC/DMC/DEC	3.0–5.0	Stable discharge curves based on the low resistance of the separator	[51]
PVDF	NMP	Phase separation	35–70	$LiPF_6$-EC/DMC	0.7–2.1	Good long-term stability and decent capacity at high charge rate	[52]
PVDF/PAN blend	DMSO	Phase separation	146–149	$LiPF_6$-EC/DMC/DEC	2.0–2.9	High thermal stability and good C-rate performance	[53]
PVDF-HFP/PVAc blend	THF	Solvent evaporation	140	$LiClO_4$-EC/PC	2.3	Good electrochemical stability and high thermal stability	[54]
PVDF/PMMA blend	THF	Solvent evaporation	N/V	$LiClO_4$-EC/PC	0.4	Robust mechanical strength	[55]

(Continued)

Table 5.2 (Continued)

Materials	Solvents	Processing method	Separator thickness (μm)	Electrolyte	Ion conductivity ($mS\ cm^{-1}$)	Performance	References
Multilayer separator							
PP + PE substrate		Dry process	12–38			Enhanced puncture strength and thermal resistance	[56]
PP + PE substrate		Dry process	31			Enhanced mechanical and thermal properties	[57]
PVDF + PE substrate	NMP	Phase separation	33	$LiPF_6$-EC/DEC/PC	0.89	Enhanced C-rate performance and electrochemical stability	[58]
PVDF/PMMA/ PEGDA blend + PET substrate	Electrolyte	Dip coating	100	$LiPF_6$-EC/DMC/EMC	4.5	Good cycle life and C-rate capacity	[59]
PVDF-HFP/ PEGDMA blend + PE substrate	Acetone	Dip coating	>12	$LiClO_4$-EC/DEC	0.38	Improved ion conductivity and thermal stability	[60]
Modified separator							
PE/AN	Electrolyte	Plasma	23	$LiPF_6$-DEC/EMC/EC	1.4	Good wettability and stable columbic efficiency	[37]
PP/AA, DEGDM	Electrolyte	Electron beam	50	$LiAsF_6$-PC/EC/DME	1.0	Good mechanical and wettability	[39]
PE	Electrolyte	Gamma ray		$LiPF_6$-EMC/EC	0.9	Improved thermal resistance and good capacity retention	[38]
PVDF/DMAM graft	Electrolyte, THF	Gamma ray grafting	80–100	$LiPF_6$-PC/EC	N/V	Increased C-rate performance and stable cycle life	[61]
PVDF-HFP/PEGDA blend	Electrolyte	UV curing	50	$LiPF_6$-EC/DEC	4.0	Good cycle performance based on improved electrolyte retention	[62]

Nonwoven separator							
PVDF	Acetone, DMAc	Electrospinning	30	$LiPF_6$-EC/DMC/DEC	1.0	Stable charge and discharge curve	[63]
PVDF-CTFE/Sb_2O_3 blend	DMF	Electrospinning	42	$LiPF_6$-EC/DEC	2.88	Excellent cycling stability and rate capability	[40]
PPESK/PVDF core-shell nanofiber	NMP, THF	Coaxial electrospinning	40–60	$LiPF_6$-EC/DMC/EMC	1.62	Robust mechanical strength and superior rate capability	[64]
Composite separator							
Al_2O_3/PP substrate/PEO-PVDF-HFP binder	Acetone	Casting and drying	26	$LiPF_6$-EC/DMC/EMC	3.8	Good rate capacity and cycle stability	[65]
Al_2O_3 particle/PE substrate/PLSS binder	DMAc	Casting and drying	30	$LiPF_6$-EC/DMC	0.7–0.8	Better capacity retention and cycling performance	[66]
Al_2O_3/PE substrate/PVDF-HFP binder	Acetone	Phase separation	30	$LiPF_6$-EC/DEC	0.51	Superior discharge C-rate capacity	[36]
$Al(OH)_3$ particle/PVDF/Cellulose blend	DMAc	Phase separation	50	$LiPF_6$-EC/DMC/EMC	N/V	Improved thermal stability and decent C-rate capability	[41]
MnO_2 particle/PVDF-HFP/PMIA blend	DMAc	Electrospinning	39–43	$LiPF_6$-EC/DEC/EMC	2.27	Excellent cycle performance and superior rate capability	[46]
SiO_2 particle/PVDF-HFP blend	Acetone	Casting and drying	50–70	$LiPF_6$-EC/DMC	0.09–0.31	Good rate capability	[67]
$LiAlO_2$ particle/PVDF-HFP/PVA blend	DMF	Casting and extraction of PVA	50–80	$LiPF_6$-EC/DEC	8.1	Low interfacial resistance and good long-term stability	[68]
Al_2O_3 particle/PVDF-HFP binder/PP substrate	Acetone	Dip coating	26	$LiPF_6$-EC/DMC/EMC	3.8	Enhanced mechanical strength, good rate capacity, and cycle stability	[65]
SiO_2 particle/PVDF binder/PE substrate	Acetone	Dip coating	30	$LiPF_6$-EC/DEC	N/V	Excellent thermal stability and cycle performance	[69]

PLSS: poly(lithium 4-styrenesulfonate)

excellent cycling stability, and rate capability. **Composite separators** are made of coating inorganic particles, such as Al_2O_3 or SiO_2, on top of PVDF or PVDF copolymer separators and polyolefin separators [41, 46, 65, 67–69]. Composite separators are widely used now in actual application of Li-ion batteries due to the high thermal stability of the composite separator over non-coated separators. SiO_2 or Al_2O_3 nanoparticle-coated separators show excellent cycle performance and superior rate capability, low interfacial resistance, and good long-term stability.

Regardless of the type of separators, a homogeneous hybrid film can be obtained with PVDF when EC and or PC with lithium salt of proper proportions [80]. Choe et al. [81] reported PVDF-based electrolytes plasticized with a solution of $LiN(CF_3SO_2)_2$ in PC that offered conductivity of $1.74 \times 10^{-3}\ S\,cm^{-1}$ at 30 °C and oxidation potential limits between 3.9 and 4.3 V versus Li^+/Li.

A novel polymer electrolyte comprising of PVDF-EC-PC and a lithium salt ($LiCF_3SO_3$, $LiPF_6$ or $LiN(SO_2CF_3)_2$) was prepared by thermal extrusion [82]. The mechanical strength of the polymer electrolyte film varied widely and depended on the PVDF content.

A rechargeable all-plastic battery can be fabricated with PVDF-PC-$LiClO_4$ membrane as electrolyte using polyacetylene film as an active electrode material. This battery system has been found to be feasible for small power source for low current electronic devices. However, the problems associated with ionic conductivity polymer electrolyte and the degradation of battery performance and adhesion between the acetylene films and polymer electrolyte are to be rectified [83].

The ionic conductivity of new ternary and quaternary solvent mixtures containing significant amount of toluene was reduced upon the addition of PVDF host matrices [84]. The addition of toluene in the quaternary mixture solvents not only increased the stability of the anodic film but also reduces the solubility and swelling of PVDF. The electrochemical and dielectric properties of PVDF polymer hosts have been reported [85].

5.5.3 Poly(vinylidene fluoride-hexafluoro propylene)

One example of the lithium-ion gel polymer cells, also known as plastic lithium-ion cell (PLIOJN), was developed by Bellcore [86]. Tarascon et al. developed a microporous plasticized PVDF-HFP-based polymer electrolyte which served both as separator and electrolyte. In PLION cells, the anode and cathode are laminated onto both side of the gellable membrane. Good adhesion between the electrodes and the membranes is possible because all three sheets contain large amounts of PVDF copolymer that can be melted and bonded during the lamination step. Due to the thickness limitation of PVDF copolymer, Bellcore process has proven to be an elegant laboratory process but is difficult to implement in large-scale production.

PVDF-HFP separator can be prepared in monolayer, multilayer, modified separator with PERDA, and composite separator by coating inorganic nanoparticles on top of the separator. Gel polymer electrolytes based on a copolymer of PVDF/HFP and a solution of EC, DEC, and $LiN(SO_2CF_3)_2$ as salt were prepared by Capiglia et al.

[87], by changing the content of the polymer. The change in salt concentration was reflected in the conductivity change between 10^{-2}–10^{-8} S cm^{-1}. The ionic conductivity, thermal stability, and compatibility of PVDF-HFP gel electrolytes with a combination of EC and PC as plasticizers and three different lithium salts ($LiCF_3SO_3$, $LiBF_4$, and $LiClO_4$) have been examined [88]. The ionic conductivity of the gel electrolytes increased with the polymer in the gel. Among the salts studied, the film, which possess $LiBF_4$ as salt exhibited maximum conductivity attributed to the low lattice energy of the salt. Saikia and Kumar [89] systematically compared the ionic conductivity transport properties of polymer electrolytes made of PVDF-HFP (or PVDF) + PC + DEC + $LiClO_4$. The film possessing PVDF-HFP host exhibited higher conductivity and transport number than the film with PVDF. High conductivity of PVDF-HFP copolymer came from its amorphous nature compared with higher degree of crystallinity of PVDF, which results in lower conductivity.

5.6 Next-Generation Li Battery

LIBs have changed portable electronics and transformed the electrification of transportation. However, the highest energy storage possible for LIBs is not sufficient for the long-term needs of society, for example, extended range electric vehicles. To go beyond the horizon of LIBs is a formidable challenge; there are few options such as Li-air (O_2) and Li-S (Figure 5.10) [90]. Li-air batteries use air as a cathode without an electrode, thus the specific energy is about, nine times higher than that of the conventional Li-ion batteries (Table 5.3). Similarly, Li-S batteries also exhibit a specific energy of 2567 Wh kg^{-1} through the theoretical redox reaction, offering the potential for electric vehicles with a target driving range of ~500 km. In general, Li-O_2 and Li-S batteries both adopt Li metal as anode because Li possesses the largest capacity (3860 mAh g^{-1}) and the lowest negative electrochemical potential (−3.04 V) versus standard hydrogen electrodes. Although both batteries have active cathode components (O_2 and S) that are nearest neighbors in group 16 of the periodic table, there are

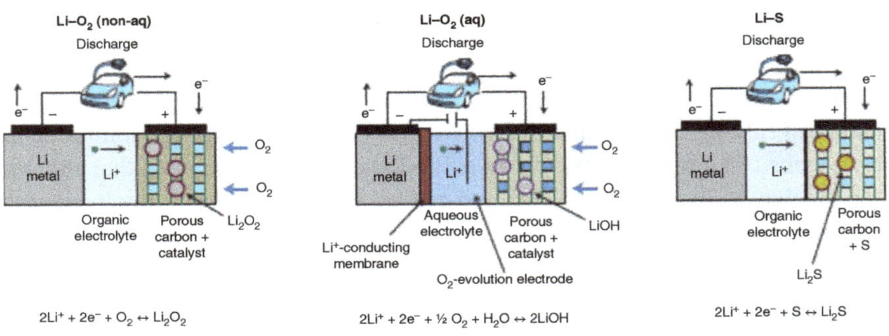

Figure 5.10 Schematic representation of Li-ion, non-aqueous and aqueous Li-O_2 and Li-S cells. Source: Reproduced from Bruce et al. [90]/with permission from Nature Publishing Co.

Table 5.3 Data for several electrochemical reactions in Li-based batteries.

Battery	Cell voltage (V)	Specific energy ($Wh\,kg^{-1}$)	Energy density ($Wh\,l^{-1}$)	References
Conventional Li-ion: $½C_6Li + Li_{0.5}CoO_2 \leftrightarrow 3C + LiCoO_2$	3.8	387 (Theoretical)	1015 (Theoretical)	[90]
$Li–O_2$ (non-aqueous): $2Li + O_2 \leftrightarrow Li_2O_2$	3.0	3505 (Theoretical)	3436 (Theoretical)	[90]
$Li–O_2$ (aqueous): $2Li + ½O_2 + H_2O \leftrightarrow 2LiOH$	3.2	3582 (Theoretical)	2234 (Theoretical)	[90]
Li-S: $2Li + S \leftrightarrow Li_2S$	2.2	2567 (Theoretical)	2199 (Theoretical)	[90]

important differences related to the different chemistry of O and S and the different states of matter of their cathodes (Eqs. (5.6)–(5.8)).

$$\text{Li} - \text{air (non} - \text{aq)} : 2\text{Li}^+ + \text{O}_2 + 2\text{e}^- \Longleftrightarrow \text{Li}_2\text{O}_2, \quad E^0 = 2.96\,\text{V versus Li}^+/\text{Li} \tag{5.6}$$

$$\text{Li} - \text{air (aq)} : 2\text{Li}^+ + \frac{1}{2}\text{O}_2 + \text{H}_2\text{O} + 2\text{e}^- \Longleftrightarrow 2\text{LiOH}, \quad E^0 = 3.35\,\text{V versus Li}^+/\text{Li} \tag{5.7}$$

$$\text{Li} - \text{S} : 2\text{Li}^+ + \text{S} + 2\text{e}^- \Longleftrightarrow \text{Li}_2\text{S}, \quad E^0 = 2.15\,\text{V versus Li}^+/\text{Li} \tag{5.8}$$

However, despite the advantage of Li-air and Li-S batteries, they have shown the problems related with the use of liquid electrolyte such as liquid leakage, ignitibility, and toxicity. Furthermore, the repeat of charging process leads to the accumulation of side products, inducing overpotential or reduction of energy capacity in a cell. In non-aqueous Li-air battery, non-ionized solid LiO_2 and side products caused by decomposition of organic electrolyte cover the electrode and finally detriment cycle performance of the cell. When non-aqueous Li-air battery is operated in ambient air, corrosion of Li metal is also an issue because water or other gases can penetrate open structure of cathode and organic electrolyte [91–93]. To overcome the limitation of non-aqueous Li-air battery, aqueous Li-air battery has been developed that contain water capable of dissolving LiO_2. An ion conductive and electronically insulating membrane was also introduced on the anode to prevent the explosion of Li-metal due to contact with water. However, because aqueous Li-air battery requires a relatively large amount of water to dissolve solid LiO_2, its low energy density (energy per unit volume) and ionic conductivity are still problems to be solved [90, 94]. Li-S battery also has a similar phenomenon to Li-air battery. The polysulfide produced as an intermediate by-product of the redox reaction is dissolved in the organic electrolyte and transferred to the anode, where they are reduced to solid lithium sulfide.

This insoluble lithium sulfide repeatedly crosses the cathode and anode, interfering with the redox reaction (shuttle effect). This problem induces poor cycle and C-rate performances in a cell, hindering the practical application of Li-S battery [95, 96].

A method to solve these problems is to use a solid electrolyte. Solid electrolyte not only solves volatilization or leakage of the electrolyte, but also significantly reduces the risk of fire and explosion. Moreover, solid electrolyte prevents water and carbon dioxide from entering the open structure of Li-air cathode to avoid corrosion of Li metal. It also effectively inhibits the shuttle effect in Li-S battery. Important advances have been made to solid electrolytes in recent years, but significant challenges remain such as low ionic conductivity and interfacial resistance between electrodes and electrolytes [94].

The next generation of lithium-ion battery fully based on nanomaterials will soon be here, followed by all solid lithium-air batteries and others using organic materials [11].

5.6.1 Li-Air Battery Separator

Elia et al. investigated the gel polymer membrane in a Li-O_2 cell using a nanostructured LixSn-C alloy anode [97]. The protective effect of the gel polymer membrane avoids excessive increase of the interphase resistance, as demonstrated by the study of the time evolution of open circuit voltage and by electrochemical impedance spectroscopy. The use of the gel polymer membrane as well as the replacement of lithium metal by a safe Li-alloying electrode are considered valid strategies suitable for the practical improvement of Li-O_2 battery with increased safety content and enhanced electrochemical performance.

A novel oxygen selective membrane consisting of Teflon-coated fiberglass cloth is used to prevent corrosion of the lithium metal negative electrode by water vapor during discharge in ambient air [92]. Water vapor from air must be prevented from corroding the Li metal negative electrode during discharge under ambient conditions, i.e. in humid air. One method of protecting the Li metal from corrosion is to use an O_2-selective membrane that allows O_2 into the cell while stopping or slowing the ingress of water vapor. The O_2-selective membrane has the additional advantage of significantly decreasing the evaporation rates of volatile solvents. This is important because volatile solvents tend to have higher O_2 solubility leading to better cell performance.

The performance of the cell was strongly dependent on the atmosphere; furthermore, it was found that the performance degradation of the Li metal anode comes from moisture contamination from the feed lines [91]. In an ambient atmosphere, the cells showed higher first discharge capacity, higher impedance, and significant increase of the cell weight owing to contamination of the oxygen by moisture. Post-mortem analysis revealed that the deterioration of lithium metal anode leads to the cell failure mechanism, and this comes from the moisture contamination. It was found that the performance of Li-O_2 batteries is very sensitive to even trances of moisture contamination and every single part of the cell design including the choice of the fitting parts and water permeability of the fitting material should be verified in order to obtain credible and reproducible results. This finding supports the idea that protection of the Li metal electrode is indispensable to realize the practical application of Li-O_2 or Li-air batteries.

Further research on the protection of water vapor while transporting O_2 in polymer electrolyte should be continued in the future.

5.6.2 Li-S Battery Separator

One of the major drawbacks of Li-S batteries is the dissolution of polysulfides into liquid electrolytes. In order to overcome this difficulty, polymer/silicate nanocomposite electrolytes composed of polyolefin or PVDF-*co*-HFP and different types of nano-layered silicates were fabricated and analyzed [98, 99]. All the electrolyte membranes showed ionic conductivity of 5–7 $mS\,cm^{-1}$ at room temperature and electrochemical stability window up to 4.8 V. Tested by two different sulfur composite cathodes, PVDF-HFP/organically modified silicate nanocomposite electrolyte delivered a higher discharge capacity and showed an improved cyclability compared to the other cells. Such electrochemical performance is attributed to the small uniformly distributed pores within the structure of the polymer nanocomposite membrane, which promotes immobilizing the electrolyte solution and prevents dissolution of polysulfides.

Huang et al. demonstrated a strategy of introducing an ion selective membrane to improve the stability and Coulombic efficiency of Li-S batteries [100]. The sulfonate-ended perfluoroalkyl ether groups on the ionic separators are connected by pores or channels that are around several nanometers in size. These SO_3^- groups-coated channels allow ion hopping of positively charged species (Li^+) but reject hopping of negative ions, such as polysulfide anions (Sn^{2-}) in this specific case due to the Coulombic interactions. Consequently, this cation permselective membrane acts as an electrostatic shield for polysulfide anions and confines the polysulfides on the cathode side. An ultra-low decay rate of 0.08% per cycle is achieved within the initial 500 cycles for the membrane, which is less than half of the routine membranes.

The same group reported the use of ultrathin graphene oxide (GO) membrane for high stability of Li-S batteries [95]. The oxygen electronegative atoms modified GO into a polar plane, and the carboxyl groups acted as ion-hopping sites of positively charged species (Li^+) and rejected the transport of negatively charged species (Sn^{2-}) due to the electrostatic interactions. Such electrostatic repulsion and physical inhibition largely decreased the transference of polysulfides across the GO membrane in the Li-S system. Consequently, the GO membrane with highly tunable functionalization properties, high mechanical strength, low electric conductivity, and facile fabrication procedure is an effective permselective separator in Li–S batteries. By incorporation of a permselective GO membrane, the cyclic capacity decay rate is also reduced from 0.49% to 0.23%/cycle. As the GO membrane blocks the diffusion of polysulfides through the membrane, it is also with advantages of anti-self-discharge properties.

Highly porous PAN/GO membrane separator exhibited excellent anti-self-discharge feature for Li-S batteries [101]. Porous carbon nanowire (N-PCNW) was designed to modify the separator [102]. The N-PCNW with high N-doped level (7.12 wt%) not only possesses excellent electron conductivity, but also traps the soluble polysulfides effectively with both physical and chemical interactions.

Lithiated Nafion membrane was used as a cation-selective electrolyte for Li-S batteries to suppress the polysulfide diffusion [103]. An advanced cathode structure containing a high-surface-activated carbon nanofiber (AC-CNF) current collector filled with dissolved lithium polysulfides and a thin AC-CNF upper current collector is used to improve the sulfur utilization. The Li-S battery system with the lithiated Nafion membrane and the sandwich cathode exhibits significantly enhanced cyclability relative to the cells with the traditional liquid-electrolyte integrated porous separator. Highly porous glass fibers were demonstrated to be a separator in Li-S batteries [104].

Metal-organic framework (MOF)/GO-based battery separator mitigated the shuttling problem in Li-S batteries and acted as an ionic sieve, which selectively sieves Li^+ ions while efficiently suppressing undesired polysulfides migrating to the anode side [96]. When a sulfur-containing mesoporous carbon material (approximately 70 wt% sulfur content) is used as a cathode composite without elaborate synthesis or surface modification, a Li-S battery with a MOF-based separator exhibits a low-capacity decay rate (0.019% over cycle over 1500 cycles).

5.6.3 All Solid-State Li-Ion Battery

Conventional LIBs use liquid electrolytes and are widely used in portable electronic equipment such as laptop computers, cell phones, and EVs [105]. Liquid electrolytes possess high ionic conductivity (10^{-2}–10^{-3} S cm^{-1}) and excellent electrode wettability at ambient and but a low $t_{Li+} < 0.5$ and suffer from limited electrochemical and thermal stability. This includes limited capacity and cycle life at temperatures over 30 °C, as well as safety issues (e.g. leakage, flammability, and combustibility). The common origin of these LIB safety issues is thermal runaway events that occur when the battery is overheated above 80 °C. Once overheated, the temperature rises exponentially, leading to decomposition of the solid electrolyte interphase (SEI), melting of the separator, and reaction between the anode (lithium metal or intercalated abode) with the organic solvent; in addition, the decomposition of the cathode material releases flammable gases and oxygen. The traditional liquid electrolyte consists almost exclusively of a combination of cyclic and linear alkyl carbonates (i.e. EC + DMC in a 1 : 1 ratio by weight), a highly volatile and flammable solvent, and inorganic $LiPF_6$. The thermal decomposition of the liquid electrolyte, reduction of the electrolyte by the anode, and oxidation of the electrolyte by the cathode are all possible exothermic reactions that can initiate thermal runaway in the battery, which are often caused by thermal (overheating), electrical (overcharge), or mechanical abuse. The safety issues associated with the liquid electrolyte will become more pronounced if a Li-metal anode is used to boost the energy density of the battery. Leakage of electrolyte occurs if any holes are present in LIBs and is the main problem of conventional LIBs. Another problem inherent in liquid electrolyte LIBs is the formation of dendrites of Li, which make it prone to explosion. To circumvent these problems, a solid electrolyte can be positioned between the electrodes. All solid-state batteries utilize solid-state electrolytes to overcome the safety issues of liquid electrolytes [106]. Drawbacks for all solid-state LIBs include high resistance at ambient temperatures and design intricacies. The advantage of this type of battery is a reduction in the net weight and volume of the battery, greater energy output, and easy transfer of Li ions, which affords better efficiency. Solid-state batteries also exhibit some advantages over the other commonly known energy storage devices, capacitors. Solid-state batteries also present challenges, such as their relatively lower power density, high ionic resistance at room temperatures, and manufacturing cost.

Solid-state electrolytes are known to be the most promising option to eliminate the safety concerns with regard to the combination system of a polymer separator and liquid electrolyte [107]. Two fundamental roles of solid-state electrolytes are: (i) a separator between electrodes and (ii) a medium to transfer Li+ between cathode and anode. Solid-state electrolyte also needs to meet the basic requirements of high mechanical and electrochemical stability as well as high ionic conductivity.

Solid-state electrolytes can be classified into two major categories: (i) solid-polymer electrolyte (SPE) (see Figure 5.11) and (ii) solid-inorganic electrolyte (SIE). Even though there are many advantages of SPE, in particular poly(ethylene oxide)-based electrolyte, fundamental safety issues such as thermal runaway cannot be circumvented. Thermal runaway is a typical phenomenon occurred in liquid-electrolyte systems. In contrast, SIE can circumvent many safety problems such as thermal-mechanical stability.

Among oxide-based solid electrolytes, lithium phosphorous oxynitride (LiPON) and garnet-type lithium electrolytes are considered to be the promising solid electrolytes for practical all-solid-state-batteries (ASSBs), due to their excellent thermal stability and reasonable ionic conductivity [107] (see Figure 5.12).

LiPON is an amorphous-type oxide-based solid electrolyte. Main advantages of LiPON are its high stability against metallic lithium, a wide potential window of 0–5.5 V versus Li/Li+, and an extremely high shear modulus of 77 GPa. However, LiPON has relatively low ionic conductivity of $1\text{–}3 \times 10^{-6}\,\mathrm{S\,cm^{-1}}$ at RT, which limits its application in thin-film micro-batteries. Garnet-type lithium electrolyte is a crystalline oxide-based solid electrolyte and has a decent ionic conductivity ($0.1\text{–}1\,\mathrm{mS\,cm^{-1}}$), excellent compatibility with Li, a wide potential window, and high stability under chemical/electrochemical conditions. However, garnet-type solid electrolytes have severe issued inhibiting the practical use in commercial batteries, such as a poor solid–solid interface contract. Crystalline microstructure of SIE undermines the battery safety because it allows the growth of Li dendrites

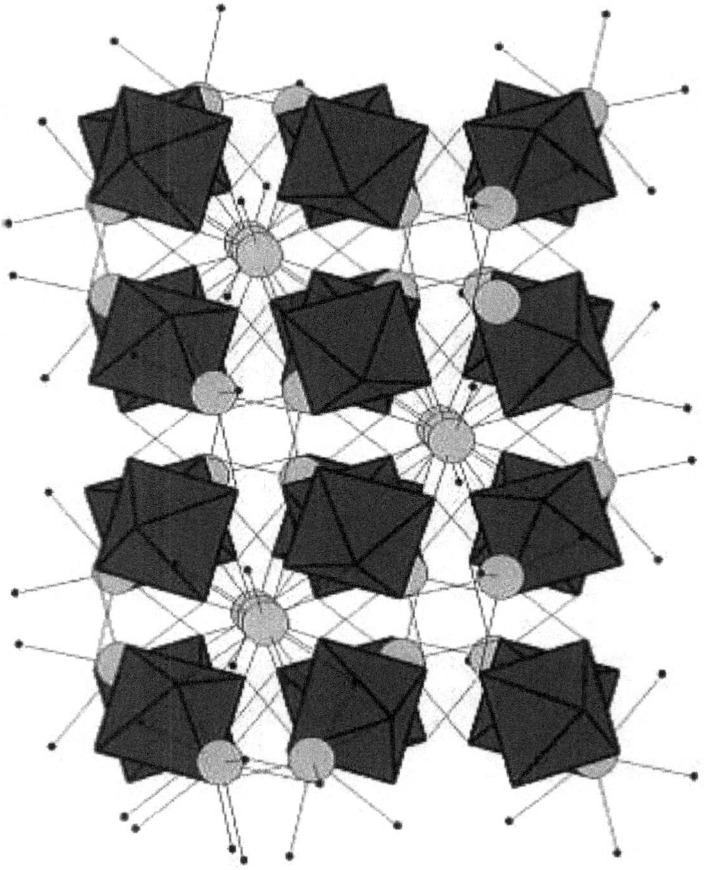

Figure 5.11 Framework of garnet-type solid electrolytes $Li_5La_3M_2O_{12}$ (M = Ta, Nb). Source: Reproduced with permission from Kim et al. [106]/Elsevier.

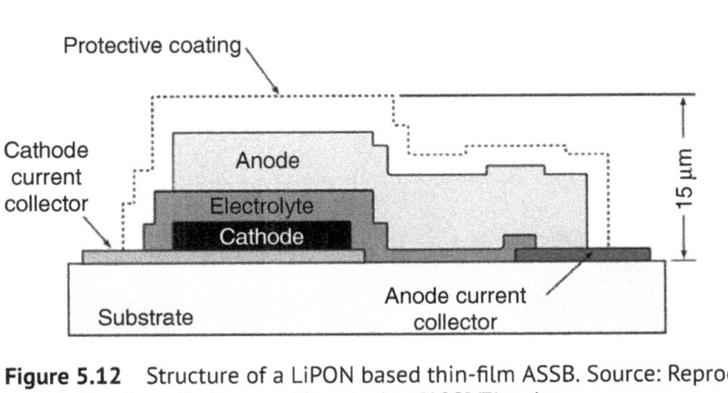

Figure 5.12 Structure of a LiPON based thin-film ASSB. Source: Reproduced with permission from Dudney and Neudecker [108]/Elsevier.

due to the structural heterogeneities. The high manufacturing cost and the multiple preparation steps also restrict the mass production of ASSBs.

Sulfide-based solid electrolytes have been regarded as the most practical solid-state electrolyte for high-power application (e.g. EVs) because of a high ionic conductivity comparable to typical liquid electrolyte and a high Li transference number. In addition, the soft and deformable feature and relatively favorable manufacturing process of sulfide-based electrolytes can overcome the inherent weaknesses of oxide-based solid electrolytes (see Figure 5.13). However, sulfide-based solid electrolytes are vulnerable to the exposure of moisture in the air, leading to the generation of toxic H_2S gas, and thus, the break-up of metal-sulfide bonds. Li dendrites formed in the sulfide electrolyte systems are also a bottleneck. All these unwanted events cause serious safety issues such as battery explosion and require a rigorous battery assemble process, i.e. processing and handling of sulfide electrolytes under a strictly controlled inert atmosphere. To address these challenges of sulfide electrolytes, further studies are inevitable (Figure 5.14).

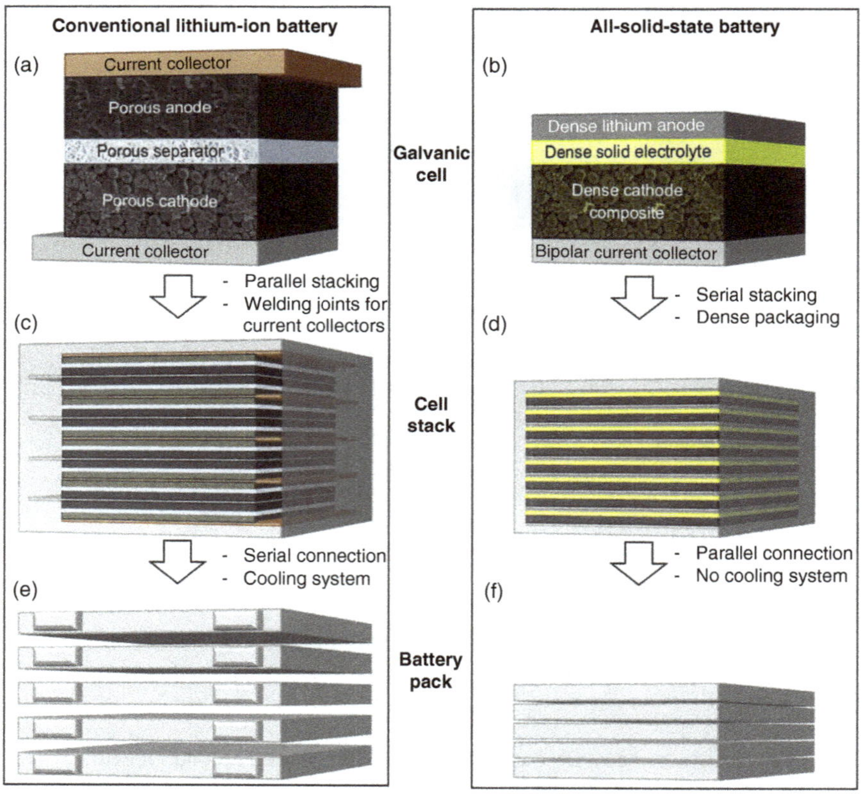

Figure 5.13 Comparison of conventional LIB and all-solid-state lithium, battery as the cell, stack, and pack levels with potentials for increased energy density. Source: Reproduced with permission from Schnell et al. [109]/Elsevier.

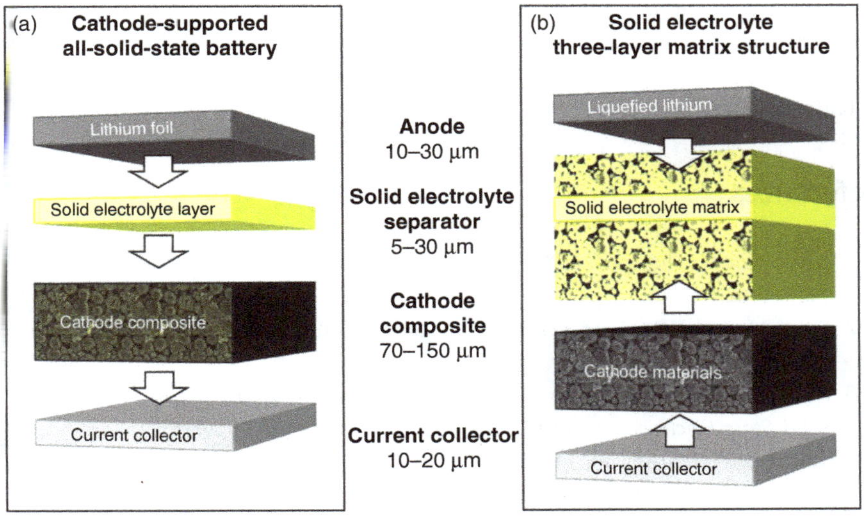

Figure 5.14 Composition and possible design concepts of high energy ASSB with desired layer thicknesses. For example, a 100 μm thick NMC811 or NCA cathode containing 30 vol% solid electrolytes, a 20 μm thick solid electrolyte layer, and a 30 μm thick lithium metal anode would result in an energy density of 1177 Wh l^{-1}, and a specific energy of 426 Wh kg^{-1} for a sulfide-based and 324 Wh kg^{-1} for a garnet-based ASSB, respectively (25) (cell level). However, porosities in the electrodes and external pressing devices will reduce the overall energy density. Source: Reproduced with permission from Schnell et al. [109]/Elsevier.

5.7 Conclusions and Future Perspectives

Spurred from the great increase in LIB market in electronic devices and electrical vehicles, separators production and sales of LIB separators have been extensively grown for the last decade (~3000 million m^2 in 2017) and expected to grow as electric vehicles sales is expected to grow from 5% market share in 2020 to ~30% in 2030 to replace cars with internal combustion engines. Separator is a crucial component in LIBs. The important role of the separator is to protect physical contact between positive and negative electrodes of the battery and to prevent the internal short while serving as the reservoir of the liquid electrolyte to enable ionic transport.

Separators are commonly microporous and are made of usually PE, PP, and fluoropolymers such as PVDF and its copolymers. Microporous membrane separators are fabricated by dry process, wet process, and phase inversion process. About 80% of separators are produced by wet process. Inorganic nanoparticles such as silica and alumina can be coated onto monolayer or multilayer separators to prepare composite separators to increase the thermal stability of the final separator. Further efforts have been focused on improving the energy density of LIBs by replacing Li-ion battery to the next-generation battery such as Li-S and Li-air battery or all solid-state battery. Theoretical energy density is doubled and tripled by replacing Li-ion battery to Li–S and LiO_2 battery, respectively. However, the volatilization of liquid electrolytes of Li-air battery and polysulfide shuttle effect in Li-S battery remains to be solved. Efforts are made to replace Li with Na ions to boost the safety issue of Li ions. Therefore, further efforts should be paid to meet the requirements of these next-generation Li batteries.

References

1 Arora, P. and Zhang, Z. (2004). Battery separators. *Chemical Reviews* 104: 4419–4462.

2 Yang, M. and Hou, J. (2012). Membranes in lithium ion batteries. *Membranes (Basel)* 2: 367–383.

3 Huang, X. (2011). Separator technologies for lithium-ion batteries. *Journal of Solid State Electrochemistry* 15: 649–662.

4 Etacheri, V., Marom, R., Elazari, R. et al. (2011). Challenges in the development of advanced Li-ion batteries: a review. *Energy & Environmental Science* 4: 3243–3262.

5 Schmid, R. and Pillot, C. (2014). Introduction to energy storage with market analysis and outlook. In: *AIP Conference Proceedings*, 3–13. USA: American Institute of Physics.

6 Sanders, M. (2017). Lithium-ion battery raw material supply and demand 2016–2025. Advanced Automotive Battery Conference, Avicenne Energy, San Francisco, USA.

7 Nykvist, B. and Nilsson, M. (2015). Rapidly falling costs of battery packs for electric vehicles. *Nature Climate Change* 5: 329–332.

8 Berckmans, G., Messagie, M., Smekens, J. et al. (2017). Cost projection of state of the art lithium-ion batteries for electric vehicles up to 2030. *Energies* 10: 1314.

9 Hannan, M.A., Lipu, M.S.H., Hussain, A., and Mohamed, A. (2017). A review of lithium-ion battery state of charge estimation and management system in electric vehicle applications: challenges and recommendations. *Renewable and Sustainable Energy Reviews* 78: 834–854.

10 Lu, L., Han, X., Li, J. et al. (2013). A review on the key issues for lithium-ion battery management in electric vehicles. *Journal of Power Sources* 226: 272–288.

11 Armand, M. and Tarascon, J.-M. (2008). Building better batteries. *Nature* 451: 652–657.

12 Curry, C. (2017). Lithium-ion battery costs and market, Bloomberg New Energy Finance, London, UK. 1–12.

13 D. Chung, E. Elgqvist, and S. Santhanagopalan (2015). Automotive lithium-ion battery supply chain and US competitiveness considerations. Clean Energy Manufacturing Analysis Center (CEMAC), CO, USA.

14 Lebedeva, N., Di Persio, F., and Boon-Brett, L. (2016). *Lithium Ion Battery Value Chain and Related Opportunities for Europe*. Luxembourg: European Commission.

15 Lee, H., Yanilmaz, M., Toprakci, O. et al. (2014). A review of recent developments in membrane separators for rechargeable lithium-ion batteries. *Energy & Environmental Science* 7: 3857–3886.

16 Chen, W., Shi, L., Wang, Z. et al. (2016). Porous cellulose diacetate-SiO_2 composite coating on polyethylene separator for high-performance lithium-ion battery. *Carbohydrate Polymers* 147: 517–524.

17 Choi, Y., Kim, J.I., Moon, J. et al. (2018). Electron beam induced strong organic/inorganic grafting for thermally stable lithium-ion battery separators. *Applied Surface Science* 444: 339–344.

18 Na, W., Koh, K.H., Lee, A.S. et al. (2019). Binder-less chemical grafting of SiO_2 nanoparticles onto polyethylene separators for lithium-ion batteries. *Journal of Membrane Science* 573: 621–627.

19 Hao, X., Zhu, J., Jiang, X. et al. (2016). Ultrastrong polyoxyzole nanofiber membranes for dendrite-proof and heat-resistant battery separators. *Nano Letters* 16: 2981–2987.

20 Gao, Y., Sang, X., Chen, Y. et al. (2019). Polydopamine modification electrospun polyacrylonitrile fibrous membrane with decreased pore size and dendrite mitigation for lithium ion battery. *Journal of Materials Science* 55: 3549–3560.

21 Patel, K.K., Paulsen, J.M., and Desilvestro, J. (2003). Numerical simulation of porous networks in relation to battery electrodes and separators. *Journal of Power Sources* 122: 144–152.

22 Costa, C.M., Silva, M.M., and Lanceros-Méndez, S. (2013). Battery separators based on vinylidene fluoride (VDF) polymers and copolymers for lithium ion battery applications. *RSC Advances* 3.

23 Luo, D., Chen, M., Xu, J. et al. (2018). Polyphenylene sulfide nonwoven-based composite separator with superior heat-resistance and flame retardancy for high power lithium ion battery. *Composites Science and Technology* 157: 119–125.

24 Zhang, H., An, X., Liu, L. et al. (2019). Preparation of cellulose-based lithium ion battery membrane enhanced with alkali-treated polysulfonamide fibers and cellulose nanofibers. *Journal of Membrane Science* 591.

25 Deimede, V. and Elmasides, C. (2015). Separators for lithium-ion batteries: a review on the production processes and recent developments. *Energy Technology* 3: 453–468.

26 Xia, M., Liu, Q., Zhou, Z. et al. (2014). A novel hierarchically structured and highly hydrophilic poly(vinyl alcohol-co-ethylene)/poly(ethylene terephthalate) nanoporous membrane for lithium-ion battery separator. *Journal of Power Sources* 266: 29–35.

27 Xu, Q., Wei, C., Fan, L. et al. (2017). A bacterial cellulose/Al_2O_3 nanofibrous composite membrane for a lithium-ion battery separator. *Cellulose* 24: 1889–1899.

28 Li, Y., Wang, X., Liang, J. et al. (2020). Design of a high performance zeolite/polyimide composite separator for lithium-ion batteries. *Polymers (Basel)* 12.

29 Zhang, J., Kong, Q., Liu, Z. et al. (2013). A highly safe and inflame retarding aramid lithium ion battery separator by a papermaking process. *Solid State Ionics* 245–246: 49–55.

30 Chen, W., Liu, Y., Ma, Y., and Yang, W. (2015). Improved performance of lithium ion battery separator enabled by co-electrospinnig polyimide/poly(vinylidene fluoride-co-hexafluoropropylene) and the incorporation of TiO_2-(2-hydroxyethyl methacrylate). *Journal of Power Sources* 273: 1127–1135.

31 Lee, M.J., Kim, J.H., Lim, H.S. et al. (2015). Highly lithium-ion conductive battery separators from thermally rearranged polybenzoxazole. *Chemical Communications* 51: 2068–2071.

32 Chung, Y.S., Yoo, S.H., and Kim, C.K. (2009). Enhancement of meltdown temperature of the polyethylene lithium-ion battery separator via surface coating with polymers having high thermal resistance. *Industrial and Engineering Chemistry Research* 48: 4346–4351.

33 M. Xu, S. Hu, J. Guan, et al. (1992). Polypropylene microporous film. Google Patents.

34 Wu, T., Wang, K., Xiang, M., and Fu, Q. (2019). Progresses in manufacturing techniques of lithium-ion battery separators in China. *Chinese Journal of Chemistry* 37: 1207–1215.

35 Ding, L., Yan, N., Zhang, S. et al. (2022). Low-cost mass manufacturing technique for the shutdown-functionalized lithium-ion battery separator based on Al_2O_3 coating online construction during the β-iPP cavitation process. *ACS Applied Materials & Interfaces*.

36 Jeong, H.-S., Kim, D.-W., Jeong, Y.U., and Lee, S.-Y. (2010). Effect of phase inversion on microporous structure development of Al_2O_3/poly (vinylidene fluoride-hexafluoropropylene)-based ceramic composite separators for lithium-ion batteries. *Journal of Power Sources* 195: 6116–6121.

37 Kim, J.Y., Lee, Y., and Lim, D.Y. (2009). Plasma-modified polyethylene membrane as a separator for lithium-ion polymer battery. *Electrochimica Acta* 54: 3714–3719.

38 Kim, K.J., Kim, Y.H., Song, J.H. et al. (2010). Effect of gamma ray irradiation on thermal and electrochemical properties of polyethylene separator for Li ion batteries. *Journal of Power Sources* 195: 6075–6080.

39 Gineste, J.L. and Pourcelly, G. (1995). Polypropylene separator grafted with hydrophilic monomers for lithium batteries. *Journal of Membrane Science* 1: 155–164.

40 Wang, L., Wang, Z., Sun, Y. et al. (2019). Sb_2O_3 modified PVDF-CTFE electrospun fibrous membrane as a safe lithium-ion battery separator. *Journal of Membrane Science* 572: 512–519.

41 Cui, J., Liu, J., He, C. et al. (2017). Composite of polyvinylidene fluoride–cellulose acetate with $Al(OH)_3$ as a separator for high-performance lithium ion battery. *Journal of Membrane Science* 541: 661–667.

42 Liu, Z., Jiang, Y., Hu, Q. et al. (2021). Safer Lithium-ion batteries from the separator aspect: development and future perspectives. *Energy & Environmental Materials* 4: 336–362.

43 Asghar, M.R., Anwar, M.T., and Naveed, A. (2019). A review on inorganic nanoparticles modified composite membranes for lithium-ion batteries: recent progress and prospects. *Membranes* 9: 78.

44 Alamgir, M., Ketkar, S., and Yoo, K. (2011). Recent progresses of LG Chem's large-format Li ion polymer batteries. In: *2011 IEEE Power and Energy Society General Meeting*, 1–4. IEEE.

45 Jung, B., Lee, B., Jeong, Y.-C. et al. (2019). Thermally stable non-aqueous ceramic-coated separators with enhanced nail penetration performance. *Journal of Power Sources* 427: 271–282.

46 Zhao, H., Deng, N., Kang, W., and Cheng, B. (2020). Designing of multilevel-nanofibers-based organic–inorganic hybrid gel electrolyte enabling an innovative lithium-ion battery with superior ionic transport capability and advanced security. *Chemical Engineering Journal* 390.

47 Djian, D., Alloin, F., Martinet, S. et al. (2007). Lithium-ion batteries with high charge rate capacity: influence of the porous separator. *Journal of Power Sources* 172: 416–421.

48 Venugopal, G., Moore, J., Howard, J., and Pendalwar, S. (1999). Characterization of microporous separators for lithium-ion batteries. *Journal of Power Sources* 77: 34–41.

49 Ihm, D., Noh, J., and Kim, J. (2002). Effect of polymer blending and drawing conditions on properties of polyethylene separator prepared for Li-ion secondary battery. *Journal of Power Sources* 109: 388–393.

50 Shi, Q., Yu, M., Zhou, X. et al. (2002). Structure and performance of porous polymer electrolytes based on P (VDF-HFP) for lithium ion batteries. *Journal of Power Sources* 103: 286–292.

51 Saunier, J., Alloin, F., Sanchez, J.Y., and Caillon, G. (2003). Thin and flexible lithium-ion batteries: investigation of polymer electrolytes. *Journal of Power Sources* 119–121: 454–459.

52 Djian, D., Alloin, F., Martinet, S., and Lignier, H. (2009). Macroporous poly(vinylidene fluoride) membrane as a separator for lithium-ion batteries with high charge rate capacity. *Journal of Power Sources* 187: 575–580.

53 Wu, Q.-Y., Liang, H.-Q., Gu, L. et al. (2016). PVDF/PAN blend separators via thermally induced phase separation for lithium ion batteries. *Polymer* 107: 54–60.

54 Choi, N.-S., Lee, Y.-G., Park, J.-K., and Ko, J.-M. (2001). Preparation and electrochemical characteristics of plasticized polymer electrolytes based upon a P (VdF-co-HFP)/PVAc blend. *Electrochimica Acta* 46: 1581–1586.

55 Nicotera, I., Coppola, L., Oliviero, C. et al. (2006). Investigation of ionic conduction and mechanical properties of PMMA–PVdF blend-based polymer electrolytes☆. *Solid State Ionics* 177: 581–588.

56 Yu, W.-C. and Hux, S.E. (1999). Method of making a trilayer battery separator. Google Patents.

57 Callahan, R.W., Call, R.W., Harleson, K.J., and Yu, T.-H. (2003). Battery separators with reduced splitting propensity. Google Patents.

58 Lee, Y.M., Kim, J.-W., Choi, N.-S. et al. (2005). Novel porous separator based on PVdF and PE non-woven matrix for rechargeable lithium batteries. *Journal of Power Sources* 139: 235–241.

59 Song, M.-K., Kim, Y.-T., Cho, J.-Y. et al. (2004). Composite polymer electrolytes reinforced by non-woven fabrics. *Journal of Power Sources* 125: 10–16.

60 Sohn, J.-Y., Gwon, S.-J., Choi, J.-H. et al. (2008). Preparation of polymer-coated separators using an electron beam irradiation. *Nuclear Instruments and Methods in Physics Research Section B: Beam Interactions with Materials and Atoms* 266: 4994–5000.

61 Jarvis, C., Macklin, W., Macklin, A. et al. (2001). Use of grafted PVdF-based polymers in lithium batteries. *Journal of Power Sources* 97: 664–666.

62 Song, M.-K., Kim, Y.-T., Kim, Y.T. et al. (2003). Thermally stable gel polymer electrolytes. *Journal of the Electrochemical Society* 150: A439–A444.

63 Kim, J.R., Choi, S.W., Jo, S.M. et al. (2004). Electrospun PVdF-based fibrous polymer electrolytes for lithium ion polymer batteries. *Electrochimica Acta* 50: 69–75.

64 Gong, W., Wang, X., Li, Z. et al. (2018). A high-strength PPESK/PVDF fibrous membrane prepared by coaxial electrospinning for lithium-ion battery separator. *High Performance Polymers* 31: 948–958.

65 Liao, Y., Li, X., Fu, C. et al. (2011). Polypropylene-supported and nano-Al_2O_3 doped poly (ethylene oxide)–poly (vinylidene fluoride-hexafluoropropylene)-based gel electrolyte for lithium ion batteries. *Journal of Power Sources* 196: 2115–2121.

66 Choi, J.-A., Kim, S.H., and Kim, D.-W. (2010). Enhancement of thermal stability and cycling performance in lithium-ion cells through the use of ceramic-coated separators. *Journal of Power Sources* 195: 6192–6196.

67 Du Pasquier, A., Warren, P., Culver, D. et al. (2000). Plastic PVDF-HFP electrolyte laminates prepared by a phase-inversion process. *Solid State Ionics* 135: 249–257.

68 Kalyana Sundaram, N.T. and Subramania, A. (2007). Nano-size $LiAlO_2$ ceramic filler incorporated porous PVDF-co-HFP electrolyte for lithium-ion battery applications. *Electrochimica Acta* 52: 4987–4993.

69 Jeong, H.-S. and Lee, S.-Y. (2011). Closely packed SiO_2 nanoparticles/poly(vinylidene fluoride-hexafluoropropylene) layers-coated polyethylene separators for lithium-ion batteries. *Journal of Power Sources* 196: 6716–6722.

70 Meyer, W.H. (1998). Polymer electrolytes for lithium-ion batteries. *Advanced Materials* 10: 439–448.

71 Song, J., Wang, Y., and Wan, C. (1999). Review of gel-type polymer electrolytes for lithium-ion batteries. *Journal of Power Sources* 77: 183–197.

72 Zhang, Q., Liu, K., Ding, F., and Liu, X. (2017). Recent advances in solid polymer electrolytes for lithium batteries. *Nano Research* 10: 4139–4174.

73 Murugan, R., Thangadurai, V., and Weppner, W. (2007). Fast lithium ion conduction in garnet-type Li(7)La(3)Zr(2)O(12). *Angewandte Chemie (International Ed. in English)* 46: 7778–7781.

74 Li, Y., Han, J.-T., Wang, C.-A. et al. (2012). Optimizing Li+ conductivity in a garnet framework. *Journal of Materials Chemistry* 22.

75 Sakuda, A., Hayashi, A., Takigawa, Y. et al. (2013). Evaluation of elastic modulus of Li2S–P2S5 glassy solid electrolyte by ultrasonic sound velocity measurement and compression test. *Journal of the Ceramic Society of Japan* 121: 946–949.

76 Zhu, Y., Wang, F., Liu, L. et al. (2013). Composite of a nonwoven fabric with poly(vinylidene fluoride) as a gel membrane of high safety for lithium ion battery. *Energy & Environmental Science* 6: 618–624.

77 Wang, Y., Travas-Sejdic, J., and Steiner, R.J.S.S.I. (2002). Polymer gel electrolyte supported with microporous polyolefin membranes for lithium ion polymer battery. *Solid State Ionics* 148: 443–449.

78 Pan, Q., Barbash, D., Smith, D.M. et al. (2017). Correlating electrode-electrolyte interface and battery performance in hybrid solid polymer electrolyte-based lithium metal batteries. *Advanced Energy Materials* 7.

79 Manuel Stephan, A. (2006). Review on gel polymer electrolytes for lithium batteries. *European Polymer Journal* 42: 21–42.

80 Watanabe, M., Kanba, M., Matsuda, H. et al. (1981). High lithium ionic conductivity of polymeric solid electrolytes. *Die Makromolekulare Chemie. Rapid Communications* 2: 741–744.

81 Choe, H., Giaccai, J., Alamgir, M., and Abraham, K. (1995). Preparation and characterization of poly (vinyl sulfone)-and poly (vinylidene fluoride)-based electrolytes. *Electrochimica Acta* 40: 2289–2293.

82 Jiang, Z., Carroll, B., and Abraham, K. (1997). Studies of some poly (vinylidene fluoride) electrolytes. *Electrochimica Acta* 42: 2667–2677.

83 Nagatomo, T. (1987). All-plastic batteries with polyacetylene electrodes. *Journal of the Electrochemical Society* 134: 305.

84 Shiao, H.-C.A., Chua, D., Lin, H.-P. et al. (2000). Low temperature electrolytes for Li-ion PVDF cells. *Journal of Power Sources* 87: 167–173.

85 Mohamed, N. and Arof, A. (2004). Investigation of electrical and electrochemical properties of PVDF-based polymer electrolytes. *Journal of Power Sources* 132: 229–234.

86 Tarascon, J.-M., Gozdz, A., Schmutz, C. et al. (1996). Performance of Bellcore's plastic rechargeable Li-ion batteries. *Solid State Ionics* 86: 49–54.

87 Capiglia, C., Saito, Y., Kataoka, H. et al. (2000). Structure and transport properties of polymer gel electrolytes based on PVdF-HFP and LiN $(C_2F_5SO_2)_2$. *Solid State Ionics* 131: 291–299.

88 Stephan, A.M., Kumar, S.G., Renganathan, N., and Kulandainathan, M.A. (2005). Characterization of poly (vinylidene fluoride–hexafluoropropylene) (PVdF–HFP) electrolytes complexed with different lithium salts. *European Polymer Journal* 41: 15–21.

89 Saikia, D. and Kumar, A. (2004). Ionic conduction in P (VDF-HFP)/PVDF–(PC+ DEC)–$LiClO_4$ polymer gel electrolytes. *Electrochimica Acta* 49: 2581–2589.

90 Bruce, P.G., Freunberger, S.A., Hardwick, L.J., and Tarascon, J.-M. (2012). Li–O_2 and Li–S batteries with high energy storage. *Nature Materials* 11: 19.

91 Cho, M.H., Trottier, J., Gagnon, C. et al. (2014). The effects of moisture contamination in the Li-O_2 battery. *Journal of Power Sources* 268: 565–574.

92 Crowther, O., Keeny, D., Moureau, D.M. et al. (2012). Electrolyte optimization for the primary lithium metal air battery using an oxygen selective membrane. *Journal of Power Sources* 202: 347–351.

93 Zhang, J., Xu, W., and Liu, W. (2010). Oxygen-selective immobilized liquid membranes for operation of lithium-air batteries in ambient air. *Journal of Power Sources* 195: 7438–7444.

94 Liu, Y., He, P., and Zhou, H. (2018). Rechargeable solid-state Li-Air and Li-S batteries: materials, construction, and challenges. *Advanced Energy Materials* 8.

95 Huang, J.-Q., Zhuang, T.-Z., Zhang, Q. et al. (2015). Permselective graphene oxide membrane for highly stable and anti-self-discharge lithium–sulfur batteries. *ACS Nano* 9: 3002–3011.

96 Bai, S., Liu, X., Zhu, K. et al. (2016). Metal–organic framework-based separator for lithium–sulfur batteries. *Nature Energy* 1.

97 Elia, G.A. and Hassoun, J. (2016). A gel polymer membrane for lithium-ion oxygen battery. *Solid State Ionics* 287: 22–27.

98 Jeddi, K., Zhao, Y., Zhang, Y. et al. (2013). Fabrication and characterization of an effective polymer nanocomposite electrolyte membrane for high performance lithium/sulfur batteries. *Journal of the Electrochemical Society* 160: A1052–A1060.

99 Deng, N., Kang, W., Liu, Y. et al. (2016). A review on separators for lithium-sulfur battery: progress and prospects. *Journal of Power Sources* 331: 132–155.

100 Huang, J.-Q., Zhang, Q., Peng, H.-J. et al. (2014). Ionic shield for polysulfides towards highly-stable lithium–sulfur batteries. *Energy & Environmental Science* 7: 347–353.

101 Zhu, J., Chen, C., Lu, Y. et al. (2016). Highly porous polyacrylonitrile/graphene oxide membrane separator exhibiting excellent anti-self-discharge feature for high-performance lithium–sulfur batteries. *Carbon* 101: 272–280.

102 Zhou, X., Liao, Q., Tang, J. et al. (2016). A high-level N-doped porous carbon nanowire modified separator for long-life lithium–sulfur batteries. *Journal of Electroanalytical Chemistry* 768: 55–61.

103 Yu, X., Joseph, J., and Manthiram, A. (2015). Polymer lithium–sulfur batteries with a Nafion membrane and an advanced sulfur electrode. *Journal of Materials Chemistry A* 3: 15683–15691.

104 Zhu, J., Yanilmaz, M., Fu, K. et al. (2016). Understanding glass fiber membrane used as a novel separator for lithium–sulfur batteries. *Journal of Membrane Science* 504: 89–96.

105 Kim, K.J., Balaish, M., Wadaguchi, M. et al. (2021). Solid-state Li–metal batteries: challenges and horizons of oxide and sulfide solid electrolytes and their interfaces. *Advanced Energy Materials* 11: 2002689.

106 Kim, J.G., Son, B., Mukherjee, S. et al. (2015). A review of lithium and non-lithium based solid state batteries. *Journal of Power Sources* 282: 299–322.

107 Zheng, F., Kotobuki, M., Song, S. et al. (2018). Review on solid electrolytes for all-solid-state lithium-ion batteries. *Journal of Power Sources* 389: 198–213.

108 Dudney, N.J. and Neudecker, B.J. (1999). Solid state thin-film lithium battery systems. *Current Opinion in Solid State and Materials Science* 4: 479–482.

109 Schnell, J., Günther, T., Knoche, T. et al. (2018). All-solid-state lithium-ion and lithium metal batteries–paving the way to large-scale production. *Journal of Power Sources* 382: 160–175.

6

Membranes in Hydrogen Production by Water Electrolysis

6.1 Introduction

As carbon neutral society needs green hydrogen to combat with the global climate issues, economic ways to produce hydrogen is inevitable. Using cation exchange perfluoro sulfonic acid membranes and iridium oxide, for example, hydrogen can be produced in large scale based on proton exchange membrane water electrolysis (PEMWE). As PEMWE requires expensive catalysts and membranes, as well as the corrosion issues, cheaper anion exchange membrane water electrolysis (AEMWE) is appealing as a competing technology to produce green hydrogen [1]. This chapter introduces the differences between alkaline water electrolysis (AWE), PEMWE, and AEMWE, advantages and disadvantages of each technology, and their recent advances and opportunities in the future.

Hydrogen is mainly produced by steam reforming natural gas or other fossil fuels, such as propane, gasoline, diesel, methanol, or ethanol [2], with an efficiency of around 70–75%, or by coal gasification, which is about 45–67% and has high CO_2 emission. This is alleviated in a processing device called a reformer in which water vapor at high temperature (~700–1000 °C) is reacted with a fossil fuel, in the presence of a metal-based catalyst (usually nickel) [2]. However, steam reforming of fossil fuels produces low purity hydrogen with a high concentration of carbonaceous species such as carbon monoxide. Generally, these processes are suitable for large-scale hydrogen production in central plants, which also offer a favorable cost structure. More importantly, steam reforming does not relieve dependencies on scarce fossil fuels or reduce their pollutants, ultimately, not contributing to the establishment of a "carbon-balanced" energy matrix.

High-quality hydrogen (~100% hydrogen) can be produced by the electrochemical conversion of water to hydrogen and oxygen through a process known as water electrolysis. The reaction with the thermodynamic energy values is described in Eq. (6.1):

$$H_2O_{(l)} + \underset{\text{Electricity}}{237\ \text{kJ mol}^{-1}} + \underset{\text{Heat}}{48.6\ \text{kJ mol}^{-1}} \rightarrow H_2 + \uparrow \frac{1}{2}O_2 \qquad (6.1)$$

For the last decades, the increasing price of electricity has hampered and/or postponed the production of electrolytic hydrogen. This can be changed with the recent growth in energy capacity based on renewable sources like photovoltaics and wind

Membranes for Energy Applications, First Edition. Young Moo Lee.

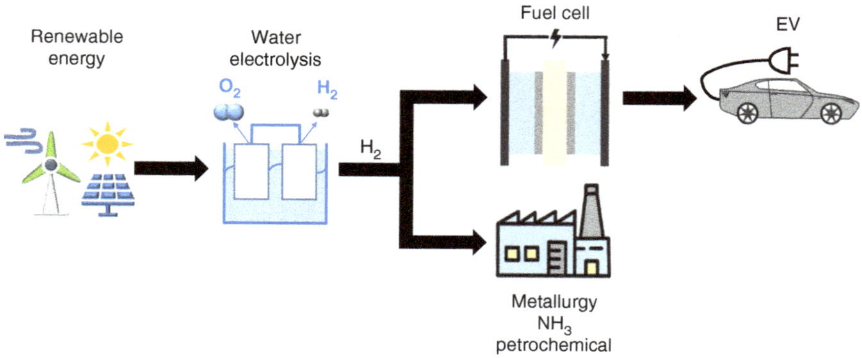

Figure 6.1 The schematic of green hydrogen energy system.

turbines, as well as the global carbon neutral economy policy. In this case, electrolytic hydrogen can not only be produced from a renewable energy source, work as an energy carrier, and work as an energy storage medium, but also overcome the intermittency of typical renewable energy resources.

Some European countries are targeting 100% renewable energy in the energy and transport sectors by 2050. Figure 6.1 depicts the overall scheme of using renewable energy for production of so-called "green hydrogen" from water for industrial sectors and mobility.

Hydrogen gas is used as a feedstock for vital industrial processes including ammonia-based fertilizer production and accounts for 1% of global energy use, with 50 billion kg of gas produced globally each year (~53% for fertilizer) [3]. It is of significant interest for future potential energy storage and transportation and central in plans for reducing, and eventually eliminating, CO_2 emissions. Such technology, however, is only "green" if powered using renewably produced hydrogen, for example, through water electrolysis powered by wind or solar. Of the 70 Mt of hydrogen used globally each year, only 1% is produced through electrolysis. The rest is essentially produced from fossil fuels by coal gasification and natural gas reforming and contributes more than 830 Mt of CO_2 emission annually. Although the cost of H_2 produced by water electrolysis is currently dominated by electricity prices, electrolyzer capital costs will become increasingly important in the future [4]. Hydrogen is mainly used in petroleum refining, ammonia production, and to a lesser extent, metal refining such as nickel, tungsten, molybdenum, copper, zinc, uranium, and lead, amounts to more than 50 million metric tons worldwide in 2006. The large-scale nature of such hydrogen consumptions requires large-scale hydrogen production to match them. For this end, hydrogen production is dominated by reforming of natural gas and gasification of coal and petroleum coke, as well as gasification and reforming of heavy oil [5]. As steam reforming of methane or other hydrocarbons results in high emissions of carbon dioxide, water electrolyzers that electrochemically split water into hydrogen and oxygen have gained much interest.

The development of water electrolysis can be classified into four stages as described in Table 6.1. The first stage is between 1800s–1920s where the first

Table 6.1 Development of water electrolysis technology.

Stages	Years	Events
First	1800s ~ 1920s	Electrolytic water splitting was found
Second	1920s ~ 1970s	100 MW alkaline water electrolyzer was built for ammonia production and petroleum refining
Third	1970s ~ present	PEMWE was established
Fourth	Present ~	WE can be combined with renewable energy

electrolytical splitting of water was found [6]. The phenomenon of water electrolysis was studied, and the gases produced by electrolysis were finally identified to be hydrogen and oxygen, and gradually recognized for its potential applications. As the electrochemistry was gradually studied, the relationship between the electrical energy consumption and the number of gases produced was established through the Faraday's law of electrolysis, reaching the concept of water electrolysis in scientific ways.

The second stage was from 1920s till 1970s where several large 100 MW size plants were built worldwide to produce hydrogen for industrial uses such as ammonia production and petroleum refining. During this period, the first pressurized electrolyzer was built by Zdansky and Lonza in 1948 [7]. The first solid polymer electrolyte system was built by General Electric Company in 1966 [8]. Therefore, this stage is called as the golden age for water electrolysis technology. Driven by the industrial need for hydrogen and oxygen, information known in the first stage was applied to the industrial electrolysis technologies. Some important technology components are still being used today.

The third stage is from 1970s to present where systematic innovative developments were initiated to improve the net efficiency due to the fear of energy shortage and environmental issues. The advances in space exploration spurred the development of PEMWE, which is the reverse of the PEMFC operations, and military needs provoked the development of high-pressure compact alkaline water electrolysis for submarine applications.

The fourth stage is ongoing now for the rapid conceptual development of combining water electrolysis technology with the renewable energy technologies as a means for a distributed energy production and storage of hydrogen in remote communities.

Although water electrolysis to produce hydrogen and oxygen has been known for around 200 years and has the advantage of producing extremely pure hydrogen, its applications are often limited to small scale and unique situations where access to large-scale hydrogen production plants is not possible or economical, such as marine, rockets, space crafts, electronic industry, and food industry as well as medical applications. Water electrolysis represents only 4% of the world hydrogen production [9, 10].

Fossil fuels and nuclear fuels have been massively utilized that causes serious environmental issues such as global climate changes, natural resources exhaustion, pollutant gases emission, and waste generation. The development of renewable energies, represented as hydraulic, wind, and photovoltaic, has been enormous recently. The existing worldwide renewable power capacity reached about 1230 GW in 2009, 7% more than in 2008, and a quarter of the global power-generating capacity, estimated at 4800 GW. In 2009, these renewable energies, other than hydropower rose to a total of 305 GW in 2011, a 22% increase over 2008. Global wind power capacity increased by 38 GW, resulting in the global total to 158 GW. On the other hand, hydropower supplied 15% of the total global electricity production in 2008. Approximately, 31 GW were added during 2009. Concerning the photovoltaic solar energy, an estimated 7.2 GW were added in 2009, increasing the existing total by 53% up to about 21 GW. Global share of renewables were about 22 TWh estimating about 20% of global renewable energy while it was 27 TWh in 2020 estimating about 29% of global power generation and is expected to grow to 48 TWh (60% of global energy) in 2035. Hydrogen demand is projected to grow fivefold by 2050, driven primarily by road transportation, maritime, and aviation. Growth to 2035 is driven by fuel cell electric vehicles that will likely displace conventional diesel trucks. Hydrogen supply is projected to shift from nearly 100% grey hydrogen to 60% clean production by 2035 according to McKinsey & Company report in 2022.

Hydrogen production from water receives an increasing interest as water, including the seawater, is the most abundant natural resources in the globe. Hydrogen economy considers a future scenario in which hydrogen is produced through low-cost environmentally clean processes by means of renewable energy sources. Final use technologies, such as fuel cells, will be widely spread.

Hydrogen is the lightest gas molecule. Its density as a gas ($0.0899\ kg\,Nm^{-3}$) is 15 times lighter than that of air. Hydrogen is a fuel with a wide inflammability range both in air, from 4 to 75 vol%, and in oxygen, from 4 to 95 vol%. It is also the fuel with the highest energy content per mass unit, being its higher heating value (HHV) $3.54\ kWh\,Nm^{-3}$ ($39.42\ kWh\,kg^{-1}$), that is, 2.5 and around three times more energetic than methane and gasoline, respectively.

Although hydrogen is the most abundant chemical element in the universe, it cannot be found in its pure state in the nature. Up to 96% of hydrogen is produced from fossil fuels, and the rest 4% from water. Steam reforming and partial oxidation of hydrocarbons, especially natural gas (methane) prevails, even though coal gasification is also utilized in those countries rich in coals. Renewable hydrogen might also be obtained from biomass through these same thermochemical processes. Hydrogen can be produced from biomass by fermentation. Lastly, water is another interesting renewable resource for hydrogen production by electrolysis.

Regarding the hydrogen economy, hydrogen production methods based on fossil fuel are considered transitional technologies. Water receives a wide attention due to the fact that it is an interesting and clean source of sustainable hydrogen of the future, mainly because the process is easy to be integrated with renewable energy sources and processes.

Electrolysis is the most important way to produce hydrogen from water. It is a well-known electrochemical technology because water generates hydrogen and oxygen upon applying electric current to water to dissociate it. Hydrogen obtained by electrolysis has a high purity of 99.999 vol% when the produced hydrogen is dried and oxygen impurities are removed. This level of hydrogen purity can be used in low-temperature fuel cells, and such purity level is much advantageous over those from fossil fuels or biomass. Electrolytic hydrogen production cost ranges from US$2 to 13 depending on the size of the plant.

Currently, two commercial low-temperature water electrolyzer technology exist: alkaline water electrolyzer (AWE) and proton exchange membrane water electrolyzer (PEMWE). However, major drawbacks of these two technologies are: AWE exhibits low productivity and efficiency, while PEMWE uses a significant amount of critical raw materials. AEMWE has been proposed to overcome the limitations of the current commercial systems.

AWE is the most widely used industrial production of hydrogen that uses aqueous KOH (30 wt%) at 353 K applying a cathodic potential of −1.47 V to overcome the enthalpy change [10]. Classical electrolyzers, using a KOH aqueous solution at −2.0 V give current densities of c. 100 mA cm^{-2}. For this system, nickel alloys as electrode materials with several depositions on the electrode surface have been successfully used as electrocatalysts to increase the current density and to enhance its life span.

Water electrolysis hydrogen produced in conjunction with renewable energy sources are attracting an increasing interest because it is considered only one without raising environmental issues such as generating pollutant gases or nuclear wastes. Hydrogen can be stored in the form of ammonia. Ammonia storage is considered to be an efficient way compared with the electricity storage system (ESS) using lithium-ion battery due to the safety reasons. Combination of reverse electrodialysis (RED) and water electrolysis is also an efficient way to produce hydrogen from water. Note that RED can treat sewage water and seawater to generate electricity. Treated water from RED can be fed to water electrolysis system for hydrogen production.

In this chapter, we will mainly deal with hydrogen production from polymer electrolyte membrane water electrolysis (PEMWE) and alkaline exchange membrane water electrolysis (AEMWE).

6.2 Alkaline Water Electrolysis

6.2.1 History of Water Electrolysis

Since the electrolysis phenomenon was first discovered by Troostwijk and Diemann in 1789 [6], AWE had become a well matured technology for hydrogen production up to the megawatt range, and had become the most welcome technology at a commercial level worldwide [11]. It is characterized by having two electrodes in a liquid alkaline electrolyte consisting of a caustic potash solution at a level of 20–30% KOH. The two electrodes are separated by a diaphragm (Figure 6.2), which has the function of keeping the product gases apart from one another for the sake of efficiency and safety. The diaphragm must also be permeable to the hydroxide ions and water molecules. Table 6.2 lists the state-of-the-art for the specifications of an alkaline electrolyzer and compares the differences between AWE, PEMWE, and AEMWE.

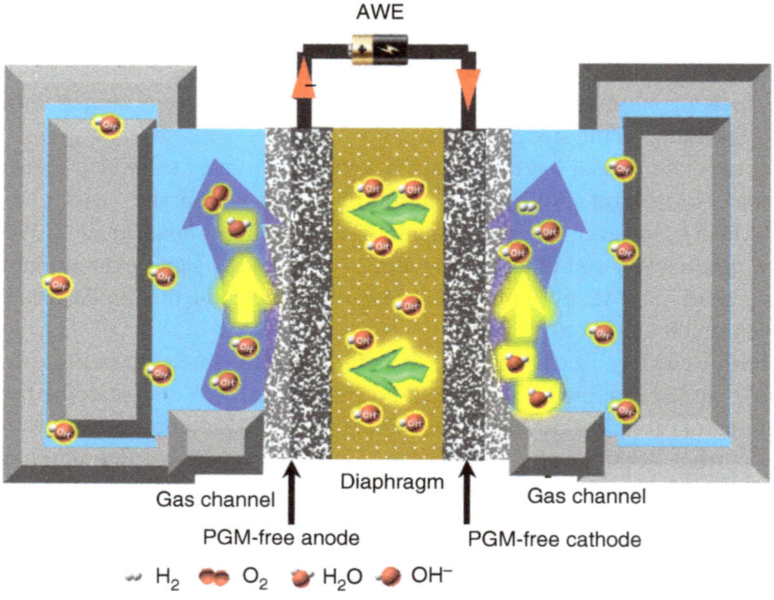

Figure 6.2 KOH-circulating alkaline electrolyzer consisting of a PGM-free electrode (Ni and Fe), diaphragm membrane, and KOH electrolyte.

Table 6.2 Comparison of water electrolysis technologies.

	AWE	PEMWE	AEMWE
Electrolyte	20–30% KOH	PFSA	AEM
Charge carrier	OH^-	H^+	OH^-
Temperature (°C)	65~100	70~90	50~70
Typical discharge H_2 pressure (bar)	25~30	30~80	~30
Separator	Asbestos	PFSA (Nafion)	AEM
OER catalyst	$NiCoO_4$, La-Sr-CoO_3, Co_3O_4	Ir oxide/Ru Oxide	Co_3O_4
HER catalyst	Ni	Platinum	CeO_2–La_2O_3
Typical current collector	Ni	Titanium	Ni
Conventional current density (mA cm^{-2})	200–500	800–2,500	200–500
Demonstrated durability (h)	100,000	100,000~50,000	12,000
Hydrogen purity (vol%)	99.3~99.9	99.9999	99.99
Technology states	Mature	Mature for small scale	R&D

Source: Vincent and Bessarabov [12]/with permission from Elsevier.

Membrane water splitting technology is a general-purpose unit operation for converting water soluble salts to their corresponding acids and bases. The process uses bipolar ion exchange membranes in conjunction with conventional cation and/or anion exchange membranes and the separation and rearrangement of ions is affected by a direct current driving force. The absence of electrochemical transformations enables the water splitting process to be energy efficient, as well as permits direct processing of oxidation sensitive chemicals and fluoride salts. The technology has many applications in the areas of pollution control/resource recovery and chemical processing [13].

6.2.2 Alkaline Electrolysis

The principle of an alkaline electrolyzer is shown in Figure 6.2. The electrodes are immersed in a liquid electrolyte separated by a diaphragm. The electrolyte is usually a 25–30 wt% KOH solution. It is circulated for the removal of product gas bubbles and heat either by pumps or by natural circulation due to temperature gradients and buoyancy of the gas bubbles. The electrolyte is stored in two separate 4d drums for each product gas (O_2 and H_2) which serve also as gas–liquid separator. The product gas quality after drying is typically in the range of 99.5–99.9% for H_2 and 99–99.8% for O_2 which can be increased to above 99.999% by catalytic gas purification. The partial reaction at the electrodes is given by:

$$2H_2O + 2e^- \rightarrow H_2 + 2OH^- \text{ for cathode} \tag{6.2}$$

$$2OH^- \rightarrow \tfrac{1}{2}O_2 + H_2O + 2e^- \text{ for anode} \tag{6.3}$$

$$H_2O \rightarrow H_2 + \tfrac{1}{2}O_2 \text{ for overall reaction} \tag{6.4}$$

Water is consumed on the cathode side while water is produced on the anode side. As a result, the streams of both sides must be mixed before entering the electrolyzer streams. This results in a contamination of product gases due to the electrolyte streams carrying dissolved gases, as the separators can only remove the gas bubbles. A critical contamination (flammable mixture) in part-load must be avoided by a proper control of the circulation.

6.2.3 Major Issues

Three major issues that are normally associated with alkaline electrolyzers are low partial load range, limited current density, and low operating pressure. First, the diaphragm does not completely prevent the product gases from cross-diffusing through it. The diffusion of oxygen into the cathode chamber reduces the efficiency of the electrolyzer since oxygen will be catalyzed back to water with the hydrogen present on the cathode side. Additionally, extensive mixing (particularly, hydrogen diffusion to the oxygen evolution chamber) also occurs and must be avoided to preserve the efficiency, as well as safety. This is particularly severe at a low load (<40%) where the oxygen production rate decreases, thus drastically increasing the hydrogen concentration to unwanted and dangerous levels (lower explosion limit >4 mol% H_2) [14]. The second drawback for alkaline electrolyzers is the low maximum achievable current density, due to the high ohmic losses across the liquid electrolyte and diaphragm. The third problem, also attributed to the liquid electrolyte, is the inability to operate at high pressure, which makes for a bulky stack design configuration.

6.3 Proton Exchange Membrane Water Electrolysis

Proton exchange membrane water electrolysis (PEMWE) is a viable alternative for generation of hydrogen from renewable energy sources [15], overcoming the drawbacks of AWE. Ion exchange polymers have been used in electrochemical application from the late 1950s in the United States at the dawn of the US space program [16]. The concept was mostly used to develop H_2–O_2 acidic fuel cells, the solid electrolyte being considered as a significant advantage for energy management in zero-gravity environments. Later, the discovery of perfluoro sulfonic acid membrane which were able to sustain the very oxidizing conditions found at oxygen evolving electrodes opened the way for PEMWE applications. General Electric Co. developed the first PEMWE in the 1960s and small-scale systems were commercialized for laboratory-scale applications. The first electrolyzers using this polymer began to be commercialized in 1978 [17]. The polymer electrolyte membrane (Nafion) is responsible for providing high proton conductivity, low gas crossover, compact system design, and high-pressure operation. The low membrane thickness (~20–300 μm thick) is in part the reason for many of the advantages of the solid polymer electrolyte.

The basic schematic of PEMWE is shown in Figure 6.3. A proton exchange membrane (Nafion in most cases) separates the two half-cells, and the electrodes are usually directly mounted on the membrane forming the MEA. The corrosive acidic regime provided by the PEM requires the use of metal catalysts like iridium for the anode and platinum for the cathode. Water is supplied at the anode (partly transported to the cathode side due to the electroosmotic grad) and the following partial reactions take place:

$$2H^+ + 2e^- \rightarrow H_2 \text{ for cathode} \tag{6.5}$$

$$H_2O \rightarrow \tfrac{1}{2}O_2 + 2H^+ + 2e^- \text{ for anode} \tag{6.6}$$

$$2H_2O \rightarrow H_2 + \tfrac{1}{2}\,O_2 \text{ for overall reaction} \tag{6.7}$$

The polymer electrolyte membrane requires a very low cross-over of gases, yielding a higher purity than AWE of typically greater than 99.99% H_2 after drying. PEMWE shows a compact module design due to the solid electrolyte and high current density compared with AWE. This supports the high-pressure operation of PEMWE. The structural properties of the solid electrolyte also allow a high differential pressure between the hydrogen and the oxygen side.

PEMWE is well suited for water splitting using intermittent power sources, especially when hydrogen is produced at close-to-atmospheric pressures since

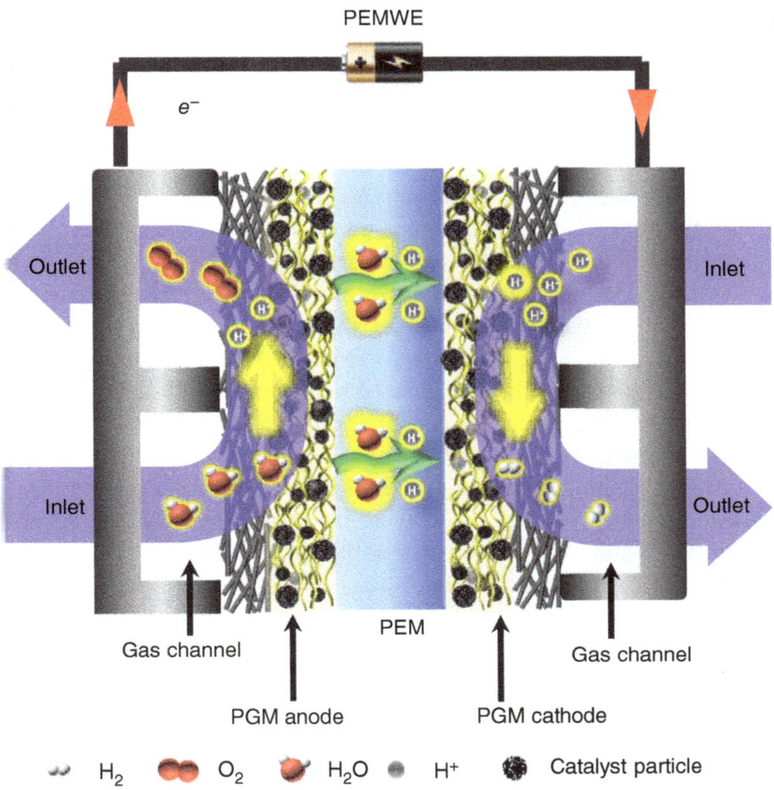

Figure 6.3 Schematic of PEMWE. Electrodes use precious-metal group catalysts in both anode and cathode which is different from AWE and AEMWE processes. PEM is proton exchange membrane.

cross-permeation phenomena is less significant. With conversion efficiencies of 60% and more at high current density (>1 A cm^{-2}), PEMWE is identified as a key process for transforming zero-carbon electricity sources into the supply of zero-carbon hydrogen and oxygen for miscellaneous end-uses [6]. Currently, small-scale systems (<200 kW) can be found on the market with unit prices in the range of 10^3–10^4 \$/kW for units producing 10^{-1} Nm^3H_2/h, depending on the size and operating conditions. To further bring down unit costs to meet market requirements, new materials, and improved manufacturing techniques are still needed. Higher conversion efficiencies are also required. For example, a conventional PEMWE cell operating at 1 A cm^{-2} with a cell voltage of 1.8 V (common good practice at the laboratory) has an efficiency of 70% (based on the lower heating value of 3.0 kWh/Nm^3 H_2). But, potential exists for further process development and optimization of this technology. If the cell voltage is reduced to 1.7 V and the current density is doubled, the efficiency will reach 75% LHV and the investment cost will be reduced by a factor of two.

6.3.1 Advantages of PEMWE

PEMWE can operate at much higher current densities, capable of achieving values above 2 A cm^{-2}, this reduces the operational costs and potentially the overall cost of electrolysis (Table 6.2). Ohmic loss limits maximum achievable current densities. As thin membrane is capable of providing good proton conductivity (0.1 ± 0.02 S cm^{-1}) [18], and is achieving high current densities. Therefore, solid polymer electrolyte allows a thinner electrolyte for PEMWE compared with diaphragm in AWE.

The low gas crossover rate of the polymer electrolyte membrane yields high purity hydrogen, allowing the PEMWE working under a wide range of power input. As discussed above, in AWE operating at low load, the rate of hydrogen, and oxygen production is reducing while the hydrogen permeability through the diaphragm remains constant, yielding a highly concentrated hydrogen on the anode (oxygen) side, and thus creating a hazardous and less efficient conditions [15].

The high-pressure operation of an electrolyzer brings the advantage of delivering hydrogen at a high pressure for the end user, thus requiring less energy to further compress and store the hydrogen. It also reduces the volume of the gaseous phase at the electrodes thus significantly improving product gas removal. In a differential pressure configuration, only the cathode (hydrogen) side is under pressure, this can eliminate the hazards related to handling pressurized oxygen and the possibility of self-ignition of Ti in oxygen [19, 20]. The pressure increase minimizes the expansion and dehydration of the membrane, preserving the integrity of the catalytic layer [19].

6.3.2 Disadvantages of PEMWE

High operational pressure in PEMWE causes problems, such as cross-permeation phenomenon which increases with pressure [21, 22]. Pressures above 100 bar will require the use of thicker membranes and internal gas recombiners to maintain the critical concentrations (mostly H_2 in O_2) under safety threshold (4 vol% H_2 in O_2) [21]. Lower gas permeability (crossover) through the membrane is obtained by adding fillers into the membrane materials, but this is normally leading to less conducting materials.

The corrosive acidic environment provided by the proton exchange membrane requires the use of distinct materials. These materials must not only resist the harsh corrosive low pH condition (pH ~ 2), but also sustain the high applied over voltage (~2 V), especially at high current densities. Corrosion resistance is required for the catalysts, current collectors, and separator plates.

Electrolytic cells require 1.23 V, thermodynamically, to drive the water-electrolysis reaction. The overall cell potential (E_{cell}) also includes electrode overpotentials and other losses that are summarized in Eq. (6.1):

$$E_{cell} = 1.23\text{V} + \eta_{anode} + \eta_{cathode} + iR_{sol} + E_M + E_{pH} \tag{6.8}$$

where η_{anode}, $\eta_{cathode}$, iR_{sol}, E_M, and E_{pH} are, respectively, the potential losses due to anode and cathode kinetics, the solution resistance, the membrane resistance, and pH gradients that can develop across the membrane or near the electrode surfaces.

Most research on electrocatalyst development for photoelectrochemical water splitting has focused on η_{anode}, and $\eta_{cathode}$, and with notable exceptions, has not considered the system-level losses iR_{sol} and, E_M, and E_{pH}. Indeed, in strongly acidic or basic electrolytes, the effects of iR_{sol} and E_{pH} are greatly diminished. This is the case for commercial water electrolyzers, which typically operate at high current density in strongly acidic or basic electrolytes.

6.3.3 Membranes

A thin PFSA membrane with a thickness of ~100 μm is widely used in PEMWE as a solid polymer electrolyte [11, 23–25]. The first commercial PFSA membrane, Nafion from DuPont, is used due to its excellent chemical stability, mechanical strength, and high proton conductivity (see Figure 6.4) [11, 24]. Please refer to Chapter 4 for further information on Nafion. However, cost and disposal of Nafion are the disadvantages. Disposal is an issue as Nafion contains fluorine groups in the polymer and it is expensive to dispose the membrane after use.

Nafion membranes have been extensively investigated in PEM fuel cells [27, 28]. The hydration state of the membrane differs between fuel cell operation and electrolysis operation. The membrane is humidified by a humidified gases and equilibrated with water vapor in PEM operation, while the electrolyte membrane is exposed to liquid phase of water and fully hydrated during water electrolysis [29]. Despite these differences, the water uptake and performance of the membrane may or may not be affected as there is contradictory evidence in the literature [26, 30, 31]. Figure 6.5 compared the polarization curves for the two different single cells equipped with the Nation and Aquivion membranes [32] measured at 80 °C. Nafion 115 MEA showed about 1.5 A cm^{-2} while Aquivion showed 2.4 A cm^{-2} at 1.8 V and 80 °C. At a current density of 1.0 A cm^{-2}, the voltage efficiency referring to the electrolysis thermoneutral potential (~1.48 V) was about 87% for

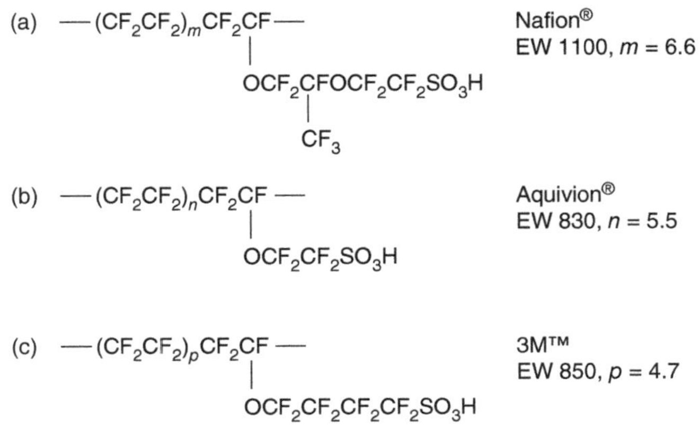

Figure 6.4 Chemical structure of Nafion, Aquivion, and 3M perfluorosulfonic acid membranes and ionomers from DuPont, Solvay Specialty Polymers, and 3M, respectively.

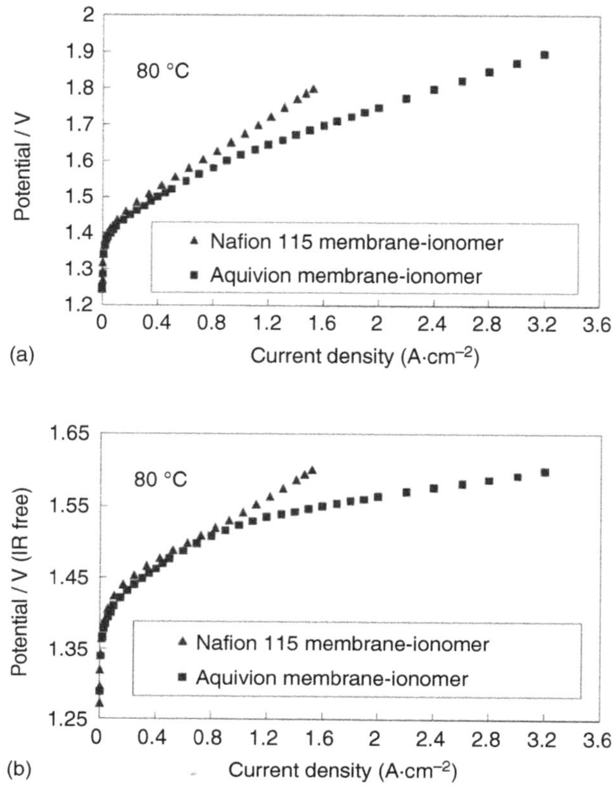

Figure 6.5 (a) Comparison of the water electrolysis polarization behavior for Aquivions- and Nafions-based MEAs at 80 °C. (b) iR-free electrolysis polarization curves. Source: Siracusano et al. [26]/with permission from Elsevier.

the Nafion-based MEA compared to 92% for the Aquivion-based MEA. The voltage efficiency referring to the reversible potential for water splitting (~1.23) was about 72% for the Nafion-based MEA compared to about 77% for the Aquivion-based MEA. The better performance of the Aquivion-based MEA is recorded at high current density in the ohmic-drop corrected polarization curves (Figure 6.5b).

Siracusano et al. [33] investigated on the extruded and recast short side chain Aquivion membranes for PEMWE. The 90 μm thick PFSA membranes provide resilience to hydrogen crossover under differential pressure operation and conductivity to reduce ohmic losses at high current densities. Membranes showed a voltage efficiency >85% and 80% at 1.85 V and 3 and 4 A cm^{-2}, respectively, at 90 °C. Hydrogen crossover analysis at a differential pressure of 20 bar showed low gas permeation through both membranes allowing for a wide load range (15–100%) and high Faradaic efficiency >99% at practical current densities (1–4 A cm^{-2}).

Many studies on the development of hydrocarbon-based membranes have done intending to seek for an alternative for PEM in water electrolysis. Smith et al. [34]

reported on the sulfonated multiblock polyetheretherketone as a proton conducting membrane for production of high purity hydrogen. Electrolysis testing of these ionomers showed good energy efficiency (1.67 V at 1 A cm^{-2} and 80 °C, corresponding to 4 kWh/Nm^3 of H_2). Durability of the cell was tested for 60 hours. Klose et al. [35] studied on the hydrocarbon ionomers that bear the potential to lower the material cost and increase the efficiency of PEMWE. They used sulfonated poly(phenylene sulfone)s (sPPSs) as membrane and electrode binder reaching 3.5 A cm^{-2} at 1.8 V and outperforming state of the art Nafion MEAs (N115 as membrane, 1.5 A cm^{-2} at 1.8 V) due to a significantly lower high-frequency resistance (57 ± 4 mΩ cm^2 versus 161 ± 7 mΩ cm^2). Pristine sPPS membranes show a three times lower gas crossover (<0.3 mA cm^{-2}) than Nafion N115 membranes (>1.1 mA cm^{-2}) in a fully humidified test. Durability test of sPPS MEA showed 80 hours at a constant current density hold at 1 A cm^{-2} at 80 °C. Han et al. [36] reported on sulfonated poly(arylene ether sulfone) (BPSH) with ion exchange capacity (IEC) from 1.2 to 2.0 meq g^{-1}. BPSH membranes show much lower H_2 permeability (20–45 Barrer) than that of perfluorosulfonic acid (PFSA) membranes (~115 Barrer for Nafion) at 80 °C and 100% RH conditions. Random BPSH membranes show slightly higher selectivity of proton to hydrogen than the membranes from multi-block BPSH copolymers at a similar IEC due to the enhanced hydrogen barrier property by less-developed hydrophilic phase. The best performing random BPSH with the IEC ~ 1.9 meq g^{-1} shows 5.3 A cm^{-2} than the similarly thick (~50 μm) Nafion 212 (4.8 A cm^{-2}) at 1.9 V, but the higher degradation rate (951 mV h^{-1}) than Nafion 212 (613 mV h^{-1}) at an accelerated stress test with the 360 consecutive alternating current densities of 3 and 0.02 A cm^{-2}, simulating on and off a water electrolyzer. Park et al. [37] reported similarly on the sulfonated poly(arylene ether sulfone) with degree of sulfonation of 50 mol% (SPAES50) as membrane and ionomer for PEMWE intending to replace Nafion. SPAES50 membrane with a thickness of 20 μm displayed the 1.1 A cm^{-2} at 1 V and 90 °C, which is similar to the performance of Nafion membrane (25 μm). It is attributed to the low ohmic resistance caused by the high proton conductivity of SPAES50 and the small membrane thickness. There is no report on the durability test using these membranes and ionomers. Bender et al. [38] studied proton conducting aromatic multi-block copolymer and blend membranes for PEMWE as an alternative to common PFSA-based polyelectrolytes. They synthesized sulfonated polyarylsulfones and polybenzimidazoles and their blend membranes. The 48 μm thick ternary blend membrane showed a current density of 4.1 A cm^{-2} at 2.0 V and 80 °C, which is comparable with 50 μm Nafion membrane and its ionomer. There is no durability data. Choi et al. [39] studied on the sulfonated poly(arylene ether sulfone) as a membrane and Nafion as an ionomer for PEMWE. They added CeO_2 as a radical scavenger to interlock the interfacial layer that addresses the interfacial delamination issue. The ball-socket joint structure of the IIL (Figure 6.6) leads to mechanical interlocking at the interface, and the presence of CeO_2-containing IIL-based PEMWE operates more than 500 h with low polarization at a voltage decay rate of 48 mV h^{-1}, which is lower than Nafion-based PEMWE (53 mV h^{-1}).

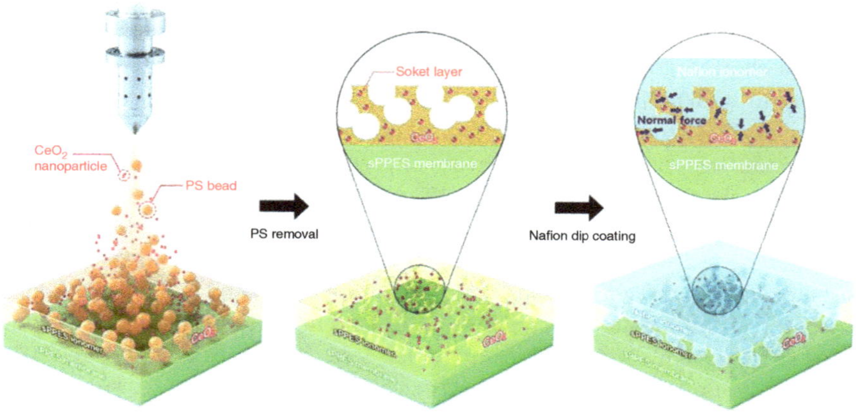

Figure 6.6 Schematic diagram of the structure and fabrication process for the radical scavenger-containing interlocking interfacial layer for PEMWE. Source: Adapted from Park et al. [37].

Wan et al. [40] reported on a series of H_2SO_4-doped polybenzimidazole (PBI) membranes for hydrogen production with acid-alkaline amphoteric water electrolysis. m-PBI membranes doped in 3.0 M H_2SO_4 attains a current density of 0.8 A cm^{-2} at 2.0 V at 60 °C. A long-term stability for 40 hours operation at 0.1 A cm^{-2} was obtained with an energy consumption of 3.35 kWh m^{-3} H_2.

As a proof-of-concept study, reverse osmosis membranes were studied to replace high-cost ion exchange membranes for water electrolysis [41]. In an applied electric field, RO membranes have a substantial capacity for proton and hydroxide transport through the active layer while excluding salt anions and cations. By comparing two types of membranes (Brackish Water [BW] and Saline Water [SW] membranes), BW membrane has acceptable performance over the membrane resistance and electrolysis over potential. Overall, there are remaining challenges for using ion excluding thin film composite membranes such as RO membranes compared with the ion exchange membranes that facilitate transport of all charged ions. RO membranes possessed promising selectivity of protons over cation salts when using high concentrated electrolytes and the polyamide-based TFC membrane.

Nevertheless, these alternative hydrocarbon-based membranes have shown rather low current densities and low durability compared with standard Nafion membranes [24]. Even for PFSA membranes, another drawback of PFAS membranes is known to loose water, and thus ionic conductivity, at high temperature above 100 °C [42], which prohibits them from being used for high temperature water electrolysis. High-temperature operation offers several advantages from thermodynamics, kinetics, and engineering point of view. When the temperature is increased, the electrode kinetics will be enhanced and therefore the over potentials will be reduced [38, 43]. If water is reacted above 100 °C in the gaseous form, the electrolysis process will be thermodynamically less energy demanding [44]. The reversible voltage of the electrolysis cell is 1.23 V at 25 °C, considerably higher than at 200 °C for steam water (1.14 V). Better membrane hydration and proton conductivity can be obtained when the electrolysis system is pressurized. A higher operation pressure is favorable for PEMWE since it reduces the gas pressurization constraints for storage purpose [11]. Therefore, it is important to develop membranes that resist at high-temperature operation and high pressure (>350 bar) [45].

Membranes blended with inorganic particles such as SiO_2, TiO_2, or WO_2 allow the operation at high temperature and pressure of the electrolyzer [46]. These composite membranes decrease the gas crossover. Hygroscopic inorganic fillers inside of the composite membrane allow the high-temperature operation as water is retained due to the presence of hydrophilic fillers. Phosphoric acid doping of PBI membranes have emerged as a promising PEM material for PEMFC operation [47–54]. PA-doped PBI was blended with PEEK and tested for hydrogen production in high-temperature copper chloride electrolysis [55].

6.3.4 Ionomers

Ionomers play a role as a binder providing a three phase boundary (TPB) between catalysts and ionomers, giving mechanical stability and durability to the electrode [24, 56]. The other role is to promote the proton transport from the bulk of the catalyst layer to the membrane, enhancing the overall efficiency by decreasing the ohmic losses. Ionomer, however, decreases the electronic conductivity of the catalyst because the polymer is electron resistant. The ionomer in the catalyst layer also makes the catalyst layer more hydrophilic, potentially reducing mass transport out of the catalyst layer. Therefore, an optimization of the TPB and ionomer layer characteristics is required.

Zhang et al. [57] studied on the effect of membrane electrode assembled (MEA) for PEMWE using catalyst coated membrane (CCM) process. Additional Nafion layer was sprayed between the membrane and catalyst layer, and this results in enlarging the roughness of membrane surface and the interface between membrane and catalyst layer without lowering the mechanical strength of membrane. Carmo et al. [58] studied on the effect of grafting the surface of carbon black with polymeric chains with proton conducting properties that enhances a better interaction of proton conducting phase with the metallic catalyst particles and hinders posterior support particle agglomeration. Xu et al. [59] studied on the effects of Nafion ionomer content on PEMWE using MEA performance. Using CCM method, an optimum ionomer content in the anode was 25% while it is 20% for cathode. They [60] also studied that the short-side-chain (SSC) Aquivion perfluorosulfonic acid (PFSA) ionomer in catalyst layer of PEMWE instead of long-side-chain (LSC) Nafion ionomer. In a single cell test, the MEA with SSC-PFSA Aquivion ionomer exhibited better thermal stability than the one with LSC-PFSA Nafion ionomer at 90 °C. The cell voltage at a current density of 1 A cm^{-2} was 1.63 V at 90 °C using the SSC-PFSA

Aquivion ionomer binder, Nafion 117 membrane, and without back pressure. In a continuous operation, the cell voltage degradation rate of the MEA using Aquivion ionomer binder was about 0.82 mV h^{-1}. Xu et al. [61] added the PTFE emulsion in Nafion binder and improved the cell performance. Bernt and Gasteiger [62] studied on the influence of ionomer content in IrO_2/TiO_2 electrodes on PEMWE performance (Figure 6.7). The best performance is obtained for an ionomer content of 11.6 wt% and a cell voltage of 1.57 V at 1 A cm^{-2} and less than 2 V at 6 A cm^{-2} (ambient pressure and 80 °C). Performance losses at lower ionomer contents are the results of a higher proton conduction resistance. For higher ionomer contents, on the other hand, performance losses can be related to a filling of the electrode void volume by ionomer, leading to a higher O_2 mass transport resistance, an increased electronic contact resistance, and the electronic insulation of parts of the catalyst

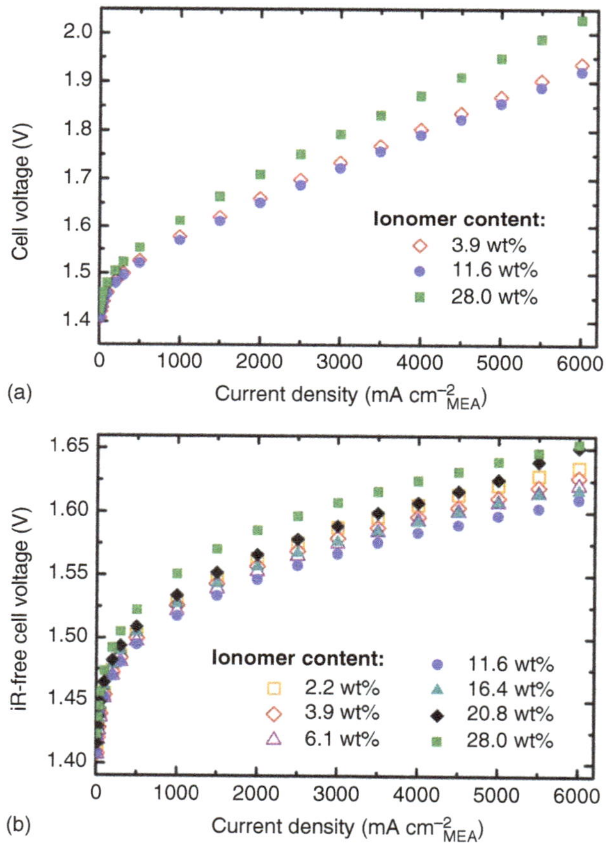

Figure 6.7 Effect of Nafion ionomer loadings in wt% using 50 mm Nafion 212 membrane. (a) Measured cell voltage; (b) cell voltage corrected by HFR. Source: Bernt and Gasteiger JES [62]/Institute of Physics Publishing/CCBY 4.0.

by ionomer. Siracusano et al. [63] reported on the effect of short-side chain perfluorosulfonic acid (PFSA) ionomer on the performance and durability of PEMWE. Using Aquivion SSC-PFSA membrane and various cathode and abode noble metal loadings, Aquivion membranes showed 3.2 A cm^{-2} at 1.8 V (~80% efficiency) at 90 °C in the presence of a total catalyst loading of 1.6 mg cm^{-2}. Durability test for 1000 hours at 1 A cm^{-2} showed stability for the MEA with noble catalyst loading of 1.6 mg cm^{-2} (cell voltage increase ~5 μV h^{-1}). Moderate degradation rate (cell voltage increase ~15 μV h^{-1}) was recorded for the low loading, 0.5 mg cm^{-2}, MEA. These were attributed to the enhanced proton conductivity and good stability of the Aquivion membrane, and the optimized structural properties of the Ir and Ru oxide catalyst [64].

6.3.5 Technical Achievements and Applications

While AWE used today offer production capacities up to several thousand Nm^3 H_2/h, commercial PEM systems are limited to several tens Nm^3 H_2/h. PEMWE can be used for industrial applications, when either hydrogen or oxygen of electrolytic grade is required. They can be used to produce hydrogen from renewable energy sources in view of the hydrogen society, but further cost reductions are still required. An automated GenHy 5000 unit which can be operated between 600 and 1000 mA cm^{-2} and can deliver gases up to 6 bar is shown in Figure 6.8 [17].

While the large-scale production of electrolytic hydrogen in the industry is performed using the more developed AWE, PEMWE technology provides a safer and more efficient way to produce hydrogen and oxygen from renewable energy sources. Stack efficiencies of up to 65% have been obtained with 1 Nm^3 H_2/h capacity electrolyzers operating at 1 A cm^{-2} and a potential exists to further increase this efficiencies [17]. Hydrogen is safely produced in the 1–10 bar pressure range, offering the possibility of direct storage in hydride tanks at moderate pressure. PEMWE can be used to develop auxiliary power units (APU) as well as hydrogen refueling stations, although higher operating pressures are still required. In the future, higher operating temperature (in the range of 100–150 °C) and higher operating current densities (in the 1–5 A cm^{-2} current density range) are still being investigated to further bring down the price of such systems.

The small-scale electrolyzed water hydrogen production equipment of PEMWE is used for hydrogen cooling in power plants [65]. In large thermal power plants, generators sets are an important part. To avoid the motor burning accident caused by the high temperature of the motor, coolant is used to cool the generator set. The commonly used coolants include air, water hydrogen, and oil. Among them, hydrogen has high thermal conductivity (0.182 W m^{-1}, K^{-1} at 25 °C) which is 8.4 times the thermal conductivity of air, small specific gravity, fast diffusion, easy to transport, and can be recycled. Therefore, hydrogen is a very effective cooling medium. For this reason, many current 330–660 MW thermal power units use hydrogen as the cooling medium of the generator set. Because of the technical advantages of

Figure 6.8 GenHy5000 PEM water electrolyzer (5.0 Nm^3 H_2/h) for the storage of renewable energy sources (CETH). Source: Grubb [16]/IOP Publishing.

PEMWE hydrogen production, this technology has been widely used in 35 power plants in United States, 6 power plants in Spain, and 6 power plants in Romania have adopted this hydrogen production technology. Application in civil and aerospace has attracted an attention all over the world.

6.4 Alkaline Exchange Membrane Water Electrolysis

Alkaline water electrolysis is one of the easiest methods for hydrogen production, offering the advantage of simplicity. The challenges for widespread use of water electrolysis are to reduce energy consumption cost and maintenance and to increase reliability, durability, and safety. AEMWE could become the cornerstone to achieve an intense, safe, and resilient green hydrogen production to fulfill the hydrogen targets to achieve the 2050 decarbonization goals. Here, we will focus on the development of AEMs for WE application and highlighting the future perspectives on the AEMWE research.

These state-of-the-art AEMWEs demonstrate that current densities have been advanced dramatically over the past several years, while most of their cell durability is limited at below 200 hours at low current density [66] except a few cases. The current density of most AEMWEs is far lower than that of state-of-the-art PEMWEs (6 A cm^{-2} at 2.0 V) [62].

AEMWE is a relatively new technology that seeks to combine the advantages of AWE and PEMWE to offer a low-cost option with compact design and high current density operation. To produce H_2, water is reduced at the cathode to form H_2 and OH^- anions through the hydrogen evolution reaction (see Figure 6.9).

$$4H_2O + 4e^- \rightarrow 2H_2 + 4OH^- \text{ for cathode} \tag{6.9}$$

$$4OH^- \rightarrow O_2 + 2H_2O + 4e^- \text{ for anode} \tag{6.10}$$

$$2H_2O \rightarrow 2H_2 + O_2 \text{ for overall reaction} \tag{6.11}$$

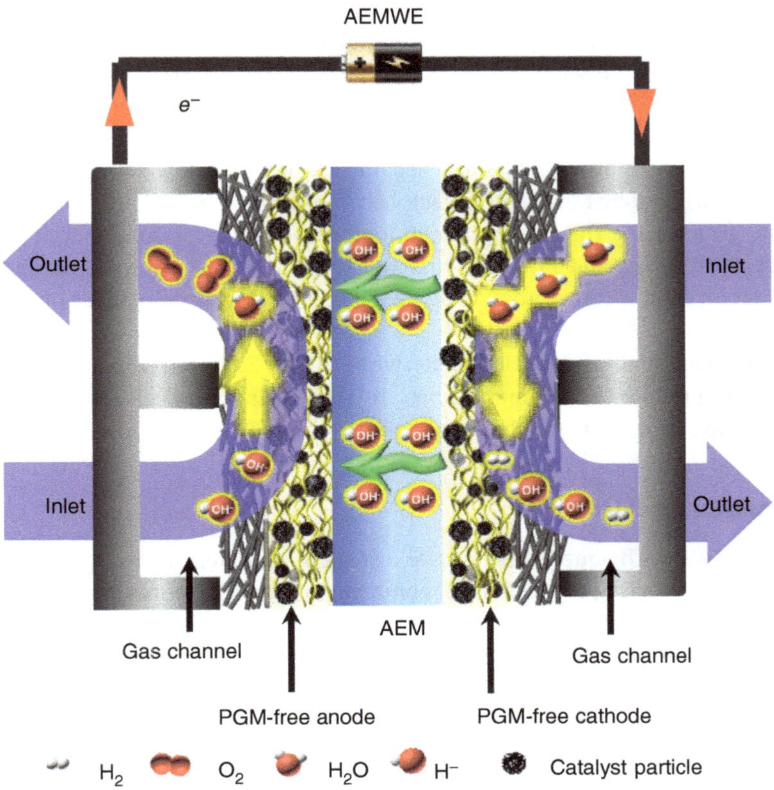

Figure 6.9 Schematic of AEMWE. Anion exchange membrane (AEM) separates the PGM-free electrodes.

The overall reaction requires 1.23 V theoretical thermodynamic cell voltage to split water into hydrogen and oxygen at 25 °C. However, in practice, the requirement of the cell voltage for efficient hydrogen production is >1.23 V. The additional voltage is required to overcome both the kinetics and ohmic resistance of the electrolyte and components of the electrolyzer. Typical operational cell voltages for AEMWE and PEMWE are 1.85–2.05 V and 1.75 V at 70–90 °C.

6.4.1 Difference Between AWE and AEMWE

A water electrolysis technology that combines the advantages of AWE and PEMWE is the anion exchange membrane water electrolysis (AEMWE). The difference between AWE and AEMWE is that the electrode separator in the AWE has no intrinsic ionic conductivity, and its conductivity is provided by KOH that fills the pores of a porous separator (see Figure 6.3) whereas in the latter, the ion-exchange polymer membrane is non-porous and possesses the intrinsic anionic conductivity (see Figure 6.9). The key advantages of AEMWE over AWE are: (i) they operate at high current densities due to lower ohmic resistance and consequently has a smaller volumetric footprint while utilizing fewer materials that in turn lead to a smaller carbon footprint; (ii) they allow pressurizing hydrogen electrochemically, thus facilitating its subsequent storage, distribution, and final utilization; (iii) using the non-porous polymeric membranes, the safety of the system is significantly improved.

Moreover, AEMWE has key advantages over PEMWE: (i) reducing the high costs of the electrocatalysts, such as iridium oxide and platinum, and the corrosion-resistant current collectors and separator plates in acidic environments that limit the large systems as the cell stack becomes a larger contributor to total system cost; (ii) reducing the environmental footprint over the entire life cycle of the product. Both AWEs and PEMWEs are considered matured technologies and have been deployed at a commercial scale depending on the specific needs of the applications.

AWEs use concentrated KOH as electrolyte, while PEMWEs are fed with deionized water (DI), using only solid polymer electrolyte. AEMWEs have been tested both with DI water and aqueous electrolyte feed typically diluted KOH or K_2CO_3 [67].

6.4.2 Liquid Electrolytes

Operating AEMWE with aqueous electrolyte instead of DI water causes significant drawbacks [68]. By making the feed water conductive, shunt currents will occur in the anode manifolds of a bipolar electrolyzer stack. Shunt currents are a well-known major problem in bipolar AWEs that have led to either very complicated stack designs or adoption of unipolar systems [68]. Shunt currents cause efficiency loss, corrosion, and possible flow distribution or safety concerns related to hydrogen and oxygen gas generation in the manifold. The less concentrated electrolytes often used in AEMWEs may reduce the magnitude of the shunt current problem compared to AWEs, but they will not eliminate it. Aqueous electrolytes increase the potential issues for corrosion among the balance of plant components. Also, the use of carbonate or bicarbonate electrolytes instead of hydroxide, while advantageous from a corrosion point of view, will cause mass transport losses due to the low pH gradient between cathode and anode, particularly in the case of bicarbonate. All these factors increase the cost and complexity of AEMWEs with aqueous electrolytes to a very significant extent and therefore DI water operation in AEMWE is recommended in the future AEMWEs as the current PEMWEs work with DI water. However, running AEMWEs with DI water is very challenging. Even the best AEMs produced so far degrade significantly when they are tested in AEMWEs running with DI water. This degradation may occur by oxidation of the ionomer at high potentials, delamination of ionomer with gas generation, or dissolution of ionomer into the circulating water, but the cause has not been investigated in detail. Running AEMWEs with aqueous electrolytes makes and postpones the effects of the degradation of AEMs on the performance of AEMWE and it does not address the intrinsic cause of the problem. Durability issues of AEMWE will be discussed in Section 6.4.5.

In AEMWE, the addition of an alkali metal solution (e.g. 1 M KOH) improves the performance. However, the specific mechanism of this performance improvements is currently unknown. Liu et al. [69] elucidated the role of hydroxide electrolyte on AEMWE performance using experiments and mathematical models. The results show that additional hydroxide plays a key role not only in ohmic resistance of the membrane and catalyst layer but also in the reaction kinetics. The modeling suggests that the added liquid electrolyte creates gas bubbles; the total effective electrochemical active surface area in the cell with 1 M KOH is five time higher than that of the cell with DI water. In the cell with 1 M KOH, more than 80% of the reaction current is associated with the liquid electrolyte. These results indicate the importance of high pH of electrolyte and catalyst/electrolyte interface in AEMWEs (Figure 6.10).

6.4.3 Anion Exchange Membranes

The role of the AEM is to transport hydroxide ions from cathode to the anode, and to act as a barrier for electrons and gases that are produced by the electrochemical reactions. AEMs are made of cationic polymers that have the following characteristics: (i) high OH^- conductivity, (ii) high chemical stability (e.g. alkaline, oxidative, and thermal stability), (iii) excellent mechanical properties under wet conditions and high differential pressures, and (iv) low fuel gas (hydrogen) permeability. To make them more attractive, they need be low cost, easy to process, and produced by sustainable ways.

Unfortunately, a compromise between mechanical properties and ionic conductivity of AEMs must be made. While an increase in the number of cationic groups in AEMs can enhance the hydroxide conductivity, their water uptake increases, tending to lower the mechanical stability. In contrast, a lower loading of cationic functional groups in polymers causes lower ionic conductivity, and therefore, a low AEMWE performance.

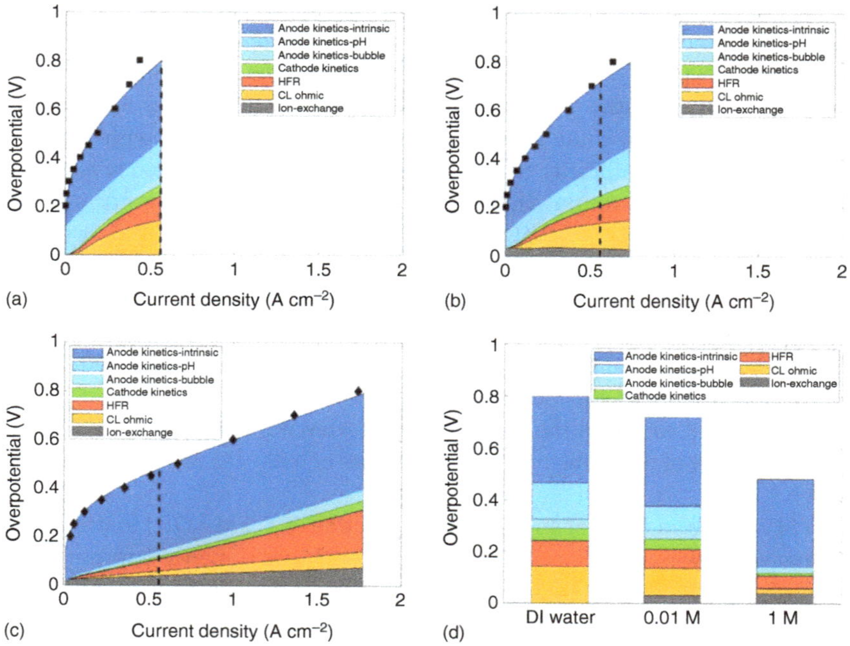

Figure 6.10 The applied voltage breakdown for (a) DI water, (b) 0.01 M KOH, and (c) 1 M KOH. The dashed line shows the location corresponding to the largest current density in (a). The bar graph in (d) shows the applied-voltage breakdowns at the same current density of 0.56 A cm^{-2}, i.e. the current density indicated by the dashed lines. The cell overpotential is broken down into anode kinetic losses (blue), cathode kinetic losses (green), HFR loss (red), CL ohmic loss (yellow), and ion-exchange loss (grey). The anode kinetic losses are further broken down into three parts: anode kinetic losses due to bubble coverage (light blue), anode kinetic losses due to low pH (medium blue), and intrinsic kinetics loss (dark blue). Source: Adapted from LeRoy [68].

Common polymer backbones used for AEMs are: polysulfone [70], poly(ether ether ketone) (PEEK) [71], poly(phenylene oxide) (PPO) [72], polybenzimidazole (PBI) [73, 74], polyphenylene, polyethylene (PE) [75], polystyrene [76], poly(fluorenyl aryl piperidinium) [77, 78], and polycarbazole [78].

Cationic groups that drive the transport of hydroxide ions are N-containing positively groups (e.g. ammonium or quaternary ammonium (QA) [79], piperidinium or dimethyl piperidinium (DMP) [77, 80–82], spirocyclic or N-heterocyclic 6-azonia-spiro[5.5]undecanium (ASU) [72], and imidazolium (IM) [83].

Commercial and laboratory researched cation exchange membranes incorporating the polymers and cation exchange groups will be discussed here.

6.4.3.1 Commercial Membranes

There are several commercially available AEMs such as Tokuyama (A201) (discontinued production), Fumatech (Fumasep®), Ionomer (Aemion™), Dioxide Materials (Sustanion®), Orion Polymer (Orion TM1), and Vesogen (PiperION®). As AEMs have been developed for over 70 years, most of these were optimized for chemically less aggressive environments such as desalination, electrodialysis, or electrodeionization [97]. Table 6.3 lists these membranes that have been reported for AEMWEs in the literatures. Table also shows the physical and electrochemical properties of membranes. Conductivity values reported need to pay more attention as there are two methods to test these values. First is to measure the hydroxide conductivity by immersing membranes in KOH solution to achieve full ion exchange and then changing the solation to pure water, to leach out excessive KOH. This should be done in a closed system using nitrogen purges water, or in a glove box to prevent the contact with the membrane with CO_2 [98]. Another way is to electrochemically purge carbonates from the membrane by applying a voltage in a CO_2-free atmosphere. Ion conductivities reported in this way are much higher than those measured from the first method and are more close to the conductivity observed in an operating electrolysis cell [99, 100].

FAA3 membrane is available through Fumatech in different thickness as non-supported or PEEK or PP-reinforced membrane. FAA3 is a polyaromatic polymer with ether bond in the main chain, and quaternary ammonium groups attached to the main chain. Newer membranes are slightly crosslinked. For the use as ionomer, fully soluble linear polymers are available. A201 were available from Tokuyama but is discontinued from production. The chemical structure of A201 membrane is not disclosed except that they are hydrocarbon-based membranes with quaternary ammonium groups.

Aemion can be obtained from Ionomer and has various IEC (1.4–1.7 or 2.1–2.5 meq g^{-1}) and thickness (25 or 50 μm). Aemion membrane was developed by the Holdcroft group and is based on methylated PBIs [101] (Figure 6.11), where one example is hexamethylterphenyl-polydimethylbenzimidazole (HMT-PDMBI). The best performing membrane from Aeomion is AFI-HNN8-25-X (25 μm in thickness) using FAA ionomer that showed a current density of 500 mA cm^{-2} at 1.7 V and a voltage increase of 2.4 mV h^{-1} over the 17 hour test period. This is significantly higher than the state-of-the art (5 μV h^{-1}) [102].

Sustainion membrane can be purchased from Dioxide Materials and independent distributors. It is based on poly(4-vinylbenzyl chloride-co-styrene) (Figure 6.12) [76, 103]. Sustainion 37–50 seems to indicate the molar ratio of 4-vinylbenzyl chloride in the copolymer (37%) and the thickness of 50 μm. Because it is a polystyrene-based polymer, Sustainion is brittle and cracks easily in dry state. "Classic Sustainion" contains a small amount of water-soluble plasticizer while "Class 60 Sustainion" has a reduced amount of plasticizer, which needs to be washed out from the assembled electrolysis cell with the first feed water. For operation at higher pressure, Dioxide Materials offer PTFE-supported "Class T Sustainion" and "Class TZ Sustainion" with PTFE support and zirconium oxide nanoparticles. There is a controversy that Sustainion is not a real AEM but conductive because of absorbed KOH and water.

Water electrolysis performance of Sustainion 37–50 membrane and $NiFe_2O_4$ anode and NiFeCo cathode catalysts showed a stable performance of 1 A cm^{-2} at 1.9 V for over 2000 hours [76]. With PGM catalysts, even 2 A cm^{-2} at 1.7 V were obtained [104]. Dioxide Materials reported that the voltage required to maintain 1 A cm^{-2} over time in an AEMWE using a 50 μm thick Sustainion X37-50 and PTFE

Table 6.3 Physical and electrochemical properties of commercially available AEMs and polymers from literatures.

AEM	Polymer backbone	Cationic group	Company	Country	Product code	Thickness (mm)	IEC (meq g^{-1})	Ionic conductivity (mS cm^{-1})
Commercial membranes								
Fumasep FAA 3			Fumatech	Germany	FAA 3-PE-30	20–30	1.4–1.6	0.8–1.3 Ω cm^2 (?)
Fumasep FAA 3			Fumatech	Germany	FAA-3	130	1.1–1.4 (Cl^-)	1.8–4.0 Ω cm^2 (Cl^-) ? °C
Fumasep FAA 3			Fumatech	Germany	FAA-3-50	50	1.6–2.1 (Cl^-)	0.6–1.5 Ω cm^2 (Cl^-) ? °C
Sustainion 37–50	Polystyrene	IM	Dioxide materials	USA	Sustainion 37–50	50		0.045 Ω cm^2@60 °C (1M KOH)
Sustainion 37–50	Polystyrene	IM	Dioxide materials	USA	Sustainion 37–50	50–80		0.045 Ω cm^2@60 °C (1M KOH)
PTFE reinforced Sustainion	Polystyrene	IM	Dioxide materials	USA	Grade-T			0.08 Ω cm^2@60 °C (1M KOH)
Fumasep FAS-50		QA	Fumatech	Germany	FAS-50	50	1.6–2.0	0.3783 Ω cm^2@60 °C (1M KOH)
Fumasep FAPQ-375		QA	Fumatech	Germany	FAPQ-375	75		0.83 Ω cm^2@60 °C (1M KOH)
AMI-7001	Gel polystyrene cross linked with divinylbenzene	QA	MI ULTREX	USA	AMI-7001	450	1.3	2.0 Ω cm^2@60 °C (1M KOH)
A201	Linear hydrocarbon backbone	QA	Tokuyama Corp.	Japan	A201	28	1.8	42 mS cm^{-1}@23 °C
Research Membranes								
HTMA-DAPP	Polyphenylene	QA	Los Alamos National Laboratory	USA		26	2.6	120@80°C
HTMA-DAPP	Polyphenylene	QA	Lawrence Berkeley National Laboratory	USA		50	2.6	120@80°C
QPC-TMA	Poly(carbazole)	QA	Korea Research Institute of Chemical Technology	Korea		50	2.31	125@70 °C
QPP-b-PSK-w-TMA	Poly(p-phenylene)	QA	Korea Research Institute of Chemical Technology	Korea			2.4	
m-PBI	Polybenzimidazole	PBI	Technical University of Denmark	Denmark		40–80		
PFTP-13	Poly(fluorenyl aryl piperidinium)	piperidinium	Hanyang University	Korea		30	2.82	163@80 °C

Anode			Cathode			Current density			
Ionomer	Catalyst	Porous transport layer	Ionomer	Catalyst	Porous transport layer	1 M KOH	DI water	Durability	References
FAA-3-SOLUT-10	Ir black	Titanium	FAA-3-SOLUT-10	NiMo alloy	Titanium	1.0@1.9 V, 50 °C			[84]
FAA-3	3.0 mg cm^{-2} IrOx	Sintered titanium frit	FAA-3	3.0 mg cm^{-2} Pt/C	Carbon paper (Toray 090)		0.5@2.29 V, 50 °C		[85]
FAA-3-Br	4.0 mg cm^{-2} IrO2	Titanium	FAA-3-Br	0.4 mg cm^{-2} Pt/C	Carbon-based (C-GDL)	1.5@1.9 V, 70 °C			[86]
PFTP-8	20 mg cm^{-2} Ni-Fe	Titanium felt	PFBP	20 mg cm^{-2} Ni-Fe	Carbon paper (Sigracet 39 BC)	0.62@2.0 V, 60 °C		140 h, 0.5 A cm^{-2}, 60 °C, 1600 mV h^{-1}	[77]
	1.8 mg cm^{-2} NiFe2O4	Stainless-steel		2.7 mg cm^{-2} modified Raney Ni	nickel fiber paper	0.74@1.8 V, 60 °C		10100 h, 1 A cm^{-2}, 60 °C, 0.7 mV h^{-1}	[87]
	1.8 mg cm^{-2} NiFe2O4	Stainless-steel		14.5 mg cm^{-2} modified Raney Ni	nickel fiber paper	0.83@1.8 V, 60 °C		12180 h, 1 A cm^{-2}, 60 °C, 0.7 mV h^{-1}	[87]
Nafion	2.0 mg cm^{-2} NiFe2O4	316L sintered stainless-steel	Nafion	2 mg cm^{-2} NiFeCo	Carbon paper (Sigracet 39 BC)	~2.4@2.3 V, 60 °C			[76]
Nafion	2.0 mg cm^{-2} NiFe2O4	316L sintered stainless-steel	Nafion	2 mg cm^{-2} NiFeCo	Carbon paper (Sigracet 39 BC)	~0.55@2.3 V, 60 °C			[76]
Nafion	2.0 mg cm^{-2} NiFe2O4	316L sintered stainless-steel	Nafion	2 mg cm^{-2} NiFeCo	Carbon paper (Sigracet 39 BC)	~0.25@2.3 V, 60 °C			[76]
AS-4	2.9 mg cm^{-2} IrO2	Ti foam	AS-4	3.2 mg cm^{-2} Pt/C	Carbon paper (Toray TGP-H-120)		~0.39@1.8 V, 50 °C	40 h, 0.2 A cm^{-2}, 50 °C, 17500 mV h^{-1}	[88]
Quaternized polystyrene	3 mg cm^{-2} Ni–Fe	Platinized titanium	Quaternized polystyrene	0.5 mg cm^{-2} PtRu/C	carbon paper (29BC, SGL)	5.3@1.8 V, 80 °C	2.7@1.8 V, 80 °C		[89]
HTMA-DAPP	0.75 mg cm^{-2} IrO2	Platinized titanium	HTMA-DAPP	0.36 mg cm^{-2} PtRu/C	carbon paper (29BC, SGL)	~1.75@2 V, 60 °C	0.4@2 V, 60 °C		[69]
QPC-TMA	2 mg cm^{-2} IrO2		QPC-TMA	0.4 mg cm^{-2} Pt/C		2.57@1.8 V, 70 °C	~0.2@1.8 V, 70 °C		[78]
QPC-TMA	2 mg cm^{-2} IrO2	stainless steel	QPC-TMA	0.4 mg cm^{-2} Pt/C	carbon paper	3.75@1.9 V, 80 °C 4.0@1.9 V, 90 °C		100 h, 1 A cm^{-2}, 80 °C, 1.2 mV h^{-1}	[90]
	Ni–Al alloy	Perforated nickel plates		Ni-Al-Mo alloy	Perforated nickel plates	1.7@1.8 V, 80 °C (24wt% KOH)			[74]
PFTP-8	2 mg cm^{-2} IrO2	Titanium felt	PFBP-14	0.5 mg cm^{-2} Pt/C	Carbon paper (Sigracet 39 BC)	7.68@2.0 V, 80 °C	1.8@2.0 V, 80 °C	1100 h, 0.5 A cm^{-2}, 60 °C, $<$200 mV h^{-1}	[77]

(Continued)

Table 6.3 (*Continued*).

AEM	Polymer backbone	Cationic group	Company	Country	Product code	Thickness (mm)	IEC (meq g^{-1})	Ionic conductivity (mS cm^{-1})
PTP-90	Poly [(terphenyl piperidinium)-*co*-(oxindole terphenylylene)]	piperidinium	Chinese Academy of Sciences	China		45	2.52	128.9@80 °C
PAP-TP-85	Poly(aryl piperidinium)	Piperidinium	University of Delaware (Vesogen)	USA		20	2.4	78@RT
PAP-TP-85	Poly(aryl piperidinium)	Piperidinium	University of Delaware (Vesogen)	USA		20	2.4	78@RT
PAP-TP-85-MQN	Poly(aryl piperidinium)	Piperidinium and QA	University of Delaware (Vesogen)	USA		20	3.2	150@RT
PTFE/LDH	Pores PTFE compisted with FeNi	Hydroxide NiFe	Tsinghua University	China		20		0.4 Ω cm^2@80 °C
SES-TMA-1.4	Polystyrene	QA	Los Alamos National Laboratory	USA		35	1.4	50@50 °C
SES-TMA-1.4	Polystyrene	QA	Los Alamos National Laboratory	USA		35	1.4	50@50 °C
SES-TMA-1.7	Polystyrene	QA	Los Alamos National Laboratory	USA		32	1.7	61@50 °C
HTMA-DAPP	Poly(phenylene)	QA	Los Alamos National Laboratory	USA		50	1.5	90@50 °C
x-QAPS	Polysulfone	QA	Wuhan University	China		70		
PISPVA	Polystyrene	IM	Hanyang University	Korea		55	1.65	89.7@60 °C (0.5 M NaOH)
LSCPi	Poly(2,6-dimethyl phenylene oxide)s	Piperidinium	Chinese Academy of Sciences	China			1.55	29.0@20 °C
SCPi	Poly(2,6-dimethyl phenylene oxide)s	Piperidinium	Chinese Academy of Sciences	China			1.65	24.8@20 °C
PBP-67	Poly(aryl piperidinium)	Piperidinium	Chinese Academy of Sciences	China		40	2.3	104.5@80 °C
PTP-83	Poly(aryl piperidinium)	Piperidinium	Chinese Academy of Sciences	China		40	2.3	110.4@80 °C
PQP-100	Poly(aryl piperidinium)	Piperidinium	Chinese Academy of Sciences	China		40	2.3	118.7@80 °C

Anode			Cathode			Current density			
Ionomer	Catalyst	Porous transport layer	Ionomer	Catalyst	Porous transport layer	1 M KOH	DI water	Durability	References
A-Radel	2.5 mg cm^{-2} IrO_2	porous Ni foam	A-Radel	0.5 mg cm^{-2} Pt/C	porous titanium	1.0@2.2 V, 75°C		120 h, 0.4 A cm^{-2}, 55 °C, 1.61 mV h^{-1}	[82]
PAP-TP-85-MQN	4.8 mg cm^{-2} FexNiyOOH-20F	Ni foam	PAP-TP-85	0.94 mg cm^{-2} Pt/C	TGP-H-60 Toray carbon paper		0.81@1.8 V, 80 °C 1.02@1.8 V, 90 °C	70 h, 0.5 A cm^{-2}, 80 °C, 1.81 mV h^{-1} 160 h, 0.2 A cm^{-2}, 80 °C, 0.56 mV h^{-1}	[91]
PAP-TP-85	4.8 mg cm^{-2} FexNiyOOH-20F	Ni foam	PAP-TP-85	0.94 mg cm^{-2} Pt/C	TGP-H-60 Toray carbon paper		0.54@1.8 V, 80 °C		[91]
									[91]
Nafion	2 mg cm^{-2} IrO_2	Ni foam	Nafion	2 mg cm^{-2} Pt/C	Ni foam		1.0@1.88 V, 80 °C	180 h, 0.5 A cm^{-2}, 80 °C, 0.03 mV h^{-1}	[92]
HTMA-DAPP	1 mg cm^{-2} IrO_2	Titanium felt	HTMA-DAPP	0.5 mg cm^{-2} PtRu/C	carbon paper (29BC, SGL)	~0.11@1.6 V, 60 °C (0.1 M NaOH)			[93]
FLN55	1 mg cm^{-2} IrO_2	Titanium felt	FLN55	0.5 mg cm^{-2} PtRu/C	carbon paper (29BC, SGL)	~0.23@1.6 V, 60 °C (0.1 M NaOH)			[93]
FLN55	1 mg cm^{-2} IrO_2	Titanium felt	FLN55	2 mg cm^{-2} PtRu/C	carbon paper (29BC, SGL)		1.25@1.9 V, 60 °C		[93]
FLN55	1 mg cm^{-2} IrO_2	Titanium felt	FLN55	2 mg cm^{-2} PtRu/C	carbon paper (29BC, SGL)		~1.1@1.9 V, 60 °C		[93]
x-QAPS	Ni–Fe		x-QAPS	Ni–Mo			0.24@1.8 V, 70 °C		[94]
Nafion	2 mg cm^{-2} IrO_2	Titanium felt	Nafion	0.5 mg cm^{-2} Pt/C	carbon paper (10BC, SGL)	0.55@2.0 V, 80 °C (0.5 M NaOH)		80 h, 1.8 V, 60 °C, 28% performance loss	[95]
LSCPi	8 mg cm^{-2} IrO_2	Porous Titanium plate	LSCPi	0.4 mg cm^{-2} Pt/C	carbon paper (HCP120, HESEN)		0.3@1.8 V, 50 °C		[80]
SCPi	8 mg cm^{-2} IrO_2	Porous Titanium plate	SCPi	0.4 mg cm^{-2} Pt/C	carbon paper (HCP120, HESEN)		0.2@1.8 V, 50°C		[80]
PBP-67	1.5 mg cm^{-2} IrO_2		PBP-67	1.3 mg cm^{-2} Pt/C		1.102@2.0 V, 85 °C			[96]
PBP-67	1.5 mg cm^{-2} IrO_2		PBP-67	1.3 mg cm^{-2} Pt/C		1.187@2.0 V, 85 °C			[96]
PBP-67	1.5 mg cm^{-2} IrO_2		PBP-67	1.3 mg cm^{-2} Pt/C		1.544@2.0 V, 85 °C		402 h, 0.2 A cm^{-2}, 60 °C, 0.11 mV h^{-1}	[96]

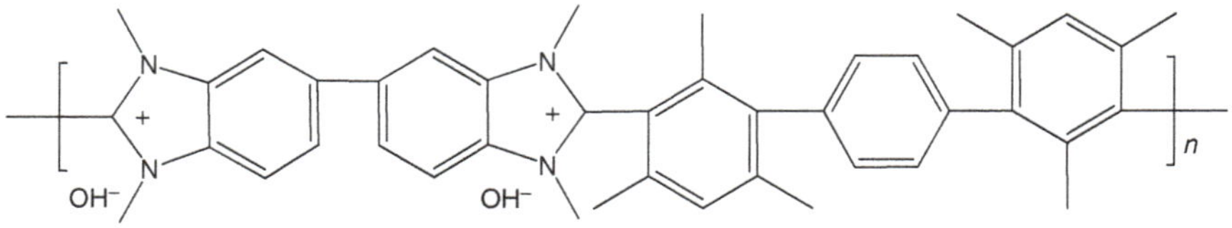

Figure 6.11 General chemical structure of Aeomion PBI polymer.

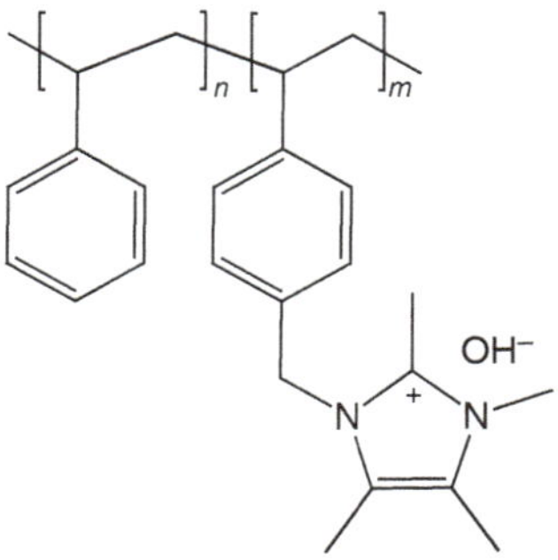

Figure 6.12 Structure of Sustainion membrane.

reinforced Sustainion (Grate T) membranes (Figure 6.13) [87]. The voltage rise of only ~8 mV in 12,000 hours (over one year), or <1 μV h^{-1} was reported. These results demonstrated that the membrane is stable for over one year in 1 M KOH at 60 °C. If the voltage rise from 1.85 to 2.035 V (>10% performance loss), then

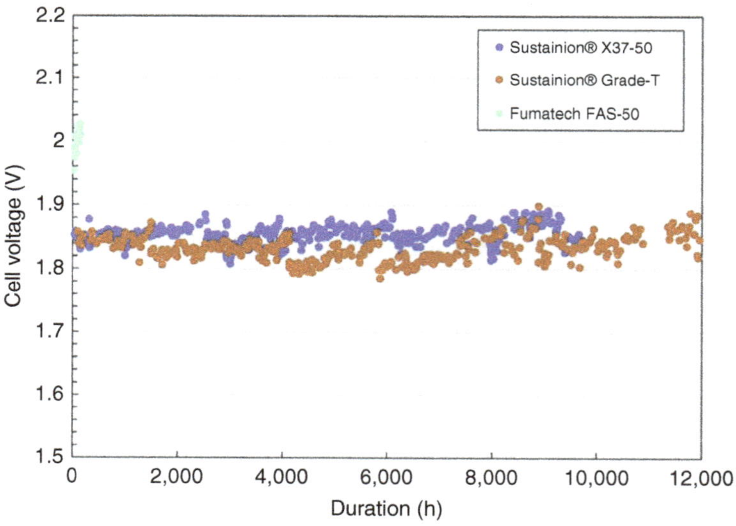

Figure 6.13 The voltage needed to maintain 1 A cm^{-2} current in an AEM electrolyzer running at 60 C in 1 M KOH. Dark green and light green points are data from two different cells using Sustainion® membranes. Blue data points are from FAS-50 running in the same conditions. Source: Adapted from Motealleh et al. [87].

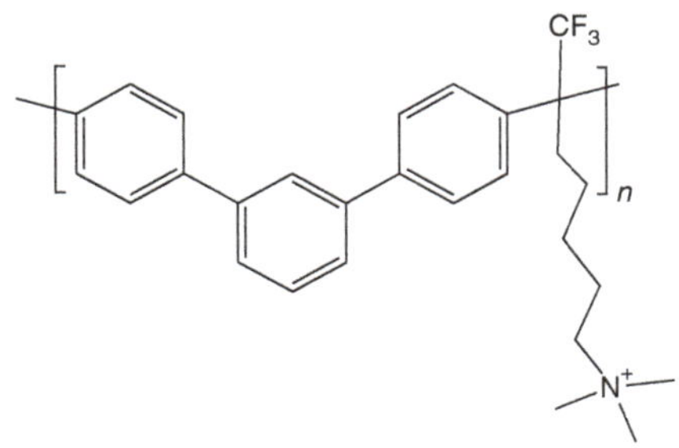

Figure 6.14 Structure of Orion TM1 polymer.

$1.0\,\mu V\,h^{-1}$ degradation would take 185,000 hours or 21.1 years, which is the best reported performance so far for commercial membrane.

TM1 (or TPN1-100) membranes are available from Orion Polymer spin off from Prof. ChulSong Bae at RPI [105]. This polymer is ether-free structure (Figure 6.14). This polymer connects the rigid aromatic structure with the flexible linkers allowing the flexibility and solubility. Orion TM1 membrane was selected as the most stable membrane and the only membrane which did not show signs of degradation among over 50 AEMs from over ten organizations [106]. In the AEMWE test, TM1 showed a stable initial performance, but suffered from low mechanical stability in the hydrated state [107].

Xiao et al. at University of Delaware studied on the water-fed AEM based on poly(aryl piperidium) PAP-TP-x as membrane or PiperION (Vesogen) polymer membrane (Figure 6.15) from Prof. Yan's group and nickel-iron oxyhydroxide catalyst in the anode [91]. They achieved a performance of $1020\,mA\,cm^{-2}$ at 1.8 V and prevented the detachment of catalysts during continuous operation (>160 hours at $0.2\,A\,cm^{-2}$) (Figures 6.16 and 6.17).

Chen et al. [77] reported on the poly(fluorenyl terphenyl piperidinium) membranes and ionomers (Figure 6.18) for AEMFC and that performed with a current density of $7.68\,A\,cm^{-2}$ at 2.0 V with a 1 M KOH anode. PGM-free AEMWEs displayed a $1.62\,A\,cm^{-2}$ at 2.0 V. PGM and PGM-free AEMWEs operated stably under a $0.5\,A\,cm^{-2}$ at 60 °C for 1000 hours (Figure 6.19). Pure water fed AEMWEs constructed based on PFTP-13 AEMs and PGM catalysts presented a current density of $1.76\,A\,cm^{-2}$ at 2.0 V and 80 °C (R_{ohm}, $0.12\,\Omega\,cm^2$). Note that most state-of-the-art

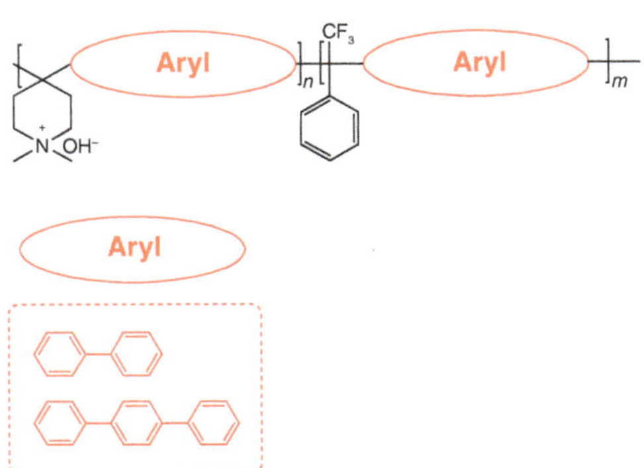

Figure 6.15 Structure of PAP-TP polymer or PiperION polymer.

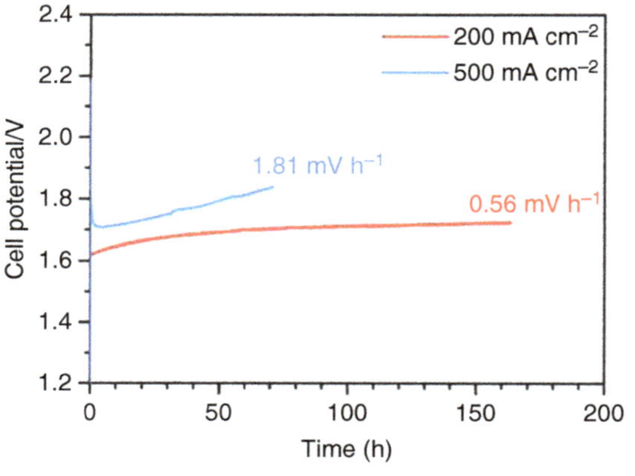

Figure 6.16 Long-term stability performance of water-fed AEMs at 200 and 500 mA cm^{-2} and 80 °C. Membrane: PAP-TP-85 (20 μm); cathode: Pt/C (47 wt%, 0.94 mgPt cm^{-2}); and anode: self-supported FexNiyOOH-20F (4.8 mg cm^{-2}). The ionomer is PAP-TP-85 in the cathode with a loading of 30 wt% and is PAP-TP-85-MQN in the anode with a loading of 0.8 mg cm^{-2}. Source: Adapted from Xiao et al. [91].

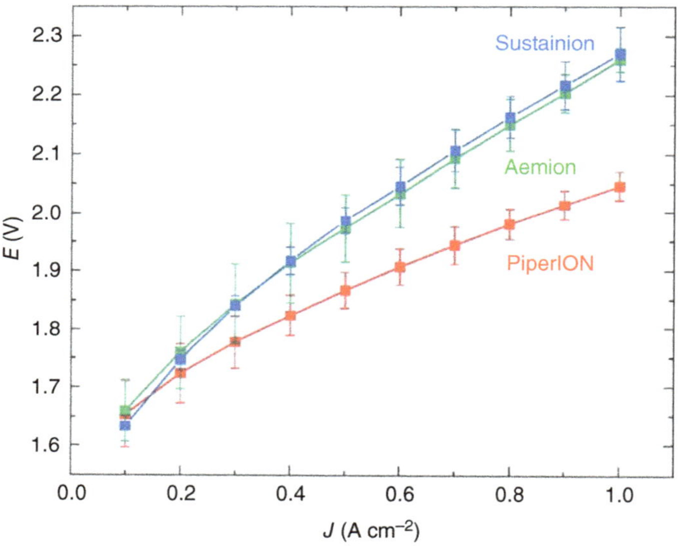

Figure 6.17 Performance comparison of electrolyzers from three commercially available AEMs with a thickness of 50 μm. Source: Adapted from Lindquist et al. [108]. All experiments were run with pure water at a cell temperature of 55 °C.

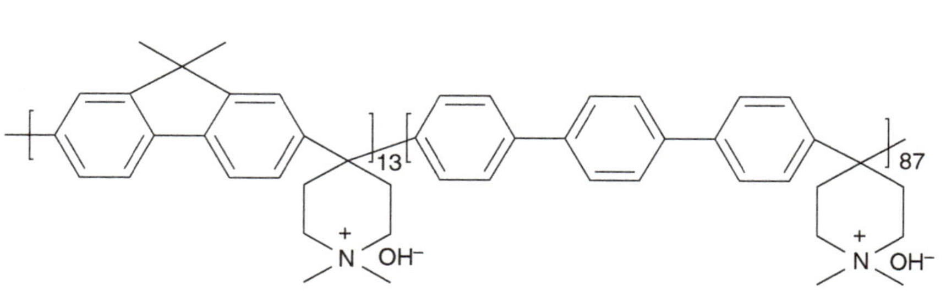

Figure 6.18 Structure of PFTP polymer.

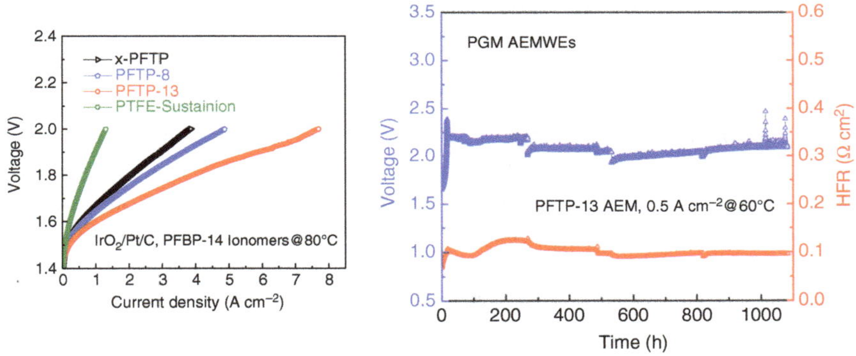

(a) (b)

Figure 6.19 The effect of AEMs on AEMWE performance. (a) Polarization curve, and (b) durability curve. Testing conditions: 1 M KOH anode, 25% PFBP cathode ionomers, PFTP-8 anode ionomers, different water diffusivities of AEMs, 2 mg cm^{-2} IrO_2 anode, and 0.5 mg cm^{-2} Pt/C cathode. *I–V* curves at 80 °C, Commercial Sustainion® AEM-based AEMWEs were measured using 10% A/C PTFE ionomers. Source: Chen et al. [77]/Royal Society of Chemistry.

AEMWEs operating under pure water conditions displayed a limited cell performance of below 1 A cm^{-2}. PGM-free AEMWEs were studied based on home-made A/C Ni-Fe composite catalysts and PFTP-13 AEMs. The A/C Ni-Fe AEMWEs reached a current density of 1.2 A cm^{-2} (R_{ohm}: 0.25 Ω cm^2, R_{CT}: 0.5 Ω cm^2) and 1.6 A cm^{-2} (R_{ohm}: 0.2 Ω cm^2, R_{CT}: 0.2 Ω cm^2) at 2.0 V at 60 and 80 °C, respectively. For comparison, Sustainion® AEM-based AEMWEs displayed limited current density of 0.62 A cm^{-2} with similar R_{ohm}: ~0.25 Ω cm^2 but much higher R_{CT}: 0.65 Ω cm^2. They concluded that the role of water on the transport of AEMs are critical for the AEMWE performance. Therefore, the search for highly efficient and durable AEMs and ionomers is crucial to the development of urgently needed advanced AEMWEs [109]. Hu et al. [82] reported on the piperidinium group-attached polyterphenylene. The MEA with highly conductive PTP-90 membrane demonstrated the current density of 910 mA cm^{-2} at 2.2 V at 55 °C. Cell durability tests showed that after operating at 400 mA cm^{-2} for 120 hours at 55 °C, performance degradation rate was 1.61 mV h^{-1}. Significant chemical degradation of the membrane was observed by analyzing the membrane where the cationic group degradation via Hofmann elimination and nucleophilic substitution reaction occurred. In contrast, no chemical degradation was found for PTP membrane even in ex situ 10 M NaOH at 55 °C for 200 hours.

Kraglund et al. [74] achieved a current density of 1.7 A cm^{-2} at 1.8 V in 24 wt% KOH at 80 °C using ion-solvating KOH-doped poly(2,2′-(*m*-phenylene)-5,5′-bibenzimidazole) (*m*-PBI) as AEM. Aili et al. [110] reported on KOH-doped PBI membranes for AEMWE. PBI is known to be a hygroscopic polymer that can be doped with aqueous KOH to give a material with high ion conductivity, which in combination with the low gas permeability makes it an interesting electrolyte material for AEMWE. While linear PBI showed a structural instability in 6 mol l^{-1} KOH at

85 °C for 176 days, while crosslinked PBI membrane indicated mechanical stability. Crosslinked PBI membraned showed a current density of about 200 mA cm^{-2} at 2.0 V and 80 °C in 30 wt% KOH solution. Diaz et al. [111] reported KOH-doped PVA-PBI membranes for AEMWE. PVA was crosslinked inside of blend by glutaraldehyde to enhance the stability of the membrane. The crosslinked blend membrane showed a current density of 360 mA cm^{-2} at 1.9 V and at 70 °C in 15 wt% KOH solution. Wan et al. [40] studied H_2SO_4-doped PBI membranes for acid-alkaline amphoteric water electrolysis 3.0 M H_2SO_4 doped-mPBI membrane showed a current density of 800 mA cm^{-2} at 2.0 V and 60 °C. The stability test was conducted for 40 hours at 100 mA cm^{-2} with an energy consumption of 3.35 kWh m^{-3} H_2.

Commercially available polymers such as polysulfone, poly(phenylene oxide) (PPO), and low density polyethylene (LDPE) have been used as membrane materials. Polysulfone has been reported as membranes for AEMWE [88, 94, 112]. Xiao et al. [94] reported on the implementation a self-crosslinked quaternary ammonium polysulfone (xQAPS) membrane and ionomer that showed a current density of 400 mA cm^{-2} at 70 °C and 1.85 V using pure water and Ni–Fe anode/Ni–Mo cathode catalyst. Leng et al. [88] studied on the aminated Radel polysulfone membranes for AEMWE and showed a current density of 399 mA cm^{-2} at 1.8 V that lasted up to 500 hours. Parrondo et al. [112] reported on the similar polysulfone for AEMWE using pure water and obtained a current density of 400 mA cm^{-2} at 1.8 V and 50 °C. They concluded that the short-term degradation of water electrolyzer performance over time was largely a consequence of carbon dioxide intrusion into the system and could be easily remedied, while the long-term deterioration was a consequence of irreversible AEM polymer degradation.

Parrondo et al. [113] studied on the stability of PPO-based AEM and binder for WE application. Brominated PPO was treated with trimethylamine (TMA) and 1-azabicyclo[2.2.2]octane (ABCO) to impart the anion exchange capacity. PPO-ABCO showed about 600 mA cm^{-2} at 50 °C and 2.0 V while PPO-TMA exhibited about <200 mA cm^{-2}. Chu et al. [80] compared the functional groups in PPO-based AEMs. They prepared a series of piperidinium-functionalized PPO with different locations of piperidinium groups. Long-side chain-type PPO (LSCPi) membrane showed a low water uptake, swelling ration at the same IEC, and displayed an alkaline stability and showed 300 mA cm^{-2} at 1.80 V and 50 °C using pure water. Long-side-chain attaching piperidinium group in PPO positively affect the fuel cell performance and water electrolysis performance. Although the chemical stability of PPO in alkaline solution is not attractive, these studies indicated an importance of cation functional groups and the location of these groups.

Faraj et al. [67] synthesized an AEM by radiation grafting of LDPE with vinylbenzyl chloride followed by conversion into quaternary ammonium sites with 1,4-diazabicyclo[2.2.2]octane (DABCO). The electrolysis of a potassium carbonate-doped water solution (1 wt% carbonate was used to boost the ionic conductivity in the electrodes) resulted in a cell voltage of 2.1–2.2 V at 0.5 A cm^{-2}. LDPE-g-VBC-DABCO MEA showed a 150 mA cm^{-2} at 45 °C. The average increase in cell potential over the period of the test was 2 mV d^{-1} (benchmark) and 6 mV d^{-1} or 250 mV h^{-1} (LDPE-g-VBC-DABCO) of degradation. LDPE membrane electrolytic cells showed a constant hydrogen production rate of about 30 cc min^{-1} over more than 500 hours.

Piperidinium functionalized SEBS (SEBS-Pi) was used for AEMs [81]. The membrane showed 400 mA cm^{-2} at 2.0 V and 50 °C that lasted for 105 hours.

Cha et al. [78] reported alkylammonium poly(carbazole) AEM-based AEMWEs, and the cells reached a high current density of 3.5 A cm^{-2} at 1.9 V (1 M KOH at 70 °C) but displayed a poor cell durability of ~3 hours.

Kim and coworkers [114] reported high-performance AEMWEs based on alkylammonium polyphenylene AEMs (HTMA-DAPP) (see Figure 6.20) and BTMA-polystyrene ionomers, and the cells reached an excellent current density of 2.7 A cm^{-2} at 1.8 V in pure water (~5.5 A cm^{-2} in 1 M KOH) at 85 °C (Figure 6.21), while the cell durability remained limited below 170 hours under 0.2 A cm^{-2} (Figure 6.22). On the other hand, most AEMWEs are based on PGM catalysts, and research on PGM-free AEMWEs is lacking. Hu et al. [115] reported NiMo/Fe–NiMo-based AEMWEs that reached a current density of 1 A cm^{-2} at 1.57 V in 1 M KOH.

Huang et al. [116] studied on the ionomer optimization for water uptake and swelling in AEMWE using poly(norbonene) (PNB) ionomer and membranes (Figure 6.23). The AEMs were reinforced tetrablock copolymer (GT74) with 5 mol% crosslinking. As ion conductivity, water uptake and swelling ratio influence the

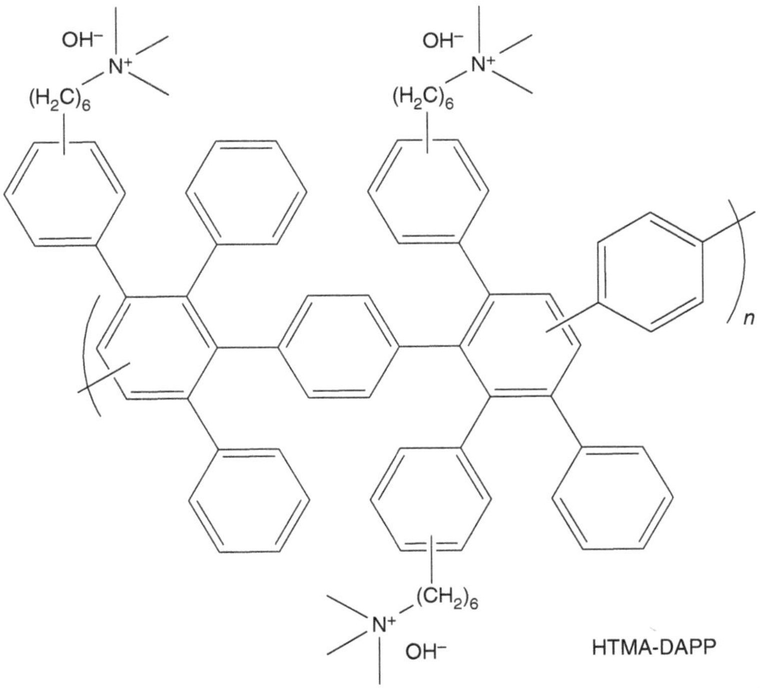

HTMA-DAPP

IEC = 2.6 meg g^{-1}

Figure 6.20 Structure of HTMA-DAPP.

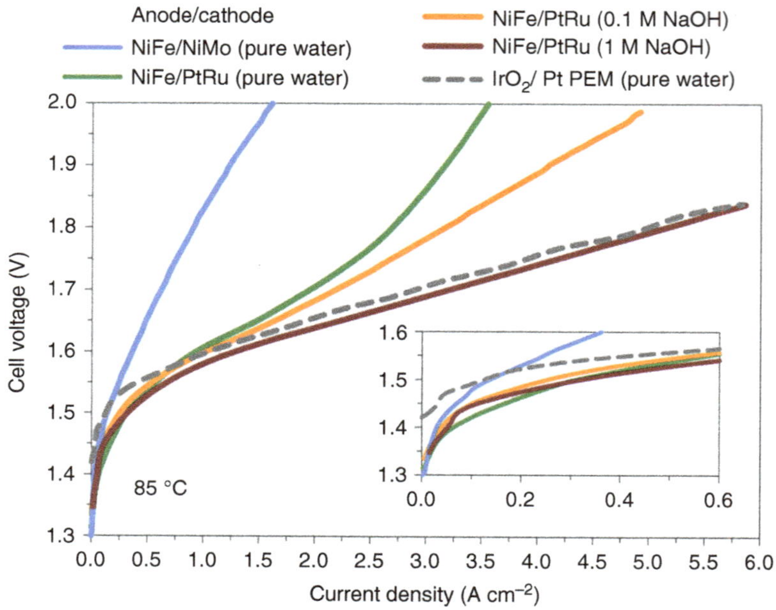

Figure 6.21 The performance of MEAs employing PGM-free catalysts (the magnified region of the current density between 0.0 and 0.6 h cm^{-2}, shown in the inset). AEM, HTMA-DAPP (26 μm thickness); anode, NiFe nanofoam (3 mg cm^{-2}); cathode, PtRu/C (50 wt% Pt, 25 wt% Ru, 2 mgPt cm^{-2}) or NiMo/C (2 mgNiMo cm^{-2}). The ionomer content at the cathode was 20 wt%. The performance was measured under ambient pressure at 60 °C for NaOH solution and 85 °C for pure water flowing in both the anode and the cathode. The performance of a state-of-the-art PEM electrolyzer11 is also shown. PEM electrolyzer, PEM 3 M 825 EW (50 μm thickness); anode, 0.25 mg cm^{-2} Ir-NSTF; cathode, 0.25 mg cm^{-2} Pt-NSTF measured at 80 °C. Source: Adapted from Leng et al. [88].

electrode performance, they used crosslinked PNB to lower water uptake of AEIs. Film form of lightly crosslinked AEIs became fine-powder ionomer from the quaternized polymer by use of a high-speed grinder with dry ice. As the crosslinked PNB AEIs are in the powder form, the ionomer detachment was observed and therefore, the electrode performance was poor. Typical crosslinked ionomer GT72-15 exhibited an IEC of 3.47 meq g^{-1}, ionic conductivity of 153 at 80 °C, and water uptake of 78%. Excessive swelling of the high IEC ionomer was identified as the cause of poor performance. They suggested that the water uptake should be lower than this value to properly manage the water content.

6.4.3.2 Chemical Stability of Cationic Groups

Regarding the stability of cation groups in ether-free polymer membranes, Kreuer and Marino have reported that DMP and ASU show superior alkaline stability at elevated temperature following the study of a significant number of different QA salts [117]. The six-membered rings in the mono- and bicyclic QAs (e.g. DMP and ASU) show high alkaline stability because the low ring strain is accompanied by a restricted conformation imposed by the ring structure. Hence, the transition

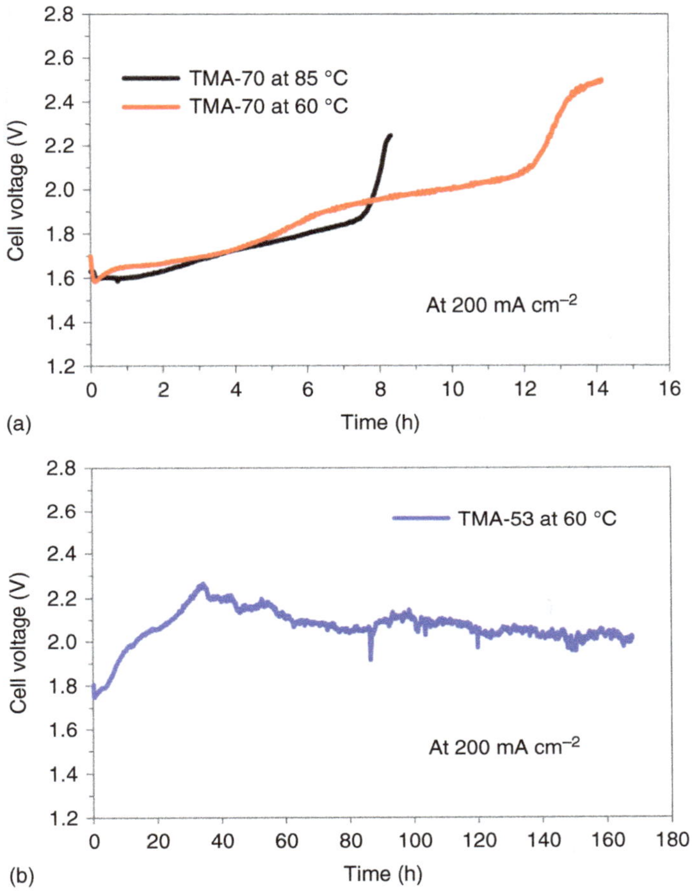

Figure 6.22 (a) Durability of the MEAs using the TMA-70 ionomer at 85 and 60 °C at a constant current density of 200 mA cm^{-2}. (b) Durability of the MEA using the TMA-53 ionomer at 60 °C at a constant current density of 200 mA cm^{-2}. In all cases, deionized water was flowing on the anode only, and HTMA-DAPP and PtRu/C were used as the AEM and the catalyst on the cathode, respectively.Source: Adapted from Leng et al. [88].

state energy increases for both elimination and substitution reactions. Chen et al. continued the study by analyzing the alkaline stability of 24 representative model compounds based on N-heterocyclic ammonium (NHA) groups for anion exchange polyelectrolytes [109]. They concluded that from the density-functional theory calculations and experimental results suggest that nucleophilic substitution is the dominant degradation pathway in NHAs, while Hofmann elimination is the primary pathway for NHA-based AEMs. Jannasch's group reported a series of studies on this topic [118–122]. Pham reported on the stability of five or six-membered ring directly attached to piperidinium and found that AEMs with six-membered ring is much more stable compared with the five-membered ring to piperidinium [118] (Figure 6.24).

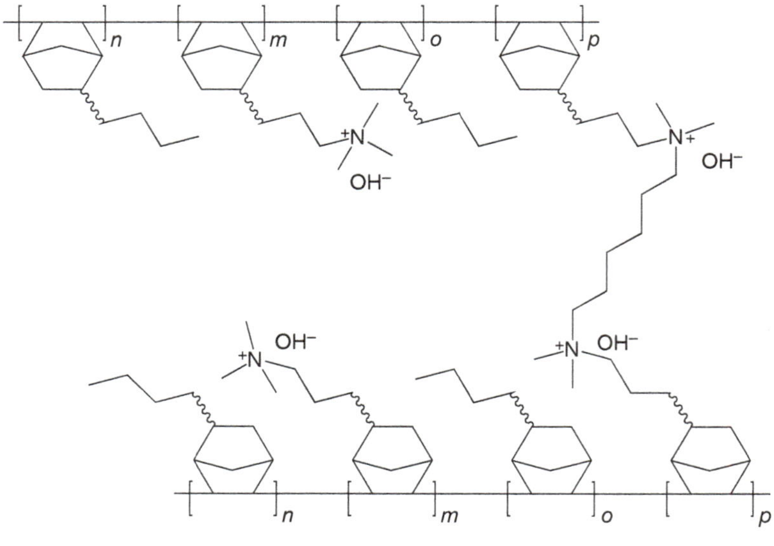

Figure 6.23 Chemical structure of crosslinked poly(norbonene).

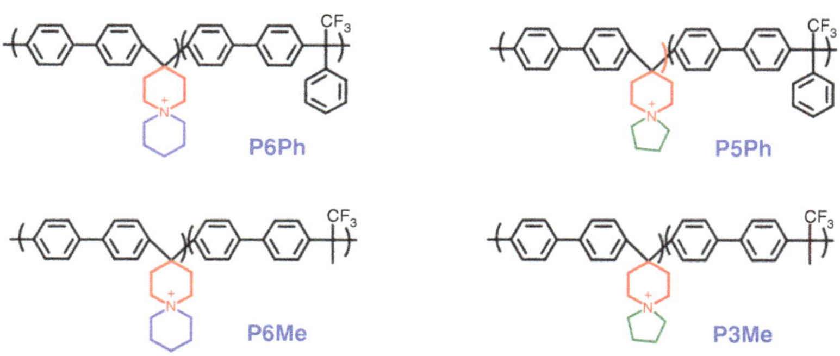

Figure 6.24 Examples of six- and five-membered ring attached to piperidinium group in polyarylpiperidinium polymer.

Piperidinium ring in six membered ring is much more susceptible to ring opening Hofmann elimination reaction than relaxed six membered or five membered ring in. Therefore, the introduction of flexible alkyl space will increase the alkaline stability of spiro-ionenes. Pham et al. reported on the alkaline stability of DMP- and ASU-based poly(terphenyl)-based AEMs [119]. AEM based on m-terphenyl backbone possess a higher ionic conductivity compared with other membranes due to its flexibility. DMP cation-based AEMs show better alkaline stability than ASU-based polyterphenyl-based AEM [119] and the polyfluorene-type polymers [120]. They also found that m-terphenyl-based AEM showed a better alkaline stability compared with the para-counterparts, because of its higher water uptake that decreases the concentration of aggressive hydroxide ions. AEMs with DMP

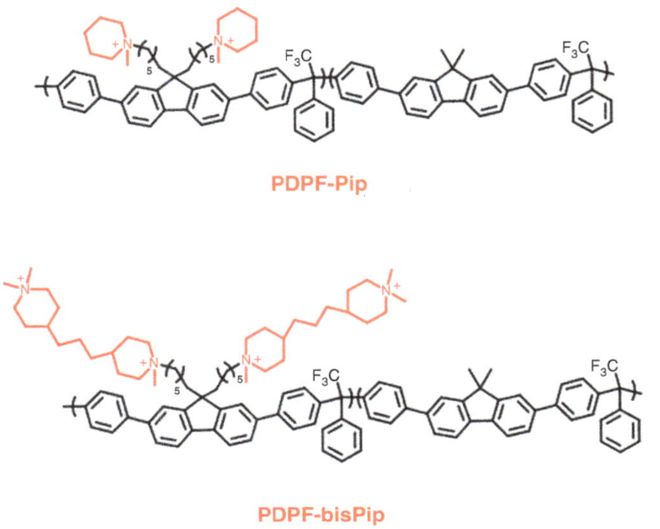

Figure 6.25 Structure of PDPF-Pip and PDPF-bisPip polymer.

group possess the higher ion conductivity compared with ASU-based AEM [120]. Pan et al. studied on the effect of mono- or bi-cationic side chains in AEMs [121]. They concluded that AEMs with mono-cationic side chain showed better alkaline stability. Long-side chain improved the alkaline stability of AEMs. Here, DMP-based AEMs possess better alkaline stability and higher ionic conductivity than ASU-based AEMs. Allushi et al. [122] studied the effect of mono-piperidium (Pip) versus bispiperidinium (bisPip) in poly(diphenyl fluorene) (PDPF) polymer backbone. They showed that AEMs with multi-cation side chain (PDPF-bisPip) possess (Figure 6.25) higher ionic conductivity and water uptake than AEMs with mono-cation side chain. BisPip AEMs show higher alkaline stability than mono-Pip AEM, indicating that increasing the local concentration of cations in polymer backbone can simultaneously enhance the conductivity and alkaline stability. They also noted that β-hydrogen in an alkyl spacer chain is about eight times more susceptible to attacks by OH- ions than a β-hydrogen in piperidinium. Therefore, the design of alkaline stable AEMs is possible to consider the importance of the stability of cationic groups in the alkaline environments and their attachments in polymer backbones for AEMWE.

Soni et al. [64] studied on the pure water PEMWE using fully aromatic and high-molecular weight poly(fluorene-alt-tetrafluorophenylene)-trimethyl ammonium AEMs and ionomers. MEAs fabricated with these polymer electrolytes showed that about 1.8 and 2 V was obtained at 1 A cm^{-2} at 80 °C using 1 M KOH solution and pure water, respectively. Durability of these MEAs showed about 140 hours.

6.4.4 Ionomers

AEIs are binders that create hydroxide ion transport pathways between the membrane and the reaction sites within electrodes and thus they drastically improve the

Table 6.4 Critical properties of typical AEIs.

AEI	Polymer backbone	Cationic group	IEC (mmol g^{-1})	Conductivity (mS cm^{-1})
AS-4	Linear hydrocarbon	QA	1.4	13 (HCO_3^-, 40 °C, wet)
qPVB/OH	Polystyrene	QA	—	16 (OH^-, 25 °C, 100% RH)
PSF-TMA/OH	Polysulfone	QA	1.8	8 (OH^-, 25 °C, water)
XQAPS	Crosslinked polysulfone	QA	1.3	15 (OH^-, 20 °C, water)
GT72-10 (2020, JES, 167, Garrett Huang)	Crosslinked polynorbonene powder	QA	3.47	69 (OH^-, 25 °C, water)
HTMA-DAPP	Polyphenylene	QA	2.6	120 (OH^-, 80 °C)
PFBP-14	Poly (fluorenyl aryl piperidinium) Copolymer	PI	3.43	146 (OH^-, 80 °C)
PAP-TP-85	Poly(aryl piperidinium)	PI	2.37	~160 (OH^-, 80 °C)
QPC-TMA	polycarbazole	QA	2.3	154 (OH^-, 80 °C)

QA: quaternary ammonium, PI: piperidinium.

utilization of the electrocatalyst particles while reducing the internal resistance. One of the most desirable properties of an ionomer for use in the electrocatalyst layer is its high solubility or dispersibility in low-boiling-point water soluble solvents such as ethanol and isopropanol, because these solvents are safe to handle and easy to remove during the electrode preparation. Ionomer should be insoluble in water, and it should have low swelling ratio, even at high temperature. Finally, the ionomer should have good hydroxide conductivity and chemical stability. So far, only there are several AEIs that have been reported in electrolyzers (Table 6.4). Like AEMs, the alkaline, oxidative, and chemical and thermal stability of AEIs are important and the degradation mechanisms of AEIs especially in membrane electrode assemblies (MEAs) should be investigated.

6.4.5 Durability

Although interest in low-cost hydrogen is growing and much attention have been pain in AEMWE because of their ability to split water using PGM-free catalysts, less expensive anode flow, and bipolar plates, large-scale realization of AEMWEs is still in its development stage because of durability limiting factors that restrict the long-term use of these devices. Li et al. extensively reviewed the durability issues of AEMWEs [123]. Similar to PEMWEs, the zero-gap configuration of AEMWEs employ a non-porous membrane that can produce hydrogen at a high rate and minimize the mechanical compression of hydrogen for storage. Recent modeling study indicated that the additional liquid electrolyte not only reduces the ohmic resistance of the membrane and catalyst layer, but also improves the reaction kinetics. By adding a liquid electrolyte to the AEMWE cell, the local pH increases at the catalyst/electrolyte interface and an additional electrochemical interface is generated. Kaczur et al. reported on the industrial AEMWEs based on Ni-based catalysts and Sustainion 37–50 membrane that produced hydrogen with ~1 A cm^{-2} at 1.85 V in 1 M KOH that approaches the performance of conventional PEMWEs under ambient pressure [87, 104]. The cell showed a voltage increase in time of only 1 μV h^{-1} and run over 10,000 hours. The longevity of pure water-fed and 1 wt% K_2CO_3-fed AEMWEs were only demonstrated for <3000 hours. The projected lifetime is over 20 years. The power consumption estimates for the AEMWE cell operating at 1 A cm^{-2} at 1.9 V was calculated to be 50.5 kWh kg^{-1} H_2. The electrolyzer energy consumption is higher compared with the electrolyzer H_2 power consumption to the DOE electrochemical hydrogen target goals which have stack efficiencies of 43 kWh kg^{-1} and hydrogen levelized cost of US$2.00 for 2020. Cycling on/off performance over the course of 30 years was simulated experimentally through accelerated voltage shock tests, resulting in a performance loss of 0.15 mV/cycle over 11,000 cycles. An improvement in performance is achieved by the addition of zirconia to the polymer matrix and mechanically reinforcing the membrane.

The stack lifetime of commercial PEMWEs is close to 20,000–60,000 hours, while the reported lifetime of AEMWEs is 12,000 hours. The liquid electrolyte provides higher hydroxide ion transport and extended catalyst-electrolyte interfacial area. It is explained that the hydroxide conductivity (σ_{bulk}) of the AEM is 18.4 mS cm^{-1} at 50 °C [123]. At the electrode, the effective conductivity ($\sigma_{effective}$) of the ionomeric binder is calculated to 1.2 mS cm^{-1} from Eq. (6.1)

$$\sigma_{bulk} = \sigma_{effective} \times \frac{\varepsilon}{\tau^2} \tag{6.12}$$

where σ_{bulk} = bulk conductivity, ε = volume fraction, and τ = tortuosity. Under the 1 M KOH solution condition, σ_{bulk} increases to 57.4 mS cm^{-1} and at the electrode, the $\sigma_{effective}$ of the ionomer binder and the liquid electrolyte increases to 27 mS cm^{-1}, respectively. Adding the highly concentrated KOH solution improves the catalyst utilization five times. The improvement of the catalyst utilization is due to the expanded catalyst-electrolyte interfacial area and the improved intrinsic kinetics of the catalysts. As a result of the low ohmic resistance, the liquid electrolyte-fed AEMWEs is much better than that of pure water-AEMWE. The current density of the pure water, 1 wt% K_2CO_3 (equivalent to 0.07 M KOH) and 1 M KOH-fed AEMWEs at 60 °C were 450, 1200, and 1700 mA cm^{-2} at 2 V, respectively [123].

There are several reasons for the gradual cell voltage increase over time (voltage decay rate) (Figure 6.26); (1) loss of cationic functional group of the AEM (Type 1), (2) gas crossover or cationic group degradation or ionomer detachment (Type 2), (3) electrochemical oxidation (Type 3), (4) electrolyte concentration changes (Type 4), (5) polymer backbone degradation (Type 5), and (6) system failure or mechanical failure (Type 6). The durability limiting factors such as ionomer detachment and poisoning under high cell voltage and current density conditions need to be considered as ionomer related degradations are accelerated under high cell

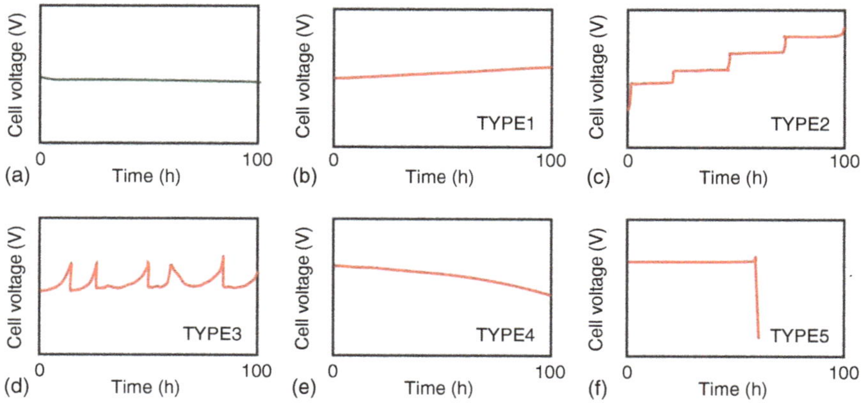

Figure 6.26 Various voltage behaviors of AEMWEs for 100 hours extended-term test at a constant current density. Source: Adapted from Pan et al. [122].

voltage and current density conditions. As the IEC of ionomers and operating temperature increase, water uptake of ionomer increases, resulting in the detachment of ionomers from the catalyst. The ionomer detachment by gas evolution occurs more severely in pure water-fed AEMWEs because the catalyst-electrolyte interfacial area of pure water-fed AEMWEs is relatively small. Since the gas permeability of polymeric materials is much lower than that of the KOH solution, it is difficult to completely remove the gas from the catalyst-ionomer interface under high current density. Compared to PEMWEs, the gas bubble-induced ionomer detachment occurs more easily in AEMWEs because of the low gas permeability of AEMs, and much lower adhesion due to their excessive swelling with water. Therefore, it is highly recommended to use low IEC ionomers and to operate at a low operating temperature although lower performance is expected.

Polymer design strategy for improved durability in AEMWEs can be focused on the low water-swollen hydrophobic ionomers [124] by introducing multi-cations [125], polar interactions [126], and crosslinking [127–133]. In particular, end group crosslinking of AEM improved the durability of AEMFC [127]. Crosslinking of ionomer prevented the catalyst particle aggregation and therefore AEMFC performance using crosslinked ionomer possesses a higher power density and more stable durability compared with AEMFC with un-crosslinked ionomer [129]. Wang et al. extended the crosslinking technology in AEMFC [133]. Chen et al. crosslinked PNB ionomers for durable low-temperature AEMWEs [132], showing the adhesion between the catalyst particles, porous transport layer, and ionomer, as determined by adhesion measurement and electrolysis performance. AEMWE performance under 1 A cm^{-2} at 1.83 V (67% voltage efficiency) for >600 hours (Figure 6.27).

6.4.6 Outlook for AEMWE

Over the past few years, AEMs and AEIs have seen a rapid growth in interest, driven by the several advantages of the AEMFC technology. Significant progress in AEMs

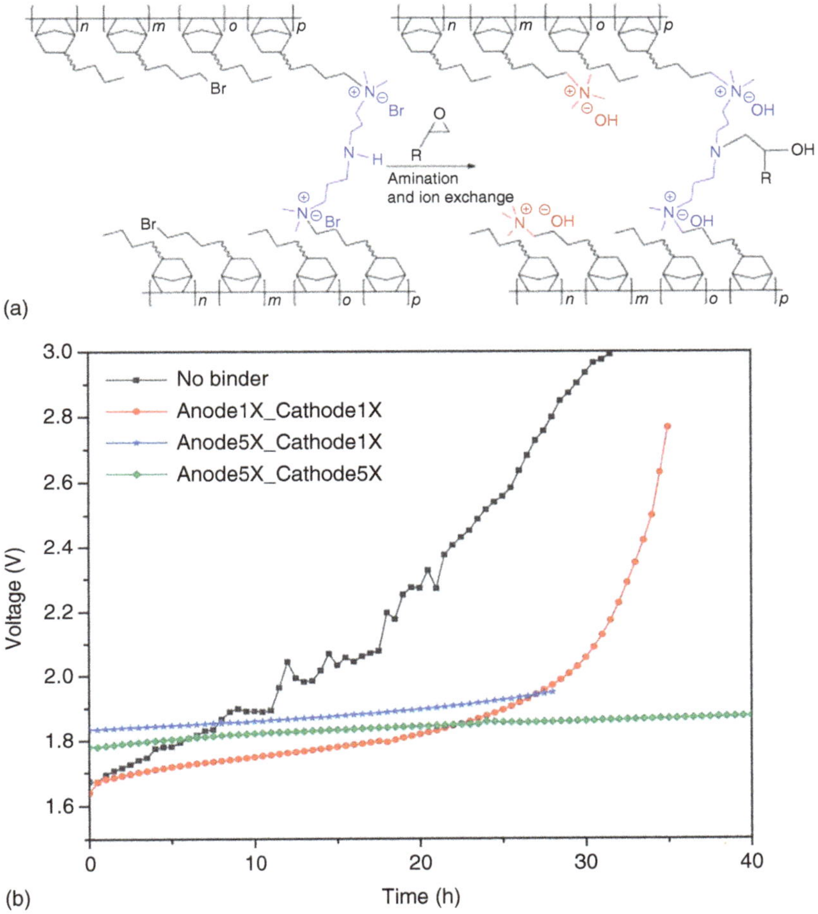

Figure 6.27 (a) Chemical structure of crosslinked PNB ionomer for self-adhesion. (b) Cell voltage versus time for MEAs with various epoxy contents at 0.5 A cm^{-2}. The AEM was GT72-10 with PTFE reinforcement (30 μm thick). The anode used GT72-10 ionomer and PbRuOx catalyst. The cathode used GT72-3 ionomer and Pt3Ni/C catalyst. The two-part epoxy binder was added to the electrode inks at a loading of 0.65 mg cm^{-2} corresponding to "1X". Source: Adapted from Lee et al. [131].

and AEIs is required for potential of AEMWEs in the future. AEMWEs meet the additional challenges over AEMFCs including gas generation, high oxidation potentials, large differential pressures, and carbon-free anode electrocatalysts, which will have to be addressed when integrating new materials.

Although a significant progress in advancement of AEM/AEIs have been made, AEMWEs need disrupting materials and breakthroughs in electrocatalysts, membranes, ionomers, and MEA. According to IRENA report on techno-economical goal by 2050 [134] to be competitive, AEMWEs need to operate at current densities above 2 A cm^{-2} (from the current 0.2–2 A cm^{-2}) with an operating voltage lower than 2 V (from the current 1.4–2.0 V). Improved system pressure would lead to reach 70 bar (from the current <35 bar) at 80 °C (from the current 40–60 °C). Higher operating temperature improve the electrocatalyst activity, the voltage efficiency (>75%), and the electrical efficiency (42 kWh kg^{-1} H_2 for the stack and 45 kWh kg^{-1} H_2 for the overall system). Current lifetime of AEMWE stack is about 12,000 hours [87], but this is expected to at least 100,000 hours to be competitive. To realize the AEMWEs, several aspects should be clarified before commercialization: (i) a fundamental understanding of in situ failure mechanisms, (ii) more investigations showing durability (>1000 hours) at current densities of >0.75 A cm^{-2} (ideally >1.0 A cm^{-2}), and (iii) durability testing with earth abundant bipolar plates and porous transport layers. High mechanical strength of AEMs is preferable because of the possible pressure differential operation in AEMWEs. With the advancement of AEM/AEI and their integration in MEAs, AEMWE device can reach to MW scale from the current 25 kW scale. If the size of AEMWE can be in MW scale, the cost of <$100/kW for the stack and <$200/kW for the complete AEMWE system can be achieved by 2050.

6.5 Conclusions and Future Perspectives

This chapter overviews the importance of hydrogen and water electrolysis using ion exchange membranes. Although hydrogen is widely used in industry, its production relies mainly on the steam reforming or industrial petroleum refining, while green hydrogen gas should be applied in the future as the climate change issue should be solved. As alkaline water electrolysis has been installed for the last several decades, ion exchange membrane water electrolysis such as PEMWE and AEMWE needs to be explored to combat with the energy crisis. PEMWE is currently available in s kilowatt scale, while AWE is now in the MW scale. AEMWE is in the infant stage. However, there are several issues for AWE and PEMWE, spurring the development of AEMWE for large-scale production of hydrogen. AEMWE in conjunction with renewable energy scenario is a viable portfolio for green hydrogen production for future carbon neutral society. Recent advances in AEMs and electrocatalysts as well as in the durability of a selected AEMWE system can produce green hydrogen in an economic way soon.

References

1 Chatenet, M., Pollet, B.G., Dekel, D.R. et al. (2022). Water electrolysis: from textbook knowledge to the latest scientific strategies and industrial developments. *Chemical Society Reviews* 51: 4583–4762.

2 Ni, M., Leung, D.Y., and Leung, M.K. (2007). A review on reforming bio-ethanol for hydrogen production. *International Journal of Hydrogen Energy* 32: 3238–3247.

3 Logan, B.E. (2019). *Ending Our Hydrogen and Ammonia Addiction to Fossil Fuels*, 257–258. ACS Publications.

4 Esposito, D.V. (2017). Membraneless electrolyzers for low-cost hydrogen production in a renewable energy future. *Joule* 1: 651–658.

5 Zeng, K. and Zhang, D. (2010). Recent progress in alkaline water electrolysis for hydrogen production and applications. *Progress in Energy and Combustion Science* 36: 307–326.

6 Trasatti, S. (1999). Water electrolysis: Who first? *Journal of Electroanalytical Chemistry* 476: 90–91.

7 Kreuter, W. and Hofmann, H. (1998). Electrolysis: the important energy transformer in a world of sustainable energy. *International Journal of Hydrogen Energy* 23: 661–666.

8 Lu, P. and Srinivasan, S. (1979). Advances in water electrolysis technology with emphasis on use of the solid polymer electrolyte. *Journal of Applied Electrochemistry* 9: 269–283.

9 Dunn, S. (2002). Hydrogen futures: toward a sustainable energy system. *International Journal of Hydrogen Energy* 27: 235–264.

10 De Souza, R.F., Padilha, J.C., Gonçalves, R.S. et al. (2007). Electrochemical hydrogen production from water electrolysis using ionic liquid as electrolytes: towards the best device. *Journal of Power Sources* 164: 792–798.

11 Ursua, A., Gandia, L.M., and Sanchis, P. (2011). Hydrogen production from water electrolysis: current status and future trends. *Proceedings of the IEEE* 100: 410–426.

12 Vincent, I. and Bessarabov, D. (2018). Low cost hydrogen production by anion exchange membrane electrolysis: a review. *Renewable and Sustainable Energy Reviews* 81: 1690–1704.

13 Mani, K.N. (1991). Electrodialysis water splitting technology. *Journal of Membrane Science* 58: 117–138.

14 Schröder, V., Emonts, B., Janßen, H., and Schulze, H.-P. (2004). Explosion limits of hydrogen/oxygen mixtures at initial pressures up to 200 bar. *Chemical Engineering & Technology* 27: 847–851.

15 Barbir, F. (2005). PEM electrolysis for production of hydrogen from renewable energy sources. *Solar Energy* 78: 661–669.

16 Grubb, W. (1959). Batteries with solid ion exchange electrolytes: I. secondary cells employing metal electrodes. *Journal of the Electrochemical Society* 106: 275.

17 Millet, P., Dragoe, D., Grigoriev, S. et al. (2009). GenHyPEM: a research program on PEM water electrolysis supported by the European Commission. *International Journal of Hydrogen Energy* 34: 4974–4982.

18 Slade, S., Campbell, S.A., Ralph, T.R., and Walsh, F.C. (2002). Ionic conductivity of an extruded Nafion 1100 EW series of membranes. *Journal of the Electrochemical Society* 149: A1556–A1564.

19 Grigor'ev, S., Khaliullin, M., Kuleshov, N., and Fateev, V. (2001). Electrolysis of water in a system with a solid polymer electrolyte at elevated pressure. *Russian Journal of Electrochemistry* 37: 819–822.

20 Bolobov, V. (2002). Mechanism of self-ignition of titanium alloys in oxygen. *Combustion Explosion and Shock Waves* 38: 639–645.

21 Millet, P., Ngameni, R., Grigoriev, S., and Fateev, V. (2011). Scientific and engineering issues related to PEM technology: water electrolysers, fuel cells and unitized regenerative systems. *International Journal of Hydrogen Energy* 36: 4156–4163.

22 Millet, P., Ngameni, R., Grigoriev, S. et al. (2010). PEM water electrolyzers: from electrocatalysis to stack development. *International Journal of Hydrogen Energy* 35: 5043–5052.

23 Russell, J., Nuttall, L., and Fickett, A. (1973). Hydrogen generation by solid polymer electrolyte water electrolysis. *American Chemical Society, Division of Fuel Chemistry* 18: 24–40.

24 Goñi-Urtiaga, A., Presvytes, D., and Scott, K. (2012). Solid acids as electrolyte materials for proton exchange membrane (PEM) electrolysis. *International Journal of Hydrogen Energy* 37: 3358–3372.

25 Wei, G., Xu, L., Huang, C., and Wang, Y. (2010). SPE water electrolysis with SPEEK/PES blend membrane. *International Journal of Hydrogen Energy* 35: 7778–7783.

26 Vallieres, C., Winkelmann, D., Roizard, D. et al. (2006). On Schroeder's paradox. *Journal of Membrane Science* 278: 357–364.

27 Hamnett, A., Vielstich, W., Lamm, A., and Gasteiger, H. (2003). *Handbook of Fuel Cells–Fundamentals, Technology and Applications*, vol. 1, 305–322.

28 Doyle, M. and Rajendran, G. (2003). Part 3: Polymer electrolyte membrane fuel cell systems (PEMFC). In: *Handbook of Fuel Cells: Fundamentals, Technology and Applications*, 351. Chichester, UK: Wiley.

29 Ito, H., Maeda, T., Nakano, A., and Takenaka, H. (2011). Properties of Nafion membranes under PEM water electrolysis conditions. *International Journal of Hydrogen Energy* 36: 10527–10540.

30 Bass, M. and Freger, V. (2008). Hydration of Nafion and Dowex in liquid and vapor environment: Schroeder's paradox and microstructure. *Polymer* 49: 497–506.

31 Onishi, L.M., Prausnitz, J.M., and Newman, J. (2007). Water–Nafion equilibria. Absence of schroeder's paradox. *The Journal of Physical Chemistry B* 111: 10166–10173.

32 Siracusano, S., Baglio, V., Stassi, A. et al. (2014). Performance analysis of short-side-chain Aquivion® perfluorosulfonic acid polymer for proton exchange membrane water electrolysis. *Journal of Membrane Science* 466: 1–7.

33 Siracusano, S., Oldani, C., Navarra, M.A. et al. (2019). Chemically stabilised extruded and recast short side chain Aquivion® proton exchange membranes for high current density operation in water electrolysis. *Journal of Membrane Science* 578: 136–148.

34 Smith, D.W., Oladoyinbo, F.O., Mortimore, W.A. et al. (2013). A microblock ionomer in proton exchange membrane electrolysis for the production of high purity hydrogen. *Macromolecules* 46: 1504–1511.

35 Klose, C., Saatkamp, T., Münchinger, A. et al. (2020). All-hydrocarbon MEA for PEM water electrolysis combining low hydrogen crossover and high efficiency. *Advanced Energy Materials* 10: 1903995.

36 Han, S.-Y., Yu, D.M., Mo, Y.-H. et al. (2021). Ion exchange capacity controlled biphenol-based sulfonated poly(arylene ether sulfone) for polymer electrolyte membrane water electrolyzers: comparison of random and multi-block copolymers. *Journal of Membrane Science* 634: 119370.

37 Park, J.E., Kim, J., Han, J. et al. (2021). High-performance proton-exchange membrane water electrolysis using a sulfonated poly(arylene ether sulfone) membrane and ionomer. *Journal of Membrane Science* 620: 118871.

38 Bender, J., Mayerhofer, B., Trinke, P. et al. (2021). H(+)-Conducting aromatic multiblock copolymer and blend membranes and their application in PEM electrolysis. *Polymers (Basel)* 13: 3467.

39 Choi, S., Shin, S.-H., Lee, D.-H. et al. (2022). Enhancing the durability of hydrocarbon-membrane-based polymer electrolyte water electrolysis using a radical scavenger-embedded interlocking interfacial layer. *Journal of Materials Chemistry A* 10: 789–798.

40 Wan, L., Xu, Z., Wang, P. et al. (2021). H_2SO_4-doped polybenzimidazole membranes for hydrogen production with acid-alkaline amphoteric water electrolysis. *Journal of Membrane Science* 618: 118642.

41 Shi, L., Rossi, R., Son, M. et al. (2020). Using reverse osmosis membranes to control ion transport during water electrolysis. *Energy & Environmental Science* 13: 3138–3148.

42 Park, C.H., Lee, S.Y., Hwang, D.S. et al. (2016). Nanocrack-regulated self-humidifying membranes. *Nature* 532: 480–483.

43 Hedegaard, K. and Meibom, P. (2012). Wind power impacts and electricity storage–a time scale perspective. *Renewable Energy* 37: 318–324.

44 Ebbesen, S.D., Jensen, S.H., Hauch, A., and Mogensen, M.B. (2014). High temperature electrolysis in alkaline cells, solid proton conducting cells, and solid oxide cells. *Chemical Reviews* 114: 10697–10734.

45 Ayers, K.E., Anderson, E.B., Capuano, C. et al. (2010). Research advances towards low cost, high efficiency PEM electrolysis. *ECS Transactions* 33: 3.

46 Antonucci, V., Di Blasi, A., Baglio, V. et al. (2008). High temperature operation of a composite membrane-based solid polymer electrolyte water electrolyser. *Electrochimica Acta* 53: 7350–7356.

47 Qingfeng, L., Hjuler, H.A., and Bjerrum, N. (2001). Phosphoric acid doped polybenzimidazole membranes: physiochemical characterization and fuel cell applications. *Journal of Applied Electrochemistry* 31: 773–779.

48 He, R., Li, Q., Bach, A. et al. (2006). Physicochemical properties of phosphoric acid doped polybenzimidazole membranes for fuel cells. *Journal of Membrane Science* 277: 38–45.

49 Kim, J.-H., Kim, H.-J., Lim, T.-H., and Lee, H.-I. (2007). Dependence of the performance of a high-temperature polymer electrolyte fuel cell on phosphoric acid-doped polybenzimidazole ionomer content in cathode catalyst layer. *Journal of Power Sources* 170: 275–280.

50 Jeong, Y.H., Oh, K., Ahn, S. et al. (2017). Investigation of electrolyte leaching in the performance degradation of phosphoric acid-doped polybenzimidazole membrane-based high temperature fuel cells. *Journal of Power Sources* 363: 365–374.

51 Tahrim, A. and Amin, I. (2019). Advancement in phosphoric acid doped polybenzimidazole membrane for high temperature PEM fuel cells: a review. *Journal of Applied Membrane Science & Technology* 23: 37–62.

52 Chen, H., Wang, S., Liu, F. et al. (2020). Base-acid doped polybenzimidazole with high phosphoric acid retention for HT-PEMFC applications. *Journal of Membrane Science* 596: 117722.

53 Skorikova, G., Rauber, D., Aili, D. et al. (2020). Protic ionic liquids immobilized in phosphoric acid-doped polybenzimidazole matrix enable polymer electrolyte fuel cell operation at 200°C. *Journal of Membrane Science* 608: 118188.

54 Rajabi, Z., Javanbakht, M., Hooshyari, K. et al. (2021). Phosphoric acid doped polybenzimidazole based polymer electrolyte membrane and functionalized SBA-15 mesoporous for elevated temperature fuel cell. *International Journal of Hydrogen Energy* 46: 33241–33259.

55 Kamaroddin, M.F.A.S., Sabali, N., Abdullah, T.A.T. et al. (2019). Phosphoric acid doped polybenzimidazole and sulfonated polyether ether ketone composite membrane for hudrogen production in high-temperature copper chloride electrolysis. *IOP Conference Series: Earth and Environmental Science* 268: 012057.

56 Carmo, M., Fritz, D.L., Mergel, J., and Stolten, D. (2013). A comprehensive review on PEM water electrolysis. *International Journal of Hydrogen Energy* 38: 4901–4934.

57 Zhang, Y., Wang, C., Wan, N. et al. (2007). Study on a novel manufacturing process of membrane electrode assemblies for solid polymer electrolyte water electrolysis. *Electrochemistry Communications* 9: 667–670.

58 Carmo, M., Roepke, T., Roth, C. et al. (2009). A novel electrocatalyst support with proton conductive properties for polymer electrolyte membrane fuel cell applications. *Journal of Power Sources* 191: 330–337.

59 Xu, W. and Scott, K. (2010). The effects of ionomer content on PEM water electrolyser membrane electrode assembly performance. *International Journal of Hydrogen Energy* 35: 12029–12037.

60 Wu, X., Scott, K., and Puthiyapura, V. (2012). Polymer electrolyte membrane water electrolyser with Aquivion® short side chain perfluorosulfonic acid ionomer binder in catalyst layers. *International Journal of Hydrogen Energy* 37: 13243–13248.

61 Xu, J., Miao, R., Zhao, T. et al. (2011). A novel catalyst layer with hydrophilic–hydrophobic meshwork and pore structure for solid polymer electrolyte water electrolysis. *Electrochemistry Communications* 13: 437–439.

62 Bernt, M. and Gasteiger, H.A. (2016). Influence of ionomer content in IrO_2/TiO_2 electrodes on PEM water electrolyzer performance. *Journal of the Electrochemical Society* 163: F3179.

63 Siracusano, S., Baglio, V., Van Dijk, N. et al. (2017). Enhanced performance and durability of low catalyst loading PEM water electrolyser based on a short-side chain perfluorosulfonic ionomer. *Applied Energy* 192: 477–489.

64 Soni, R., Miyanishi, S., Kuroki, H., and Yamaguchi, T. (2020). Pure water solid alkaline water electrolyzer using fully aromatic and high-molecular-weight poly (fluorene-alt-tetrafluorophenylene)-trimethyl ammonium anion exchange membranes and ionomers. *ACS Applied Energy Materials* 4: 1053–1058.

65 Xue, F.S., Cheng, J., Li, P.P., and Zhang, Y. (2021). Application of proton exchange membrane electrolysis of water hydrogen production technology in polwer plant. *IOP Conference Series: Earth and Environmental Science* 631: 012079.

66 Henkensmeier, D., Najibah, M., Harms, C. et al. (2021). Overview: state-of-the art commercial membranes for anion exchange membrane water electrolysis. *Journal of Electrochemical Energy Conversion and Storage* 18: 024001.

67 Faraj, M., Boccia, M., Miller, H. et al. (2012). New LDPE based anion-exchange membranes for alkaline solid polymeric electrolyte water electrolysis. *International Journal of Hydrogen Energy* 37: 14992–15002.

68 LeRoy, R. (1983). Industrial water electrolysis: present and future. *International Journal of Hydrogen Energy* 8: 401–417.

69 Liu, J., Kang, Z., Li, D. et al. (2021). Elucidating the role of hydroxide electrolyte on anion-exchange-membrane water electrolyzer performance. *Journal of the Electrochemical Society* 168: 054522.

70 Huang, T., He, G., Xue, J. et al. (2020). Self-crosslinked blend alkaline anion exchange membranes with bi-continuous phase separated morphology to enhance ion conductivity. *Journal of Membrane Science* 597: 117769.

71 Kumari, M., Douglin, J.C., and Dekel, D.R. (2021). Crosslinked quaternary phosphonium-functionalized poly (ether ether ketone) polymer-based anion-exchange membranes. *Journal of Membrane Science* 626: 119167.

72 Xue, J., Liu, X., Zhang, J. et al. (2020). Poly (phenylene oxide) s incorporating N-spirocyclic quaternary ammonium cation/cation strings for anion exchange membranes. *Journal of Membrane Science* 595: 117507.

73 Fan, J., Willdorf-Cohen, S., Schibli, E.M. et al. (2019). Poly (bis-arylimidazoliums) possessing high hydroxide ion exchange capacity and high alkaline stability. *Nature Communications* 10: 1–10.

74 Kraglund, M.R., Carmo, M., Schiller, G. et al. (2019). Ion-solvating membranes as a new approach towards high rate alkaline electrolyzers. *Energy & Environmental Science* 12: 3313–3318.

75 Kostalik, H.A. IV, Clark, T.J., Robertson, N.J. et al. (2010). Solvent processable tetraalkylammonium-functionalized polyethylene for use as an alkaline anion exchange membrane. *Macromolecules* 43: 7147–7150.

76 Liu, Z., Sajjad, S.D., Gao, Y. et al. (2017). The effect of membrane on an alkaline water electrolyzer. *International Journal of Hydrogen Energy* 42: 29661–29665.

77 Chen, N., Paek, S.Y., Lee, J.Y. et al. (2021). High-performance anion exchange membrane water electrolyzers with a current density of 7.68 A cm^{-2} and a durability of 1000 hours. *Energy & Environmental Science* 14: 6338–6348.

78 Cha, M.S., Park, J.E., Kim, S. et al. (2020). Poly (carbazole)-based anion-conducting materials with high performance and durability for energy conversion devices. *Energy & Environmental Science* 13: 3633–3645.

79 Mohanty, A.D. and Bae, C. (2014). Mechanistic analysis of ammonium cation stability for alkaline exchange membrane fuel cells. *Journal of Materials Chemistry A* 2: 17314–17320.

80 Chu, X., Shi, Y., Liu, L. et al. (2019). Piperidinium-functionalized anion exchange membranes and their application in alkaline fuel cells and water electrolysis. *Journal of Materials Chemistry A* 7: 7717–7727.

81 Su, X., Gao, L., Hu, L. et al. (2019). Novel piperidinium functionalized anionic membrane for alkaline polymer electrolysis with excellent electrochemical properties. *Journal of Membrane Science* 581: 283–292.

82 Hu, X., Huang, Y., Liu, L. et al. (2021). Piperidinium functionalized aryl ether-free polyaromatics as anion exchange membrane for water electrolysers: performance and durability. *Journal of Membrane Science* 621: 118964.

83 Zhang, F., Zhang, H., and Qu, C. (2011). Imidazolium functionalized polysulfone anion exchange membrane for fuel cell application. *Journal of Materials Chemistry* 21: 12744–12752.

84 Faid, A., Oyarce Barnett, A., Seland, F., and Sunde, S. (2018). Highly active nickel-based catalyst for hydrogen evolution in anion exchange membrane electrolysis. *Catalysts* 8: 614.

85 Xu, D., Stevens, M.B., Cosby, M.R. et al. (2018). Earth-abundant oxygen electrocatalysts for alkaline anion-exchange-membrane water electrolysis: effects of catalyst conductivity and comparison with performance in three-electrode cells. *ACS Catalysis* 9: 7–15.

86 Park, J.E., Kang, S.Y., Oh, S.-H. et al. (2019). High-performance anion-exchange membrane water electrolysis. *Electrochimica Acta* 295: 99–106.

87 Motealleh, B., Liu, Z., Masel, R.I. et al. (2021). Next-generation anion exchange membrane water electrolyzers operating for commercially relevant lifetimes. *International Journal of Hydrogen Energy* 46: 3379–3386.

88 Leng, Y., Chen, G., Mendoza, A.J. et al. (2012). Solid-state water electrolysis with an alkaline membrane. *Journal of the American Chemical Society* 134: 9054–9057.

89 Li, D., Lin, Y., Serov, A. et al. (2020). Highly quaternized polystyrene ionomers for high performance anion exchange membrane water electrolysers. *Nature Energy* 5: 375–385.

90 Cha, M.S., Park, J.E., Kim, S. et al. (2022). Oligomeric chain extender-derived anion conducting membrane materials with poly(p-phenylene)-based architecture for fuel cells and water electrolyzers. *Journal of Materials Chemistry A* 10: 9693–9697.

91 Xiao, J., Oliveira, A.M., Wang, L. et al. (2020). Water-fed hydroxide exchange membrane electrolyzer enabled by a fluoride-incorporated nickel–iron oxyhydroxide oxygen evolution electrode. *ACS Catalysis* 11: 264–270.

92 Wan, L., Xu, Z., and Wang, B. (2021). Green preparation of highly alkali-resistant PTFE composite membranes for advanced alkaline water electrolysis. *Chemical Engineering Journal* 426: 131340.

93 Motz, A.R., Li, D., Keane, A. et al. (2021). Performance and durability of anion exchange membrane water electrolyzers using down-selected polymer electrolytes. *Journal of Materials Chemistry A* 9: 22670–22683.

94 Xiao, L., Zhang, S., Pan, J. et al. (2012). First implementation of alkaline polymer electrolyte water electrolysis working only with pure water. *Energy & Environmental Science* 5: 7869–7871.

95 Park, H.J., Lee, S.Y., Lee, T.K. et al. (2020). N3-butyl imidazolium-based anion exchange membranes blended with poly(vinyl alcohol) for alkaline water electrolysis. *Journal of Membrane Science* 611: 118355.

96 Liu, M., Hu, X., Hu, B. et al. (2022). Soluble poly(aryl piperidinium) with extended aromatic segments as anion exchange membranes for alkaline fuel cells and water electrolysis. *Journal of Membrane Science* 642: 119969.

97 Merle, G., Wessling, M., and Nijmeijer, K. (2011). Anion exchange membranes for alkaline fuel cells: a review. *Journal of Membrane Science* 377: 1–35.

98 Ziv, N., Mustain, W.E., and Dekel, D.R. (2018). The effect of ambient carbon dioxide on anion-exchange membrane fuel cells. *ChemSusChem* 11: 1136–1150.

99 Ziv, N. and Dekel, D.R. (2018). A practical method for measuring the true hydroxide conductivity of anion exchange membranes. *Electrochemistry Communications* 88: 109–113.

100 Cao, X., Novitski, D., and Holdcroft, S. (2019). Visualization of hydroxide ion formation upon electrolytic water splitting in an anion exchange membrane. *ACS Materials Letters* 1: 362–366.

101 Fortin, T.K.P., Cao, X., Martinsen, S.Y. et al. (2020). High performance alkaline water electrolysis using Aemion anion exchange membranes. *Journal of Power Sources* 451: 227814.

102 Buttler, A. and Spliethoff, H. (2018). Current status of water electrolysis for energy storage, grid balancing and sector coupling via power-to-gas and power-to-liquids: a review. *Renewable and Sustainable Energy Reviews* 82: 2440–2454.

103 Kutz, R.B., Chen, Q., Yang, H. et al. (2017). Sustainion imidazolium-functionalized polymers for carbon dioxide electrolysis. *Energy Technology* 5: 929–936.

104 Kaczur, J.J., Yang, H., Liu, Z. et al. (2018). Carbon dioxide and water electrolysis using new alkaline stable anion membranes. *Frontiers in Chemistry* 6: 263.

105 Lee, W.H., Park, E.J., Han, J. et al. (2017). Poly(terphenylene) anion exchange membranes: the effect of backbone structure on morphology and membrane property. *ACS Macro Letters* 6: 566–570.

106 Meek, K.M., Antunes, C.M., Strasser, D.J. et al. (2019). High-throughput anion exchange membrane charaterization at NREL. *ECS Transactions* 92: 723–731.

107 Park, E.J., Capuano, C.B., Ayers, K.E., and Bae, C. (2018). Chemically durable polymer electrolytes for solid-state alkaline water electrolysis. *Journal of Power Sources* 375: 367–372.

108 Lindquist, G.A., Oener, S.Z., Krivina, R. et al. *ACS Applied Materials & Interfaces* 13 (2021): 51917–51924.

109 Chen, N., Jin, Y., Liu, H. et al. (2021). Insight into the alkaline stability of N-heterocyclic ammonium groups for anion-exchange polyelectrolytes. *Angewandte Chemie (International Ed. in English)* 60: 19272–19280.

110 Aili, D., Hansen, M.K., Renzaho, R.F. et al. (2013). Heterogeneous anion conducting membranes based on linear and crosslinked KOH doped polybenzimidazole for alkaline water electrolysis. *Journal of Membrane Science* 447: 424–432.

111 Diaz, L.A., Coppola, R.E., Abuin, G.C. et al. (2017). Alkali-doped polyvinyl alcohol – polybenzimidazole membranes for alkaline water electrolysis. *Journal of Membrane Science* 535: 45–55.

112 Parrondo, J., Arges, C.G., Niedzwiecki, M. et al. (2014). Degradation of anion exchange membranes used for hydrogen production by ultrapure water electrolysis. *RSC Advances* 4: 9875–9879.

113 Parrondo, J. and Ramani, V. (2014). Stability of poly(2,6-dimethyl 1,4-phenylene)oxide-based anion exchange membrane separator and solubilized electrode binder in solid-state alkaline water electrolyzers. *Journal of the Electrochemical Society* 161: F1015–F1020.

114 Li, D., Park, E.J., Zhu, W. et al. (2020). Highly quaternized polystyrene ionomers for high performance anion exchange membrane water electrolysers. *Nature Energy* 5: 378–385.

115 Chen, P. and Hu, X. (2020). High-efficiency anion exchange membrane water electrolysis employing non-noble metal catalysts. *Advanced Energy Materials* 10: 202002285.

116 Huang, G., Mandal, M., Hassan, N.U. et al. (2020). Ionomer optimization for water uptake and swelling in anion exchange membrane electrolyzer: oxygen evolution electrode. *Journal of the Electrochemical Society* 167: 164514.

117 Marino, M.G. and Kreuer, K.D. (2015). Alkaline stability of quaternary ammonium cations for alkaline fuel cell membranes and ionic liquids. *ChemSusChem* 8: 513–523.

118 Pham, T.H., Olsson, J.S., and Jannasch, P. (2018). Poly(arylene alkylene)s with pendant N-spirocyclic quaternary ammonium cations for anion exchange membranes. *Journal of Materials Chemistry A* 6: 16537–16547.

119 Pham, T.H., Olsson, J.S., and Jannasch, P. (2019). Effects of the N-alicyclic cation and backbone structures on the performance of poly(terphenyl)-based hydroxide exchange membranes. *Journal of Materials Chemistry A* 7: 15895–15906.

120 Pham, T.H., Allushi, A., Olsson, J.S., and Jannasch, P. (2020). Rational molecular design of anion exchange membranes functionalized with alicyclic quaternary ammonium cations. *Polymer Chemistry* 11: 6953–6963.

121 Allushi, A., Pham, T.H., and Jannasch, P. (2021). Highly conductive hydroxide exchange membranes containing fluorene-units tethered with dual pairs of quaternary piperidinium cations. *Journal of Membrane Science* 632: 119376.

122 Pan, D., Pham, T.H., and Jannasch, P. (2021). Poly(arylene piperidine) anion exchange membranes with tunable *N*-alicyclic quaternary ammonium side chains. *ACS Applied Energy Materials* 4: 11652–11665.

123 Li, D., Motz, A.R., Bae, C. et al. (2021). Durability of anion exchange membrane water electrolyzers. *Energy & Environmental Science* 14: 3393–3419.

124 Adhikari, S., Leonard, D.P., Lim, K.H. et al. (2022). Hydrophobic quaternized poly(fluorene) ionomers for emerging fuel cells. *ACS Applied Energy Materials* 5: 2663–2668.

125 Pan, J., Chen, C., Li, Y. et al. (2014). Constructing ionic highway in alkaline polymer electrolytes. *Energy & Environmental Science* 7: 354–360.

126 Park, E.J., Maurya, S., Martinez, U. et al. (2021). Quaternized poly(arylene ether benzonitrile) membranes for vanadium redox flow batteries. *Journal of Membrane Science* 617: 118565.

127 Lee, K.H., Cho, D.H., Kim, Y.M. et al. (2017). Highly conductive and durable poly(arylene ether sulfone) anion exchange membrane with end-group cross-linking. *Energy & Environmental Science* 10: 275–285.

128 Zhu, L., Zimudzi, T.J., Wang, Y. et al. (2017). Mechanically robust anion exchange membranes via long hydrophilic cross-linkers. *Macromolecules* 50: 2329–2337.

129 Liang, X., Shehzad, M.A., Zhu, Y. et al. (2019). Ionomer cross-linking immobilization of catalyst nanoparticles for high performance alkaline membrane fuel cells. *Chemistry of Materials* 31: 7812–7820.

130 Gao, X.L., Sun, L.X., Wu, H.Y. et al. (2020). Highly conductive fluorine-based anion exchange membranes with robust alkaline durability. *Journal of Materials Chemistry A* 8: 13065–13076.

131 Lee, S.-B., Min, C.-M., Jang, J., and Lee, J.-S. (2020). Enhanced conductivity and stability of anion exchange membranes depending on chain lengths with crosslinking based on poly(phenylene oxide). *Polymer* 192: 122331.

132 Chen, M., Mandal, M., Groenhout, K. et al. (2022). Self-adhesive ionomers for durable low-temperature anion exchange membrane electrolysis. *Journal of Power Sources* 536: 231495.

133 Wang, B., Pan, J., Zou, X. et al. (2022). UV-crosslinkable anthracene-based ionomer derived gas "Expressway" for anion exchange membrane fuel cells. *Journal of Materials Chemistry A* 10: 13355–13367.

134 Santoro, C., Lavacchi, A., Mustarelli, P. et al. (2022). What is next in anion-exchange membrane water electrolyzers? Bottlenecks, Benefits, and Future. *ChemSusChem* 15: e202200027.

7

Membranes for Power Generation

7.1 Water Energy Nexus and Membranes

In recent years, water and energy systems are interdependent [1, 2], as water is used in all phases of energy production and electricity generation [2]. Water scarcity, variability, and uncertainty lead to vulnerable concerns in the world energy system. As the future is not like the past in terms of climate, technology, and the evolving decision landscape, energy and water issues are gaining international attention. Climate change implies the future of the water-energy nexus will bring changes and variation in the availability of water resources due to increasing temperatures, changes in precipitation patterns, increasing climate variability, and more frequent extreme weather events (e.g. floods and droughts).

Energy is necessary to extract, convey, and deliver water of appropriate quality for diverse human uses, and then again to treat wastewater prior to their return to the environment. Moreover, several current trends are further increasing the urgency to address the water-energy nexus in an integrated and proactive way. First, severe climate change has already begun to affect precipitation and temperature patterns around the world and natural disasters like typhoons, floods, and drought are often occurring. Second, the population growth around the world indicates further complicating the management of both energy and water systems. Third, introduction of new technologies in the energy and the water domains could shift water and energy demands.

According to a report from Department of Energy (DOE) [1], there are six guiding strategic pillars:

- Optimize the freshwater efficiency of energy production, electricity generation, and end use systems.
- Optimize the energy efficiency of water management, treatment, distribution, and end use systems.
- Enhance the reliability and resilience of energy and water systems.
- Increase safe and productive use of nontraditional water sources.
- Promote responsible energy operations with respect to water quality, ecosystem, and seismic impacts.
- Exploit productive synergies among water and energy systems.

Membranes for Energy Applications, First Edition. Young Moo Lee.

According to published report by International Energy Agency (IEA), global renewable energy consumption increased more than 5% annually, which is three times faster than growth of total energy consumption [3]. In the next years, the share of renewables meeting global energy demand is expected to increase by a fifth, covering 40% of global energy consumption growth. There are various power generation technologies using renewable sources such as wind, solar, geothermal, and hydropower. Among them, hydropower is expected to remain the largest renewable source, meeting 16% of global electricity demand by 2023, followed by wind (6%), solar photovoltaic (PV, 4%), and bioenergy (3%) [4]. Despite the potential of renewables, these new power generation technologies have been challenged because they are not sustainable in energy production and cause various environmental issues such as large foot-prints, huge waste, and disposal problems [5, 6].

To overcome these problems and harvest renewable energy from abundant water resources, Loeb et al. designed pressure-retarded osmosis (PRO) for the first time in 1976 [7, 8]. As shown in Figure 7.1, two solutions with different concentrations are separated by a semipermeable membrane through which water moves from low to high concentration solutions (denoted as "feed" and "draw" side, respectively) [9, 10]. The osmotic-driven water flux dilutes draw solution and increase the volume flow rate of water. The resultant high-pressure draw solution is divided into a flow driving the turbine and a flow retuning to the pressure exchanger. The exchanger repressurizes the draw solution by transferring pressure energy from high to low pressure fluids. The diluted draw solution returns to the source, and the turbine generates energy.

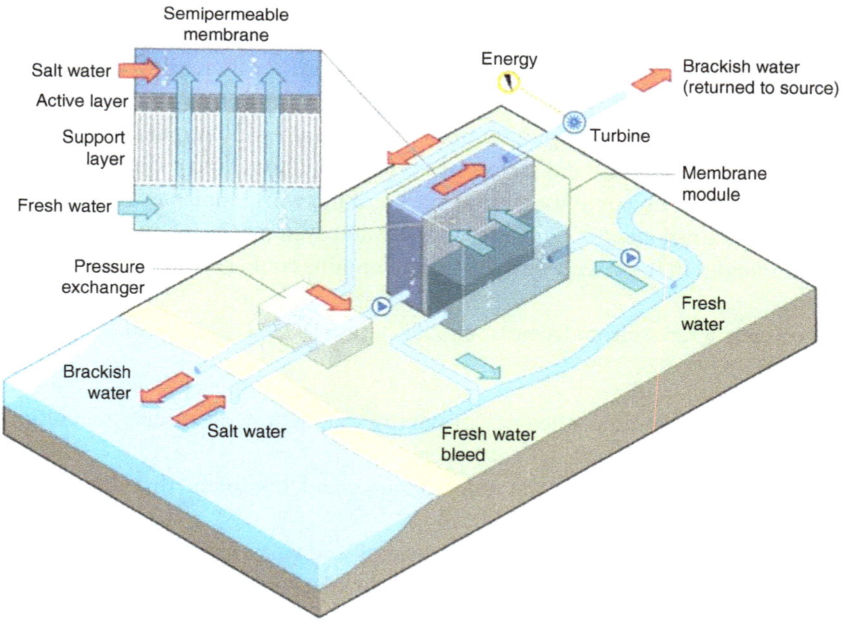

Figure 7.1 Schematic diagram of the PRO. Source: Logan and Elimelech [9]/Springer Nature.

7.2 Concept of Osmotic Power

In thermodynamic principle, useful energy can be obtained when a dilute and a concentrated solution are mixed. The issue is to convert this free energy of mixing to useful energy in an economic way. Loeb first proposed pressure-retarded osmosis [7] by mixing fresh water and saline water from a river, sea, or underground source.

Figure 7.2a presents an osmosis process where a solution or pure water is separated from its solvent or saline water by a semipermeable membrane. Under this condition, the solvent naturally flows from the pure water side of the membrane to the saline water side. If the hydraulic pressure is applied to the solution side (Figure 7.2b) to the level where $\Delta P = \Delta\pi$ osmotic pressure, the solution reaches an osmotic equilibrium (Figure 7.2c). On the contrary, when $\Delta P > \Delta\pi$, and hydraulic pressure is applied to the saline water side, water flux goes from saline water to the pure water side, which is called a reverse osmosis (RO) (Figure 7.2d). Solvent permeation rate or flux J_1 through the membrane is a function of both π and the hydraulic pressure P applied to the solution where $\Delta P > \Delta\pi$.

$$J_1 = A(\pi - P) \tag{7.1}$$

where A is the water flux constant and is characteristic of the membrane. If $\Delta P = 0$, we called it as an osmosis, for which $J_1 = A\pi$, and that if $\Delta P = \Delta\pi$, we obtain osmotic

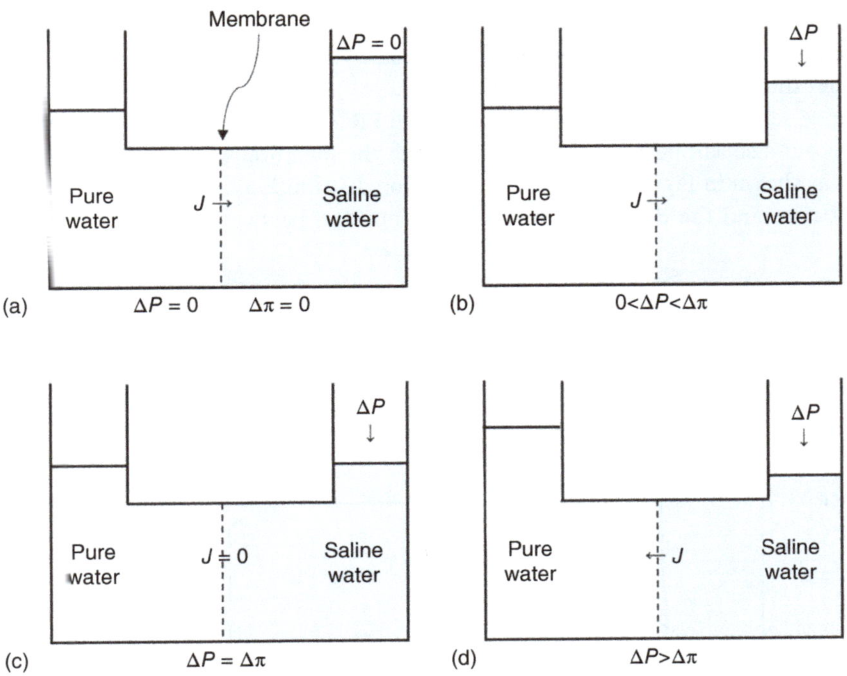

Figure 7.2 Schematics of four osmotic processes. (a) Osmosis, (b) pressure-retarded osmosis (c) osmotic equilibrium, and (d) reverse osmosis. Pure water and saline water is separated with semi-permeable membrane.

equilibrium, i.e. $J_1 = 0$. If $0 < \Delta P < \Delta\pi$, we have pressure-retarded osmosis (PRO) (Figure 7.2b) as the direction of water flux is still the same as in osmosis, but the flux is decreased as the P on the position is increased. In PRO process, solvent or pure water flows against the hydraulic pressure gradient, i.e. the flow is uphill from a hydraulic pressure standpoint. By this means, the free energy of mixing can be converted to potential mechanical energy, and therefore PRO can be used in power production. Although PRO can be the most economical osmotic power production process, it suffers from the fact that it is an intermittent or batch-type operation due to the dilution of the solution by the solvent, requiring frequent solution replacement and causing periodic reduction of power and complicating valve equipment for changing solution. Loeb have already compared the hollow fiber membrane modules vs asymmetric membrane modules for PRO and calculated the energy efficiency [7, 8]. Using Dead Sea brine and water, power cost can be calculated in a PRO power plant using hollow fiber desalination membrane module.

Later Lee et al. [11] developed a model to obtain the projected performance of membranes in PRO from direct osmosis. They concluded that for PRO to be economically feasible, significant improvement in performance will be required using seawater-fresh water as the salinity gradient resource.

Figure 7.3 represents the direction of water as a function of applied pressure in FO, PRO, and RO. FO takes place when the hydraulic pressure differential ΔP is zero and the flux is driven by the osmotic pressure differential, $\Delta\pi$, PRO occurs when the hydrostatic pressure differential is non-zero and less than the osmotic pressure differential. RO takes places when the applied hydrostatic pressure differential is greater than the osmotic pressure differential.

Since semi-permeable membrane is used in PRO system, not only water permeation but also salt permeation occurs through the membrane. Assuming the ideal case as shown in Figure 7.4a, the concentration distribution in the membrane does not occur, and the difference in bulk concentration between feed and draw will

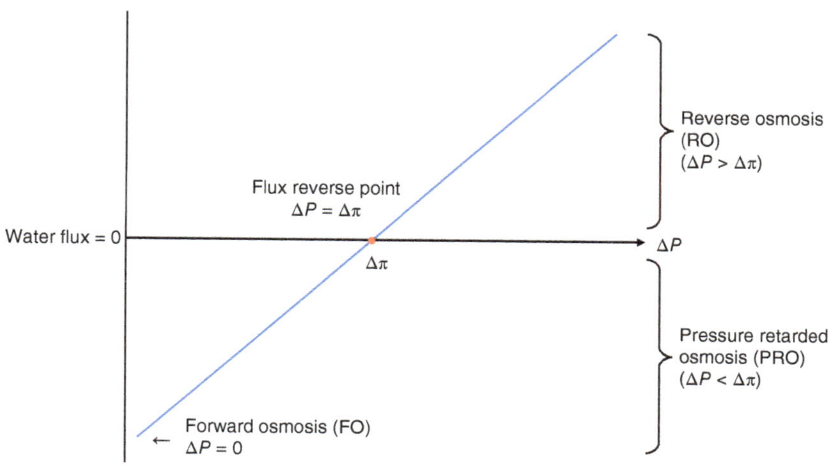

Figure 7.3 Direction of water flux as a function of applied pressure in FO, RO, and PRO.

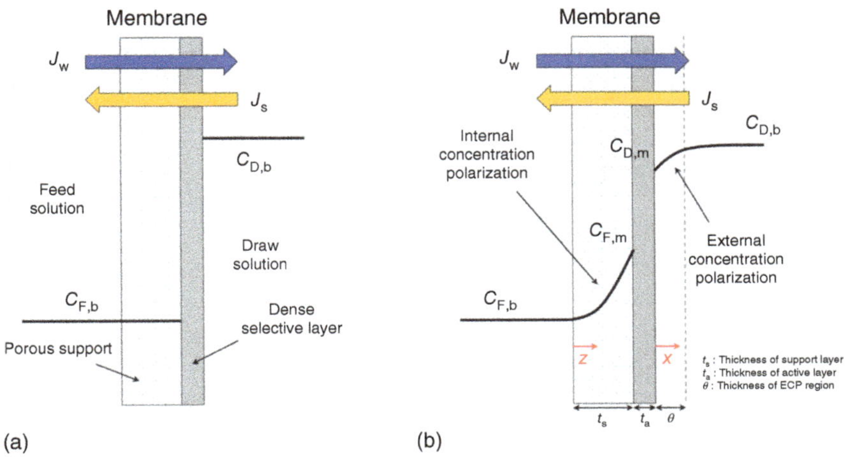

Figure 7.4 Illustration of concentration profiles for an asymmetric TFC membrane with the selective layer facing the draw solution (PRO mode); (a) Ideal case, (b) Real case.

be the driving force that directly causes water permeation. However, in practice, the solution concentration in the porous support is increased by the salt flux passing through the membrane, and the concentration of draw solution was diluted by the permeated water (Figure 7.4b). These phenomena are internal concentration polarization (ICP) and external concentration polarization (ECP), respectively [12]. For the theoretical analysis of water flux (J_w) and salt flux (J_s) in PRO system, the following equations are defined [10]:

$$J_w = A \cdot (\Delta\pi_{\text{eff}} - \Delta P_{\text{PRO}}) \tag{7.2}$$

$$J_s = B \cdot (C_{\text{D,m}} - C_{\text{F,m}}) = D_s \cdot \frac{\text{d}C(z)}{\text{d}z} - J_w \cdot C(z) \tag{7.3}$$

where A is the water permeability coefficient, $\Delta\pi_{\text{eff}}$ is the effective osmotic pressure difference on both sides of the selective layer, ΔP_{PRO} is hydraulic pressure applied to PRO draw side, B is the salt permeability coefficient, $C_{\text{D,m}}$ and $C_{\text{F,m}}$ are draw and feed concentrations at the selective layer surface, respectively, D_s is effective diffusion coefficient of salt which is expressed as:

$$D_s = D \cdot \varepsilon / \tau \tag{7.4}$$

where D is salt diffusion coefficient in water, ε is porosity of the support layer, and τ is tortuosity of the support layer. Note that J_s across the porous support is the sum of the diffusive component, driven by ΔC, and the convective component, arising from J_w. Using the following boundary conditions, Eq. (7.3) can be integrated and the equation for $C_{\text{F,m}}$ is calculated as (see Figure 7.4b):

$$\text{Boundary condition (1)}: z = 0 \rightarrow C = C_{\text{F,b}}$$

$$\text{Boundary condition (2)}: z = t_s \rightarrow C = C_{\text{F,m}}$$

$$C_{\text{F,m}} = C_{\text{F,b}} \cdot \exp\left(\frac{J_w \cdot S}{D}\right) + \frac{J_s}{J_w} \cdot \left[\exp\left(\frac{J_w \cdot S}{D}\right) - 1\right] \tag{7.5}$$

where $C_{F,b}$ is the bulk feed concentration, t_s is the thickness of the support layer, and S is the structure parameter of the support layer defined as "$t_s \cdot \tau/\varepsilon$." The equation of $C_{D,m}$ is similar to the case of $C_{F,m}$, but is obtained by integrating Eq. (7.3) using the boundary conditions related to ECP phenomenon:

$$\text{Boundary condition (3)}: x = 0 \rightarrow C = C_{D,m}$$

$$\text{Boundary condition (4)}: x = \theta \rightarrow C = C_{D,b}$$

$$C_{D,m} = C_{D,b} \cdot \exp\left(-\frac{J_w}{k_d}\right) - \frac{J_s}{J_w} \cdot \left[1 - \exp\left(-\frac{J_w}{k_d}\right)\right] \tag{7.6}$$

where $C_{D,b}$ is the bulk draw concentration, θ is the thickness of ECP region, and k_d is the mass transfer coefficient of the draw boundary layer. The theoretical equation for J_w is obtained by substituting Eqs. (7.5) and (7.6) into Eq. (7.2) [13, 14]:

$$J_w = A \cdot \left[\frac{\pi_{D,b} \cdot \exp\left(-\frac{J_w}{k_d}\right) - \pi_{F,b} \cdot \exp\left(\frac{J_w \cdot S}{D}\right)}{1 + \frac{B}{J_w} \cdot \left[\exp\left(\frac{J_w \cdot S}{D}\right) - \exp\left(-\frac{J_w}{k_d}\right)\right]} - \Delta P_{PRO}\right] \quad (\text{L m}^{-2}\,\text{hr}^{-1}) \tag{7.7}$$

The corresponding power density (W) is equal to the product of J_w and ΔP_{PRO} [15]:

$$W = J_w \cdot \Delta P_{PRO} = A(\Delta\pi_{eff} - \Delta P_{PRO}) \cdot \Delta P_{PRO} \quad (\text{W m}^{-2}) \tag{7.8}$$

The Eqs. (7.7) and (7.8) were used to analyze PRO performances for the various water sources in later chapters. By differentiating Eq. (7.8) with respect to ΔP_{PRO}, it is determined that the inverse convex function of W reaches a maximum when $\Delta P_{PRO} = \Delta\pi_{eff}/2$ (Figure 7.5). The maximum power density (W_{max}) can be calculated by substituting this value for ΔP_{PRO} in Eq. (7.8):

$$W_{max} = A \cdot \frac{\Delta\pi_{eff}^2}{4} \tag{7.9}$$

Note that Figure 7.3 can be modified to Figure 7.5 assuming the ideal case and real case. In Figure 7.5, the FO point ($\Delta P_{PRO} = 0$), the PRO region (where $\Delta P_{PRO} < \Delta\pi_{eff}$), the RO region (where $\Delta P_{PRO} > \Delta\pi_{eff}$) as well as the maximum power density (W_{max}) are indicated. Here, real values are not the same as ideal cases and in fact, are smaller than ideal values.

7.3 Energy Obtained from PRO

It is estimated that the maximum extractable energy per cubic meter of feed solution (such as river or waste water) during the mixing of two saline components ranges from 0.75 kWh m^{-3} (sea water) to 14.1 kWh m^{-3} (Dead Sea) according to draw sources [9]. Given that there are enormous amounts of rivers flowing into the sea all over the world, Achilli et al. demonstrated that the potential energy production of

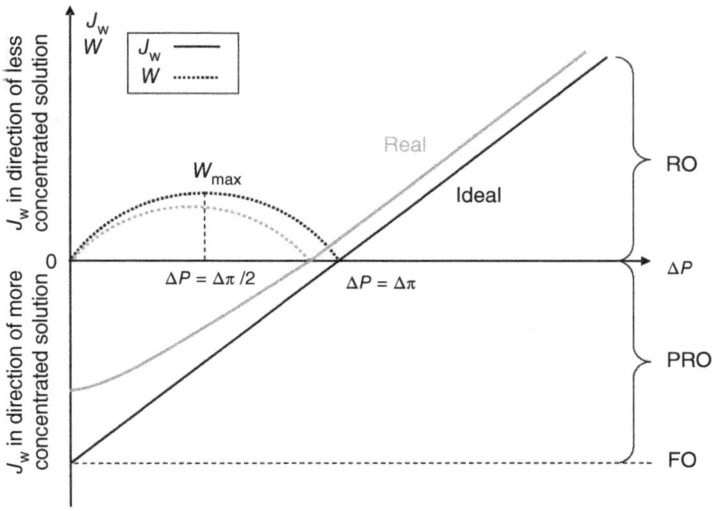

Figure 7.5 Magnitude and direction of water flux for FO, RO, and PRO. The magnitude of power density for PRO is represented as dashed lines. Source: Figure was adopted from Achilli et al. [10].

PRO can reach 2000 TWh yr^{-1}, accounting for 20% of total energy production from renewable sources [10, 16]. Furthermore, PRO can be described as a compact system requiring a relatively small footprint, and can be constructed underground (or at least partially underground) to fit the local environment [17]. Specifically, conventional hydropower generation using dams has been restricted since it interferes with the flow of water and damages enormous nature. However, the PRO only requires membrane modules and pumping system, which can be placed efficiently in a small space. Note that PRO also has the advantage of being a sustainable power generation system without any hazardous waste to the environment. Contrary to other renewable sources with intermittent supply such as wind and solar, PRO is not sensitive to changing surroundings if salinity-gradient sources are present.

Therefore, a remarkable amount of energy is available from the salinity difference resulting in osmotic difference. In fact, the large volumes of water at different salinities (river or municipal wastewater –0.1 to 1 g l^{-1}, seawater—35–40 g l^{-1}, seawater concentrate, 70–80 g l^{-1}) available at coastal urban centers make possible salinity gradient power generation. Logan and Elimelech [9] evaluated membrane processes driven by the energy of mixing between such streams. They cite a worldwide potential of 980 GW for river water mixing with seawater and 18 GW mixing municipal wastewater with seawater. In comparison, installed wind power is projected to be 500 GW by 2030. A more recent survey overviewed the different technologies available for salinity gradient power generation including capacitive processes [18], the latter being at a more preliminary stage and more likely challenged by fouling than membrane processes. The two most promising membrane techniques involve inverse operation of desalination processes to mix instead of separate streams. The inverse operation of RO is PRO, and the inverse

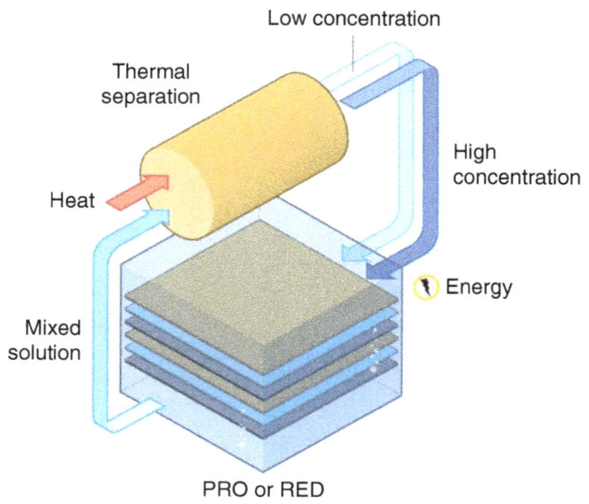

Figure 7.6 A closed-loop system in which synthetic low- and high-concentration solution pass through a PRO or RED system to generate electrical power. Source: Logan and Elimelech [9]/Springer Nature.

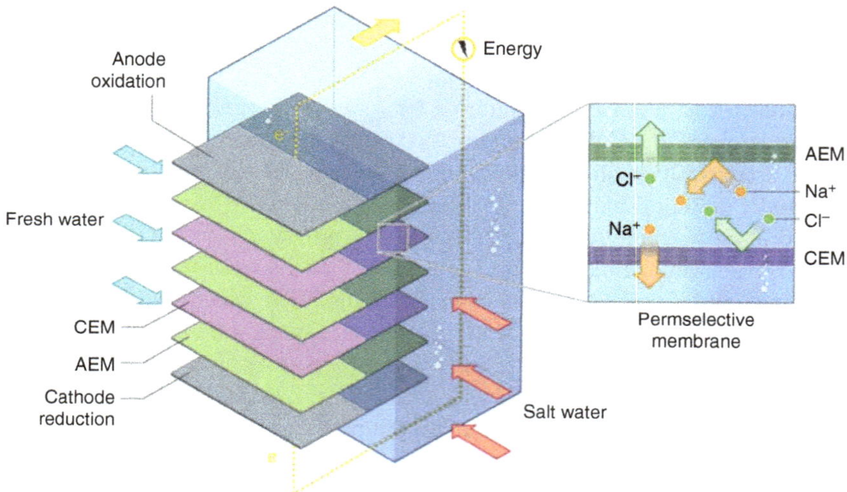

Figure 7.7 Working principle of RED. Source: Logan and Elimelech [9]/Springer Nature.

operation of electrodialysis is reverse electrodialysis (RED). These processes are presented schematically in another study comparing the two processes [19], and reproduced in Figures 7.6 and 7.7. Salinity gradient energy generates constant power instead of being variable like wind and photovoltaic (PV). A membrane-based salinity gradient plant also takes less footprint than wind and PV given the high membrane volumetric packing density. For solar PV, converted power density would range from 1.2 to 1.6 kWh d^{-1} m^{-2}, whereas a PRO plant with membranes operating at 10 W m^{-2} could have an energy density of 25–35 kWh d m^{-2} footprint, and even small RED would have 80 kWh d m^{-2} [20].

In case of the river water and seawater solution pairing, despite the remarkably high global theoretical potential, the extractable specific energy from PRO will be less than the system energy consumption. And hence, higher concentration gradients with a greater extractable energy density can improve the feasibility of PRO. For example, when the draw concentration is increased from 0.6 M NaCl (seawater concentration) to 3 M NaCl, the reversible specific energy increases nearly sixfold. To efficiently extract energy from higher concentration solutions, membrane will need to be modified for these systems. One established requirement for PRO with increased concentration gradients is a higher operating pressure in the system [21]. Both the power density and specific energy of a membrane are found to reach their maximum when the system is operated with $\Delta P_{PRO} = \Delta\pi_{eff}/2$, as shown in Figure 7.5. Meeting this condition with extremely saline solutions, such as water from the Dead Sea with an osmotic pressure around 507 bar, will be nearly impossible. Fabricating membranes that can endure high operating pressure has proven experimentally difficult. More studies will be needed to identify the structural characteristics of membranes that can operate at higher pressures and scaling-up with large membrane modules.

High concentrated solution can be produced through the conventional desalination process. The RO brine, due to its high osmotic pressure (roughly 50–59 bar in sea water reverse osmosis (SWRO) plants), is suitable as a source of high concentrated solution [22, 23]. Moreover, by incorporating with thermal separation process such as membrane distillation (MD), the closed-looped process with PRO and MD can produce electricity from low-grade heat sources, denoted as the osmotic heat engine (OHE). MD can employ low-grade heat in the thermal separation stage to regenerate the low and high concentration solutions [24]. Figure 7.6 represents a closed-loop system in which synthetic low and high concentration pass through a PRO or RED system to generate electrical power [9]. This topic will be discussed further in Section 7.5 later.

7.4 Membranes for Pressure-Retarded Osmosis

Note that the development of PRO has been hindered by the lack of membrane technologies. Because semipermeable membranes in PRO system cannot show the perfect rejection of salt ions, there is a concentration distribution in the membranes, which is called internal concentration polarization (ICP) phenomenon. Similarly, external concentration polarization (ECP) phenomenon also occurred at draw side of PRO system due to the dilution effect of water flux permeating from high to low concentration solutions. These concentration polarization (CP) phenomena significantly reduce the PRO performance by decreasing the effective osmotic pressure difference at the membrane surface [10, 25]. For example, in first PRO experiment of Loeb et al. seawater RO membrane enclosed in a "minipermeator" (Permasp® B-10) were tested [8]. The commercial RO membrane showed performances of 1.56–3.27 W m^{-2} using hyper saline draw solution, which were far below the expected results. It was determined that RO membranes were not suitable for FO and PRO applications due to their hydrophobicity and thick support layer. Until this time, the concept and theory of CP had not been established. Since 2006, Elimelech and Childress groups investigated theoretical modelling for PRO application [10, 12, 25], promoting development of FO and PRO membranes based on an understanding of CP. Their investigations that include the influence of draw solution, feed solution, CP, and hydraulic pressure on the water flux made it possible to predict the PRO power density under the certain membrane characteristics and operating conditions. Here, we will discuss about the development of membranes for PRO systems.

7.4.1 Cellulose Triacetate Membrane

In the early stage of membrane development for PRO, cellulose triacetate (CTA) membranes were specifically designed due to their robustness and resistance to ICP phenomenon [10, 26, 27]. The developed CTA membranes exhibited power density of 1.0~2.7 $W\,m^{-2}$ for river water and sea water, and when using a sea water RO brine as a draw solution, the power density reached up to 5.1 $W\,m^{-2}$ [26, 28]. However, a pilot-scale PRO plant in Norway generated less than 1 $W\,m^{-2}$ using commercial and asymmetric CTA membranes [29]. This performance was much lower than target power density of 5 $W\,m^{-2}$ required to make PRO economically feasible [30, 31]. Considering that additional energy is consumed to pretreat feed and draw solutions to prevent fouling and potential deterioration in performance, the CTA membranes with a thick selective layer fabricated by phase separation showed limitations in PRO application.

7.4.2 Thin-Film Composite Membrane

Therefore, researchers designed a thin-film composite (TFC) membrane which was composed of a porous support layer and a selective layer hundreds of nanometer thick (for example as depicted in Figure 7.8) [15, 32–35]. The selective layer of TFC membranes was fabricated through interfacial polymerization, and its low thickness promoted water permeation in PRO system. For example, Song et al. fabricated TFC membrane using polyacrylonitrile (PAN) nanofibrous support [36]. The developed TFC membrane exhibited the highest power density of 15.2 $W\,m^{-2}$ at 15 bar using river water and 1 M $NaCl_{aq}$ solution as feed and draw sources, respectively. However, theoretical calculations showed that if the developed membrane had sufficient mechanical strength to withstand high hydraulic pressure, the maximum power density could be obtained at 24 bar, half the osmotic pressure difference (theoretical equations were described in Section 7.2).

Specifically, the fabrication of TFC membrane is quite complex. Since CP phenomena reduce the effective osmotic pressure difference applied to the membrane, the support layers must have high porosity and pore connectivity [14, 32]. However, a highly porous membranes do not usually provide sufficient mechanical strength to withstand the high operating pressures in PRO application. This trade-off for the membrane properties becomes more important when hypersaline

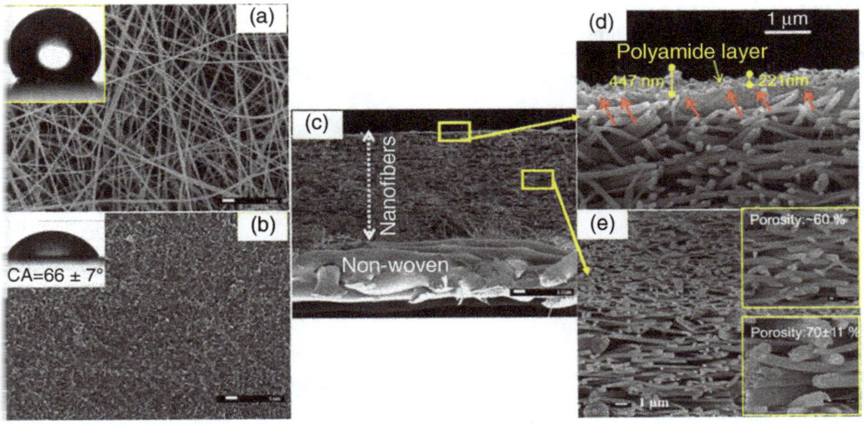

Figure 7.8 Surface morphologies of the top layer (a) before and (b) after interfacial polymerization with contact angle images; (c) cross-sectional SEM image of TFC membrane; (d) enlarged cross-sectional SEM image at the top layer of TFC membrane; (e) enlarged cross-sectional SEM image showing the rejoin between finer fiber layer and coarse fiber layer. Source: Tian et al. [15]/with permission from Elsevier.

solutions are used as draw sources because ICP becomes severe and hydro pressure should be increased in the system to maximize generated power.

7.4.3 Importance of Support Membranes

One possible solution to overcome this dilemma is using intrinsically robust polymer as the material for the support layer. Zhang et al. fabricated the polyethersulfone (PES) hollow fiber supports with diversified pore structures from macrovoid to sponge-like nanopores [37]. The developed membrane withstood under the hydraulic pressure of 20 bar, exhibiting PRO power density of $24\,\mathrm{W\,m^{-2}}$ using 1 M $NaCl_{aq}$ and deionized (D.I.) water as draw and feed solutions, respectively. Similarly, Kim et al. utilized thermally rearranged (TR) polymer having benzoxazole groups in the main chain to prepare the nanofibrous support membrane [38], which showed a tensile strength of 39 MPa with a high porosity of 62%. Given that typical nanofibrous membranes and hollow fibers only exhibited tensile strength of around 10 MPa [36, 39], the use of rigid TR polymer made it possible to control the morphology of the support layer freely without the concerns for robustness. Thus, the resultant TFC membrane showed power density of $39.5\,\mathrm{W\,m^{-2}}$ at 18 bar using 3 M $NaCl_{aq}$ and D.I. water as draw and feed solutions, respectively.

The other solution to decrease ICP phenomenon without the sacrifice of mechanical strength of support layer is hydrophilization by coating the hydrophilic polymers such as poly(vinyl alcohol) (PVA) or polydopamine (PDA) on the surface of the support. The researchers demonstrated that sufficient hydrophilicity of the support not only prevents ICP phenomenon, but also increases the water transport pathways by organizing water continuity within the membrane [40, 41]. For example, Chung's group modified surface of the polysulfone (PSf) supports with PDA through

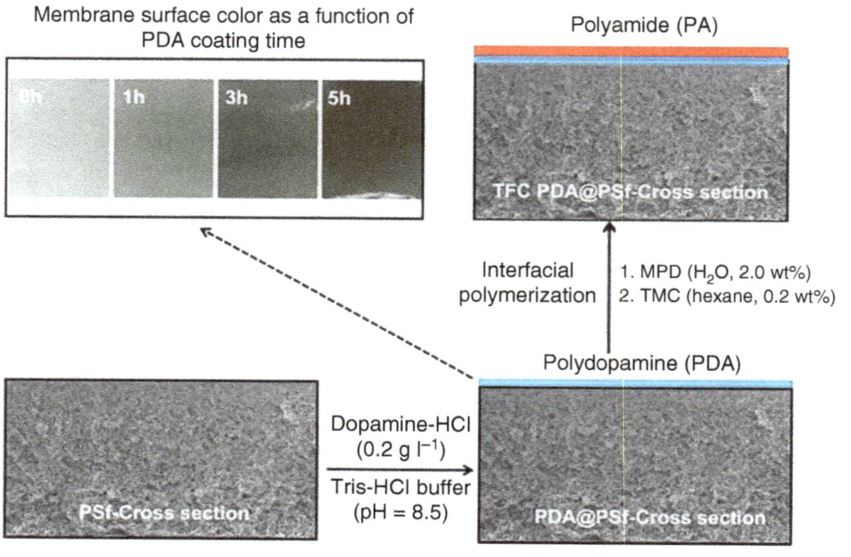

Figure 7.9 Fabrication procedure of TFC membrane by interfacial polymerization on PDA-coated PSf support. Source: Han et al. [42]/with permission from Elsevier.

the oxidant-induced dopamine polymerization in a tris-buffer solution at pH 8.5 in air (Figure 7.9) [42]. Interestingly, the PDA-coated TFC membrane improved water permeability and salt rejection properties simultaneously compared to those constructed on the pristine PSf support layer. It was attributed to hydrophilic surface of PDA-coated support, fabricating favorable polyamide layer with a high salt rejection. These results have shown that improving membrane design for osmotically driven membrane processes should focus on chemistry as well as structural characteristics of the support layer.

7.4.4 Sponge-like Porous Structure of Support

In case of membrane morphology of the support, there are two representative morphologies which are finger-like morphology and sponge-like morphology induced by non-solvent induced phase separation (NIPS). Finger-like morphology is induced by the instantaneous demixing between solvent and nonsolvent whereas sponge-like morphology is induced by delayed demixing. Instantaneous demixing means that the membrane is formed immediately after immersion in the non-solvent bath whereas in case of delayed demixing it takes some time before the membrane is formed. The finger-like morphology affects the membrane performance with excellent transport and structural properties; however, it exhibits the low mechanical strength resulting in low resistance of withstanding trans-membrane hydraulic pressure. Under high pressures, membranes with low mechanical strength are likely deformed due to various compressional, elongational, and bending stresses. However, the sponge-like morphology can withstand trans-membrane hydraulic pressure resulting in high power density.

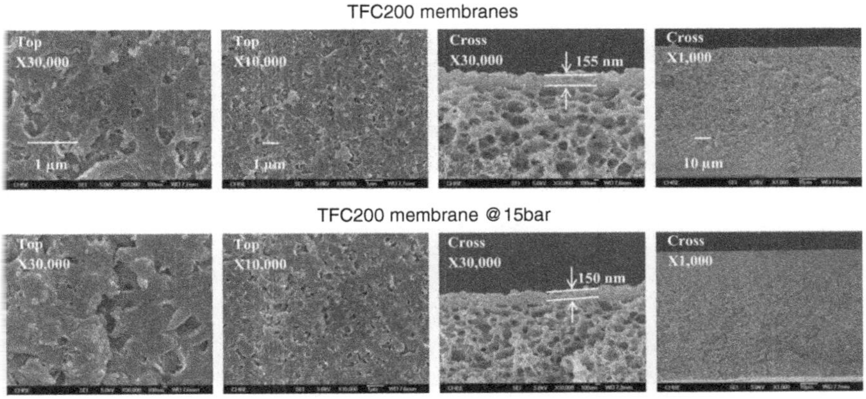

Figure 7.10 SEM images of the fabricated TFC membranes (i.e. TFC200) before and after being tested at 15 bar in the PRO process. Source: Chung et al. [13]/with permission from Elsevier.

Chung et al. firstly reported the sponge-like structure for PRO application [13]. They reported an outstanding mechanical robustness and desirable fully sponge-like morphology with a small structure parameter (Figure 7.10).

In addition, the membrane deformation was observed under the action of the high hydraulic pressure on the feed channel spacer and the resulting impact on membrane performance [43, 44]. Hollow fiber membranes are of great interest not only due to spacer free module fabrication but also high packing density. Various kinds of materials have been tried to PRO application including Matrimid [45], polyethersulfone (PES) [33], and so on, all of which are mechanically robust materials.

7.4.5 Nanofibrous Support Membrane

Nanofibrous membranes have been recently introduced for replacing the conventional support layer in TFC membranes to reduce mass transfer resistance. Nanofibrous membranes can effectively reduce ICP through the membrane, consequently maximizing the driving force and generating higher power density with PRO process. Various materials (e.g. poly(vinylidene fluoride) [PVDF], polyethersulfone [PES], Nylon 66, and poly(etherimide) [PEI], thermally rearranged [TR] polymer) exhibited excellent membrane performance due to their highly porous and interconnected structures [14, 36, 38, 46, 47] (Figure 7.11).

7.4.6 Selective Layer

The research on the selective layer of the TFC membrane has been also performed. Note that the selective layer showed a trade-off between the salt rejection and water permeation properties. If the selective layer does not sufficiently prevent salt permeation, the ICP phenomenon becomes severe and the effective osmotic pressure

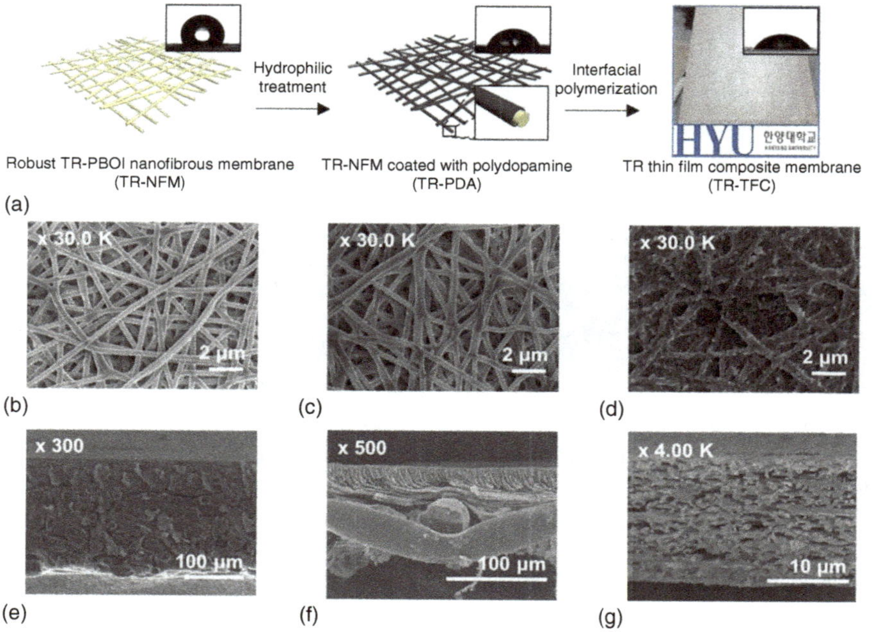

Figure 7.11 (a) Illustration of the preparation procedure of thermally rearranged thin-film composite membrane (TR-TFC). The water contact angle of the membranes was inserted in each step. Surface SEM images of (b) TR-NFM, (c) TR-NFM coated with polydopamine (TR-PDA), and (d) TR-TFC. Cross-section SEM images of conventional (e) RO (PA/PSf UF membrane, PS-20, Sepro, CA, USA) and (f) FO (HTI-TFC, Hydration Technology Innovation, OR, USA) membranes, and (g) TR-TFC. Source: Kim et al. [38]/with permission from Elsevier.

difference at the surface of the selective layer decreases. However, in the opposite case, water permeability is significantly reduced, deteriorating PRO performance despite the high osmotic driving force.

TFC polyamide membrane has been utilized because of their high intrinsic water permeabilities and low ICP. A free volume of polyamide layer synthesized by interfacial polymerization can be controlled by adjusting (i) the employment of different amines and acyl halides for the interfacial polymerization, (ii) the changes of reaction conditions such as monomer concentrations, reaction time, temperature, and solvent system, and (iii) physical and chemical post treatments. Li et al. reported that the effect of free volume of polyamide layer for PRO process [48]. The free volume of the polyamide layer is tuned by blending a bulky diamine monomer and swelling the layer by methanol treatment. Bulky diamine monomer slightly enlarges the free volume, whereas the methanol treatment greatly broadens the free volume. An increase in free volume resulted in water and salt permeabilities; however, a large free volume enlargement may lose the selectivity of the polyamide layer and make adverse impact on PRO process. In addition, various kinds of treatment can be applied to improve water flux as well as power density. Cui et al. reported that a modified polyamide layer for PRO process [49]. A surfactant of sodium dodecyl sulfate (SDS) was added into the amine interfacial polymerization solution. Also,

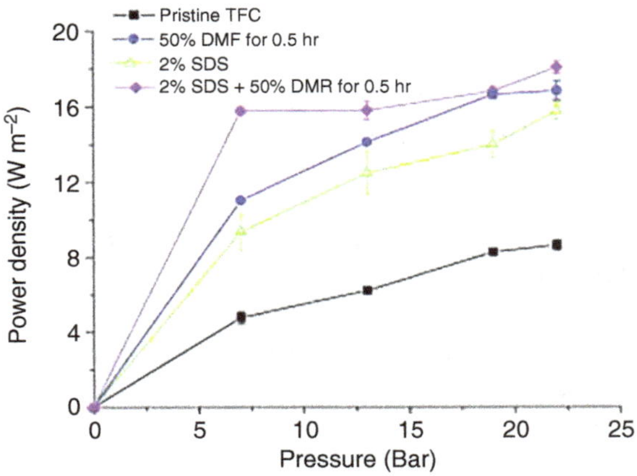

Figure 7.12 Power density of the TFC membranes with different treatments using brine (1 M NaCl) as the draw solution, and DI water as the feed solution. Source: Chung et al. [49]/with permission from Elsevier.

TFC membranes were immersed in *N*,*N*-dimethlyformamide (DMF) aqueous solution. DMF could dissolve loosely crosslinked polyamide layer without damaging the core part resulting in enhanced water flux and thus power density without losing the selectivity (Figure 7.12).

Some researchers optimized the selective properties of the membrane by introducing chlorine modification where the electron pairs of chlorine ion attack the electrophilic carbon of the polyamide (selective layer) [14, 36]. Thus, the selective layer of the TFC membrane was etched to increase the water permeability while sacrificing salt rejection property. For example, Song et al. demonstrated that the water permeability coefficient of the TFC membrane was multiplied up to three times depending on the concentration of NaOCl solution (200–2000 ppm), but the concentration of 1000 ppm was the optimal condition due to the increasing salt permeability coefficient [36]. Solvent activation method also optimized the selective properties of the membrane with a similar mechanism to chlorine modification [50, 51]. During the interfacial polymerization, unreacted and low-molecular weight oligomers exist in the dense selective layer. When the TFC membrane was immersed in mild solvent solution such as DMF or dimethylacetamide (DMAc), the oligomers and unreacted monomers were extracted out, resulting in loose selective layer. Cui et al. investigated the optimal treatment time and solution concentration for DMF activation. They reported that transport characteristics of the TFC membrane were modified in the early stages of the treatment, and no further changes occurred when the processing time exceeded 12 hours. The resultant TFC membrane exhibited higher power density of 15.8 $W\,m^{-2}$ compared to the pristine membrane (8.7 $W\,m^{-2}$) [49].

Overall, the fabrication of the TFC membrane is so complex that it is difficult to optimize the properties of membranes in PRO applications for economical energy production. In addition, feed and draw solutions require pretreatment to prevent potential deterioration in PRO performance, consuming a considerable energy of 0.2~0.3 $kWh\,m^{-3}$. Given that the PRO plant with a 50% overall efficiency using CTA membranes harvests only 0.3~0.4 $kWh\,m^{-3}$ from the mixing of sea and river waters, there is only 0.1 kWh m^{-3} of useful energy when using the most common water sources in the world [52]. This indicates that the PRO membranes should not only have high performance, but also be inexpensive, robust, and resistant to the side effects such as CP phenomenon and contamination problems.

7.5 Hybrid Systems with Membrane Distillation and Others

Note that PRO operates based on the balance between hydraulic pressure and osmotic pressure in a system where a semi-permeable membrane is employed. Therefore, PRO and conventional desalination systems (such as FO, RO, and membrane distillation [MD]) have many features in common, although these processes have fundamentally different goals. The main purpose of desalination systems is not to save energy, but to produce freshwater using energy. However, with the explosive increase in demand for freshwater, the burden on the cost of energy used in desalination plants continues to increase. In this background, an attempt has been made to increase energy efficiency by pairing the existing desalination processes and PRO. Since the water sources are usually pretreated in the desalination process, no additional pretreatment is required in PRO. This means that the energy efficiency of PRO can be further increased, and fouling of the membrane is dramatically reduced.

7.5.1 PRO-MD Hybrid System

PRO process is typically considered for operation with river water and seawater, but a far greater energy of mixing can be harnessed from high concentrated solutions [21]. In locations where fresh water naturally flows into high concentrated solution, such as the Dead Sea and the Great Salt Lake, the theoretical energy of mixing released is nearly 20 times larger than that at river water–seawater interfaces [9]. Moreover, PRO process can be applied in a closed-loop heat engine, where synthetic draw solutions are separated and regenerated using low-grade heat. High concentrated solution can be assessed through incorporating with desalination technologies such as RO and MD. In case of RO process, the RO brine which salinity is known as around 2–3 M of NaCl is produced during desalination process. MD is a thermally driven membrane process where the feed stream is heated, evaporated, and transported across the membrane. Being less sensitive to feed salt concentration compared to RO process, the MD process can produce osmotic power and re-concentrate the diluted draw solution when the PRO process is integrated with the MD process [24, 53, 54]. Straub et al. reported that operation at increased hydraulic pressure allowed unprecedented power densities, up to 60 Wm^{-2} with a 3 M NaCl draw solution [21].

However, to access the high-power density potential of high concentrated solutions, the PRO process should be able to withstand a hydraulic pressure which is theoretically half the osmotic pressure of the concentrated draw solution.

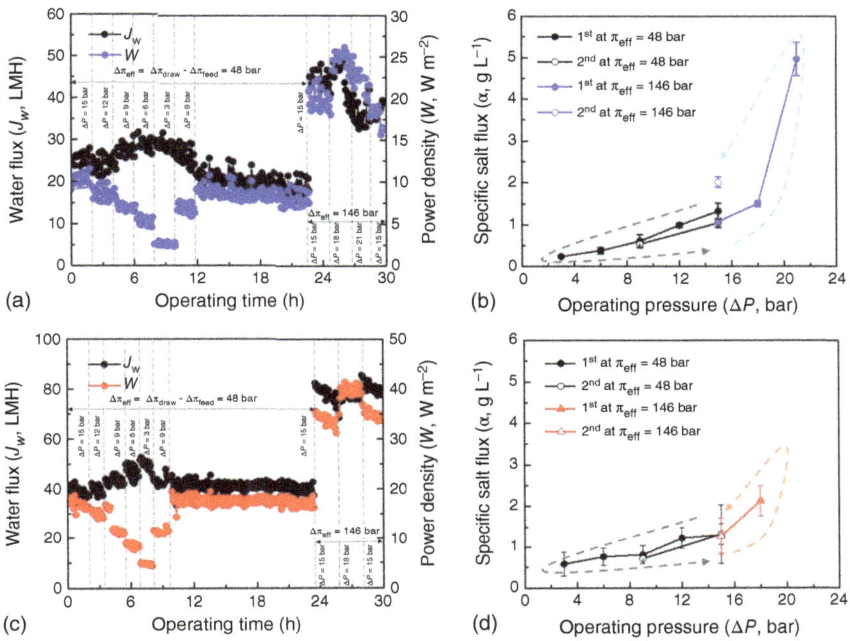

Figure 7.13 PRO performance and specific salt flux profiles of (a and b) TR-TFC and (c and d) DMF activated TR-TFC (TR-TFCDMF) were measured at different operating pressures using 1 M and 3 M NaCl aqueous solution and D.I. water as draw and feed solutions, respectively. The draw and feed solutions flowed counter-currently at 0.3 l min^{-1} (mass transfer coefficient is 52 m s^{-1}) and the PRO performance was measured with different operating pressures from 15 to 3 bar, and subsequently to 18 bar with a gap of 3 bar. Source: Kim et al. [38]/with permission from Elsevier.

In the case of a 2 M NaCl draw solution, the system should be operated at ~50 bar. However, the highest PRO operating pressure reported is ~24 bar [38, 55], with most studies reaching only ~16 bar. And hence, the development of mechanically robust membranes is required. Figure 7.13 represents the PRO performance and specific salt flux profiles of TR-TFC membranes and DMF-activated TR-TFC membranes at different operating pressures using 1 M and 3 M NaCl aqueous solution and D.I. water as draw and feed solution, respectively. Power density obtained from these membranes range from 20 to 40 W m^{-2} depending on the membrane types and the operating pressure.

7.5.2 SWRO-PRO Hybrid System

Among the PRO hybrid processes, seawater reverse osmosis (SWRO)-PRO system has been intensively researched by implementing pilot-scale plants in South Korea, Japan, and Singapore. As shown in Figure 7.14, SWRO-PRO system can be classified into turbine-based system (TBS) and energy circulation system (ECS) [56]. Since the TBS generates electricity by rotating a turbine, the produced energy can be used for various purposes, not just for RO. In the ECS, on the other hand, the produced energy

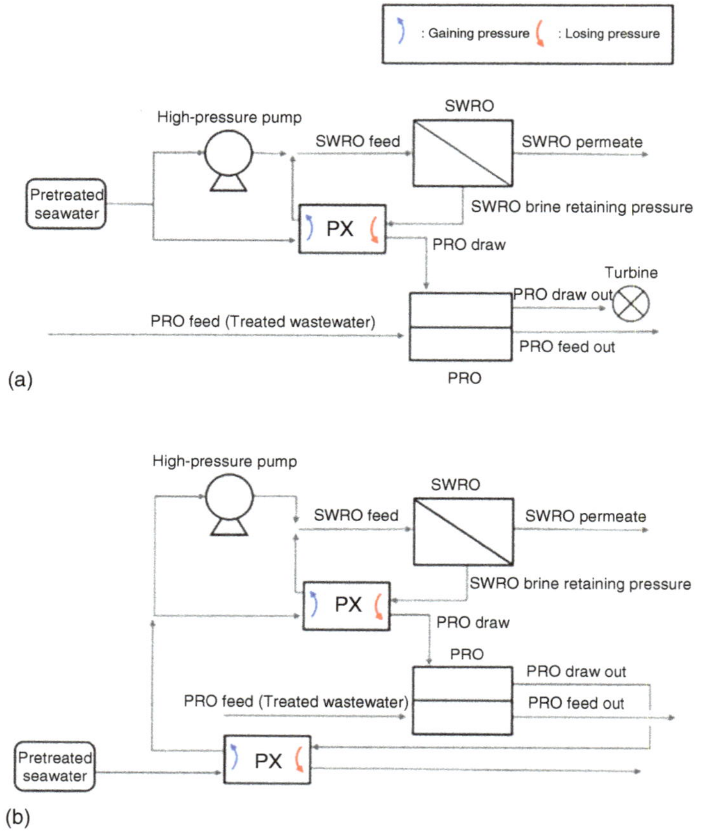

Figure 7.14 Schematic illustration representing the configuration of (a) a turbine-based system, and (b) an energy circulation system (ECS) for SWRO-PRO hybrid process. Source: Lee et al. [56]/with permission from Elsevier.

is directly transferred to the influent of SWRO in pressure exchanger as a form of pressure. The efficiency of pressure exchange is better than that of turbine in most cases, thus the ECS is preferred over the TBS. This is clearly confirmed by Wang and coworkers [57]. Wang et al. calculated the efficiencies of commercial pressure exchangers employed for SWRO-PRO system as 93.48~97.61%, which is higher than that of turbines (80~90%). Nevertheless, some researchers demonstrated that given the potential of independent power generation of TBS, the study for TBS deserves more attention [56].

The representative implementations of SWRO-PRO system are GMVP and Megaton projects that were undertaken in South Korea and Japan, respectively. During the GMVP project (2013~2018), two pilot-scale SWRO-PRO plants were implemented as the ECS, showing the desalination capacity of 20 and 240 $m^3\ d^{-1}$ in each system. The project headquarters reported that approximately 20% of energy consumed by SWRO process was recovered by PRO with a maximum power density of 8~18.3 $W\ m^{-2}$ [58]. Similarly, in the Megaton project, SWRO-PRO plant was

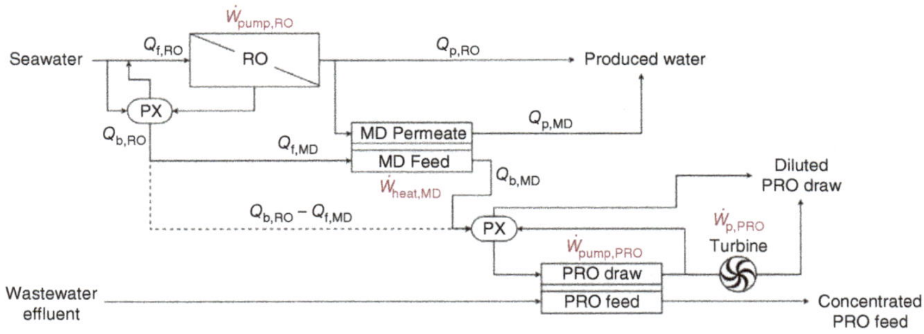

Figure 7.15 Schematic illustration of RO-MD-PRO hybrid system Source: Kim et al. [61]/with permission from Elsevier.

constructed using both TBS and ECS concepts, by dividing PRO effluent into a flow to a pressure exchanger and a flow to a turbine. Kurihara et al. demonstrated that SWRO-PRO plant in the Megaton project showed power density of 13.3 W m^{-2} using CTA hollow fiber membranes [59].

7.5.3 SWRO-PRO-MD Trihybrid System

Although SWRO-PRO is the most widely used application of PRO hybrid system, there is an important problem in the system. To produce freshwater in RO, a hydraulic pressure greater than osmotic pressure difference must be applied. If the concentration difference of the water sources is large, the desalination efficiency dramatically decreases. Thus, RO process should be used in a limited concentration range, and the concentration of RO effluent was also restricted. This means that there is an upper limit to the power density of PRO process. Therefore, some researchers designed other PRO hybrid systems using MD process, where desalination process was driven by vapor pressure (or temperature) difference rather than osmotic pressure [60]. For example, Kim et al. investigated a tri-combined process that consists of SWRO, MD, and PRO (Figure 7.15). This hybrid system outperformed SWRO-PRO in terms of energy efficiency by optimizing heat cost of MD process and the flow rates in the system [61]. It is believed that the efficiency of hybrid system using MD process will be further improved, if free thermal energy such as geothermal, solar, and industrial waste heat can be utilized.

7.5.4 Osmotic Heat Engine System

On the other hand, osmotic heat engine (OHE) was developed to make the most of waste heat [24, 54, 62, 63]. As shown in Figure 7.16, OHE is a closed-loop system consisting of MD and PRO processes. The OHE has several advantages compared to previously mentioned hybrid systems due to the characteristics of closed-loop system: (i) The pretreatment of water sources is not required; (ii) Membrane fouling is not occurred; (iii) There are no issues with the supply of water sources; (iv) Synthetic water solutions can be used in the system. Note that the objective of the

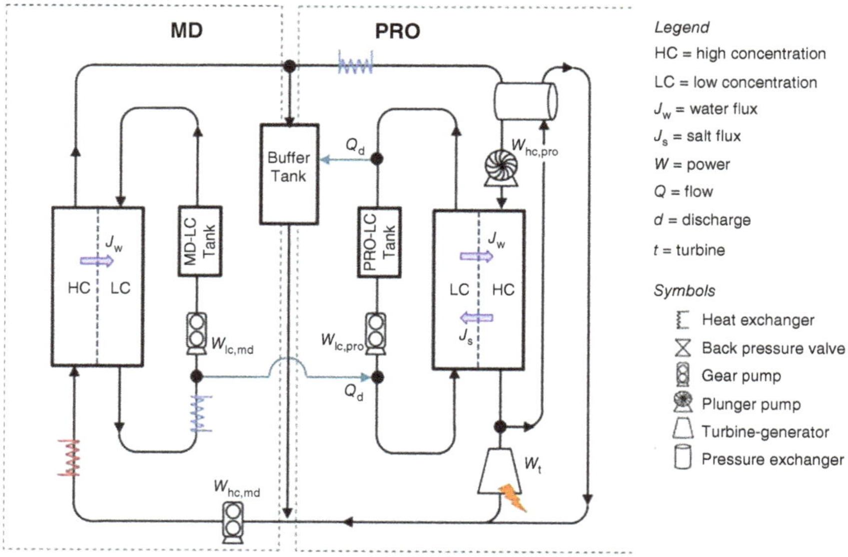

Figure 7.16 Schematic diagram of osmotic heat engine. Source: Hickenbottom et al. [63]/with permission from Elsevier.

OHE system is not to produce freshwater, but to generating electricity using waste heat (energy conversion system from heat to electricity). Once heat energy induces water flux in MD process, the corresponding amount of water flux in opposite direction would be occurred in PRO process to produce electricity. However, the PRO membrane cannot have perfect salt rejection, and PRO feed concentration increases as the OHE is driven. Therefore, there is a portion of PRO feed stream discharged to the MD feed for recovery of solutes and control of PRO feed concentration [63]. Lin et al. demonstrated that the OHE can theoretically achieve 9~10% efficiency of energy conversion (approximately 75~83% of Carnot efficiency) with low-grade heat at 60 °C [24]. This research also indicated that the OHE has the distinct advantage of being able to utilize low-grade heat sources below 80 °C, a region not easily accessible by organic rankine cycle (ORC) due to operating constraints of working fluids. Therefore, this novel concept of PRO hybrid system is expected to open a new field of green energy production.

7.6 Conclusions and Future Perspectives

Until now, we discussed about overall PRO processes and the membrane fabrication for the system. The PRO process can be used in various ways for generating renewable energy or increasing energy efficiency of desalination processes. Although PRO and its hybrid systems are promising to replace the conventional systems, there are various issues that hinder the commercialization of PRO, such as complexity of the membrane fabrication, long-term stability of the system, economic competitiveness, and absence of implementation for pilot-scale. Therefore, researchers are still required to develop high-performance membrane and to design new concept of PRO process for breakthrough of technology bottleneck.

References

1 Bauer, D., Philbrick, M., Vallario, B. et al. (2014). *The Water-Energy Nexus: Challenges and Opportunities*. US Department of Energy.

2 Spang, E.S., Moomaw, W.R., Gallagher, K.S. et al. (2014). The water consumption of energy production: an international comparison. *Environmental Research Letters* 9: 105002.

3 IEA (2018). Renewables 2018. International Energy Agency (IEA), Paris, France, pp. 1-10.

4 Moon, S.J., Kim, J.H., Seong, J.G. et al. (2020). Thin film composite on fluorinated thermally rearranged polymer nanofibrous membrane achieves power density of 87 W m−2 in pressure retarded osmosis, improving economics of osmotic heat engine. *Journal of Membrane Science* 607.

5 Alamri, B. and Alamri, A. (2009). Technical review of energy storage technologies when integrated with intermittent renewable energy. In: *2009 International Conference on Sustainable Power Generation and Supply*, 1–5. IEEE.

6 Teleke, S., Baran, M.E., Bhattacharya, S., and Huang, A.Q. (2010). Rule-based control of battery energy storage for dispatching intermittent renewable sources. *IEEE Transactions on Sustainable Energy* 1: 117–124.

7 Loeb, S. (1976). Production of energy from concentrated brines by pressure-retarded osmosis: I. Preliminary technical and economic correlations. *Journal of Membrane Science* 1: 49–63.

8 Loeb, S., Van Hessen, F., and Shahaf, D. (1976). Production of energy from concentrated brines by pressure-retarded osmosis: II. Experimental results and projected energy costs. *Journal of Membrane Science* 1: 249–269.

9 Logan, B.E. and Elimelech, M. (2012). Membrane-based processes for sustainable power generation using water. *Nature* 488: 313–319.

10 Achilli, A., Cath, T.Y., and Childress, A.E. (2009). Power generation with pressure retarded osmosis: an experimental and theoretical investigation. *Journal of Membrane Science* 343: 42–52.

11 Lee, K., Baker, R., and Lonsdale, H. (1981). Membranes for power generation by pressure-retarded osmosis. *Journal of Membrane Science* 8: 141–171.

12 Cath, T., Childress, A., and Elimelech, M. (2006). Forward osmosis: principles, applications, and recent developments. *Journal of Membrane Science* 281: 70–87.

13 Han, G., Zhang, S., Li, X., and Chung, T.-S. (2013). High performance thin film composite pressure retarded osmosis (PRO) membranes for renewable salinity-gradient energy generation. *Journal of Membrane Science* 440: 108–121.

14 Yip, N.Y., Tiraferri, A., Phillip, W.A. et al. (2011). Thin-film composite pressure retarded osmosis membranes for sustainable power generation from salinity gradients. *Environmental Science & Technology* 45: 4360–4369.

15 Tian, M., Wang, R., Goh, K. et al. (2015). Synthesis and characterization of high-performance novel thin film nanocomposite PRO membranes with tiered nanofiber support reinforced by functionalized carbon nanotubes. *Journal of Membrane Science* 486: 151–160.

16 Aaberg, R.J. (2003). Osmotic power: a new and powerful renewable energy source? *Refocus* 4: 48–50.

17 Touati, K. and Tadeo, F. (2016). Green energy generation by pressure retarded osmosis: state of the art and technical advancement—review. *International Journal of Green Energy* 14: 337–360.

18 Jia, Z., Wang, B., Song, S., and Fan, Y. (2014). Blue energy: Current technologies for sustainable power generation from water salinity gradient. *Renewable and Sustainable Energy Reviews* 31: 91–100.

19 Post, J.W., Veerman, J., Hamelers, H.V.M. et al. (2007). Salinity-gradient power: evaluation of pressure-retarded osmosis and reverse electrodialysis. *Journal of Membrane Science* 288: 218–230.

20 Post, J.W., Goeting, C.H., Valk, J. et al. (2010). Towards implementation of reverse electrodialysis for power generation from salinity gradients. *Desalination and Water Treatment* 16: 182–193.

21 Straub, A.P., Yip, N.Y., and Elimelech, M. (2013). Raising the bar: Increased hydraulic pressure allows unprecedented high power densities in pressure-retarded osmosis. *Environmental Science & Technology Letters* 1: 55–59.

22 Helfer, F., Lemckert, C., and Anissimov, Y.G. (2014). Osmotic power with pressure retarded osmosis: theory, performance and trends - a review. *Journal of Membrane Science* 453: 337–358.

23 Helfer, F. and Lemckert, C. (2015). The power of salinity gradients: An Australian example. *Renewable and Sustainable Energy Reviews* 50: 1–16.

24 Lin, S., Yip, N.Y., Cath, T.Y. et al. (2014). Hybrid pressure retarded osmosis-membrane distillation system for power generation from low-grade heat: thermodynamic analysis and energy efficiency. *Environmental Science & Technology* 48: 5306–5313.

25 McCutcheon, J.R. and Elimelech, M. (2006). Influence of concentrative and dilutive internal concentration polarization on flux behavior in forward osmosis. *Journal of Membrane Science* 284: 237–247.

26 Zhao, S., Zou, L., Tang, C.Y., and Mulcahy, D. (2012). Recent developments in forward osmosis: opportunities and challenges. *Journal of Membrane Science* 396: 1–21.

27 Zhang, S., Wang, K.Y., Chung, T.-S. et al. (2010). Well-constructed cellulose acetate membranes for forward osmosis: minimized internal concentration polarization with an ultra-thin selective layer. *Journal of Membrane Science* 360: 522–535.

28 Achilli, A. and Childress, A.E. (2010). Pressure retarded osmosis: from the vision of Sidney Loeb to the first prototype installation — review. *Desalination* 261: 205–211.

29 Nijmeijer, K. and Metz, S. (2010). Salinity gradient energy. *Sustainability Science and Engineering* 2: 95–139.

30 Gerstandt, K., Peinemann, K.-V., Skilhagen, S.E. et al. (2008). Membrane processes in energy supply for an osmotic power plant. *Desalination* 224: 64–70.

31 Skilhagen, S.E. (2010). Osmotic power—a new, renewable energy source. *Desalination and Water Treatment* 15: 271–278.

32 Park, C.H., Bae, H., Kwak, S.J. et al. (2016). Interconnection of electrospun nanofibers via a post co-solvent treatment and its open pore size effect on pressure-retarded osmosis performance. *Macromolecular Research* 24: 314–322.

33 Chou, S., Wang, R., Shi, L. et al. (2012). Thin-film composite hollow fiber membranes for pressure retarded osmosis (PRO) process with high power density. *Journal of Membrane Science* 389: 25–33.

34 Wan, C.F. and Chung, T.-S. (2015). Osmotic power generation by pressure retarded osmosis using seawater brine as the draw solution and wastewater retentate as the feed. *Journal of Membrane Science* 479: 148–158.

35 Zhu, J., Yuan, S., Uliana, A. et al. (2018). High-flux thin film composite membranes for nanofiltration mediated by a rapid co-deposition of polydopamine/piperazine. *Journal of Membrane Science* 554: 97–108.

36 Song, X., Liu, Z., and Sun, D.D. (2013). Energy recovery from concentrated seawater brine by thin-film nanofiber composite pressure retarded osmosis membranes with high power density. *Energy & Environmental Science* 6: 1199–1210.

37 Zhang, S., Sukitpaneenit, P., and Chung, T.-S. (2014). Design of robust hollow fiber membranes with high power density for osmotic energy production. *Chemical Engineering Journal* 241: 457–465.

38 Kim, J.H., Moon, S.J., Park, S.H. et al. (2018). A robust thin film composite membrane incorporating thermally rearranged polymer support for organic solvent nanofiltration and pressure retarded osmosis. *Journal of Membrane Science* 550: 322–331.

39 Chen, Y., Setiawan, L., Chou, S. et al. (2016). Identification of safe and stable operation conditions for pressure retarded osmosis with high performance hollow fiber membrane. *Journal of Membrane Science* 503: 90–100.

40 McCutcheon, J.R. and Elimelech, M. (2008). Influence of membrane support layer hydrophobicity on water flux in osmotically driven membrane processes. *Journal of Membrane Science* 318: 458–466.

41 Guo, H., Yao, Z., Wang, J. et al. (2018). Polydopamine coating on a thin film composite forward osmosis membrane for enhanced mass transport and antifouling performance. *Journal of Membrane Science* 551: 234–242.

42 Han, G., Zhang, S., Li, X. et al. (2012). Thin film composite forward osmosis membranes based on polydopamine modified polysulfone substrates with enhancements in both water flux and salt rejection. *Chemical Engineering Science* 80: 219–231.

43 Kim, Y.C. and Elimelech, M. (2012). Adverse impact of feed channel spacers on the performance of pressure retarded osmosis. *Environmental Science & Technology* 46: 4673–4681.

44 She, Q.H., Hou, D.X., Liu, J.X. et al. (2013). Effect of feed spacer induced membrane deformation on the performance of pressure retarded osmosis (PRO):

Implications for PRO process operation. *Journal of Membrane Science* 445: 170–182.

45 Han, G., Wang, P., and Chung, T.S. (2013). Highly robust thin-film composite pressure retarded osmosis (PRO) hollow fiber membranes with high power densities for renewable salinity-gradient energy generation. *Environmental Science & Technology* 47: 8070–8077.

46 Bui, N.-N. and McCutcheon, J.R. (2014). Nanofiber supported thin-film composite membrane for pressure-retarded osmosis. *Environmental Science & Technology* 48: 4129–4136.

47 Huang, L. and McCutcheon, J.R. (2014). Hydrophilic nylon 6,6 nanofibers supported thin film composite membranes for engineered osmosis. *Journal of Membrane Science* 457: 162–169.

48 Li, X., Chung, T.S., and Chung, T.S. (2013). Effects of free volume in thin-film composite membranes on osmotic power generation. *AIChE Journal* 59: 4749–4761.

49 Cui, Y., Liu, X.Y., and Chung, T.S. (2014). Enhanced osmotic energy generation from salinity gradients by modifying thin film composite membranes. *Chemical Engineering Journal* 242: 195–203.

50 Gorgojo, P., Jimenez-Solomon, M.F., and Livingston, A.G. (2014). Polyamide thin film composite membranes on cross-linked polyimide supports: improvement of RO performance via activating solvent. *Desalination* 344: 181–188.

51 Jimenez Solomon, M.F., Bhole, Y., and Livingston, A.G. (2013). High flux hydrophobic membranes for organic solvent nanofiltration (OSN)—Interfacial polymerization, surface modification and solvent activation. *Journal of Membrane Science* 434: 193–203.

52 Elimelech, M. and Phillip, W.A. (2011). The future of seawater desalination: energy, technology, and the environment. *Science* 333: 712–717.

53 Han, G., Zuo, J., Wan, C., and Chung, T.-S. (2015). Hybrid pressure retarded osmosis–membrane distillation (PRO–MD) process for osmotic power and clean water generation. *Environmental Science: Water Research & Technology* 1: 507–515.

54 Shaulsky, E., Boo, C., Lin, S., and Elimelech, M. (2015). Membrane-based osmotic heat engine with organic solvent for enhanced power generation from low-grade heat. *Environmental Science & Technology* 49: 5820–5827.

55 Zhang, S. and Chung, T.-S. (2013). Minimizing the instant and accumulative effects of salt permeability to sustain ultrahigh osmotic power density. *Environmental Science & Technology* 47: 10085–10092.

56 Lee, C., Chae, S.H., Yang, E. et al. (2020). A comprehensive review of the feasibility of pressure retarded osmosis: Recent technological advances and industrial efforts towards commercialization. *Desalination* 491.

57 Wang, Q., Zhou, Z., Li, J. et al. (2019). Investigation of the reduced specific energy consumption of the RO-PRO hybrid system based on temperature-enhanced pressure retarded osmosis. *Journal of Membrane Science* 581: 439–452.

58 Lee, S., Choi, J., Park, Y.-G. et al. (2019). Hybrid desalination processes for beneficial use of reverse osmosis brine: current status and future prospects. *Desalination* 454: 104–111.

59 Kurihara, M. and Takeuchi, H. (2018). SWRO-PRO system in "mega-ton water system" for energy reduction and low environmental impact. *Water* 10.

60 Chae, S.H., Seo, J., Kim, J. et al. (2018). A simulation study with a new performance index for pressure-retarded osmosis processes hybridized with seawater reverse osmosis and membrane distillation. *Desalination* 444: 118–128.

61 Kim, J., Park, M., Shon, H.K., and Kim, J.H. (2016). Performance analysis of reverse osmosis, membrane distillation, and pressure-retarded osmosis hybrid processes. *Desalination* 380: 85–92.

62 Straub, A.P., Yip, N.Y., Lin, S. et al. (2016). Harvesting low-grade heat energy using thermo-osmotic vapour transport through nanoporous membranes. *Nature Energy* 1.

63 Hickenbottom, K.L., Vanneste, J., Miller-Robbie, L. et al. (2017). Techno-economic assessment of a closed-loop osmotic heat engine. *Journal of Membrane Science* 535: 178–187.

Index

Membranes for Energy Applications, First Edition. Young Moo Lee.

n

o

p

q

r

www.ingramcontent.com/pod-product-compliance
Lightning Source LLC
Chambersburg PA
CBHW080932240926
51849CB00003BA/435
* 9 7 8 3 5 2 7 3 4 7 6 4 3 *

Membranes for Energy Applications